ARITHMETIC GEOMETRY AND NUMBER THEORY

Series on Number Theory and Its Applications ISSN 1793-3161

Vol. 1 Arithmetic Geometry and Number Theory
edited by Lin Weng & Iku Nakamura

Series on Number Theory and Its Applications **Vol. 1**

ARITHMETIC GEOMETRY AND NUMBER THEORY

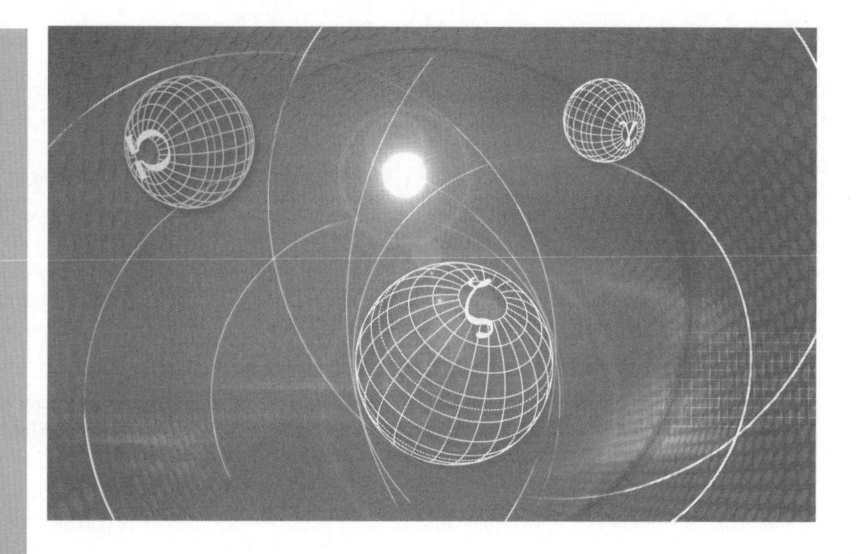

Editors

Lin Weng *Kyushu University, Japan*

Iku Nakamura *Hokkaido University, Japan*

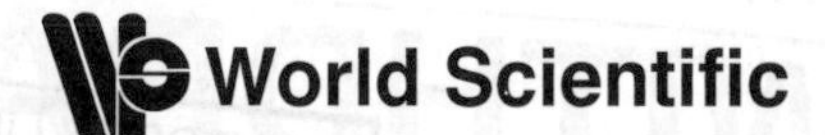

NEW JERSEY • LONDON • SINGAPORE • BEIJING • SHANGHAI • HONG KONG • TAIPEI • CHENNAI

Published by 1004949103

World Scientific Publishing Co. Pte. Ltd.
5 Toh Tuck Link, Singapore 596224
USA office: 27 Warren Street, Suite 401-402, Hackensack, NJ 07601
UK office: 57 Shelton Street, Covent Garden, London WC2H 9HE

British Library Cataloguing-in-Publication Data
A catalogue record for this book is available from the British Library.

ARITHMETIC GEOMETRY AND NUMBER THEORY

ISBN 981-256-814-X

Printed in Singapore by B & JO Enterprise

Foreword

This series aims to bring together the very many applications of number theory in a fusion of diverse disciplines such as chemistry, physics and others. It aims to provide a comprehensive and thorough coverage of the whole spectrum of (state-of-the-art knowledge of) number theory and related fields, in the form of textbooks and review volumes. Presented as an organic whole, rather than as an assembly of disjointed subjects, the volumes in the series will include ample examples to illustrate the applications of number theory. The target audience will range from the undergraduate student who hopes to master number theory so as to apply it to his or her own research, to the professional scientist who wishes to keep abreast of the latest in the applications of number theory, to the curious academic who wants to know more about this fusion of old disciplines.

Shigeru Kanemitsu
Series Editor

Preface

Mathematics is a part of our culture. As such, the works presented here serve the purposes of developing branches involved, popularizing existing theories, and guiding our future explorations.

Accordingly, the collection of this volume may be roughly divided into three categories. More precisely, first, Jiang's paper deals with local gamma factors that appeared in the theory of automorphic representations; Obitsu-To-Weng's paper investigates the intrinsic relations between Weil-Petersson and Takhtajan-Zograf metrics on moduli spaces of punctured Riemann surfaces using Deligne pairings and an arithmetic Riemann-Roch isometry; Werner's paper explains her recent works with Deninger on vector bundles on curves over $\mathbb{C}_p$; Yoshida's paper exposes his beautiful theory on CM periods; and Yu's paper studies the transcendence of special values for zetas over finite fields. All these well-prepared articles then bring us to the uppermost frontiers of the current researches in Arithmetic Geometry and Number Theory. Secondly, the lecture notes of Weng explains basic ideas and methods behind the fundamental yet famously difficult work of Langlands on the Eisenstein series and spectral decompositions. The reader will find these notes invaluable in understanding the original theory. Finally, Weng's paper of Geometric Arithmetic outlines a Program for understanding global arithmetic using algebraic and/or analytic methods based on geometric considerations – the topics touched here are a continuation of Weil's approach on non-abelian Class Field Theory using stability and Tannakian category theory; new yet genuine non-abelian zetas and Ls which are closely related with the so-called Arthur's periods; and an intersection approach to the Riemann Hypothesis.

While various important topics are selected, all papers share common themes such as the Eisenstein series, stability and zeta functions.

Jiang's paper was presented at the Conference on L-Functions (February 18-23, 2006, Fukuoka). Partial contents of the papers of

Obitsu-To-Weng, Werner, Yoshida and Yu were delivered by W.-K. To, A. Werner, H. Yoshida and J. Yu, respectively, in the (series of) lectures at our Karatsu symposium on 'Arithmetic Geometry and Number Theory', held from March 21 to March 25, 2005, immediately after the huge Fukuoka earthquake of scale M7.0 (on March 20). The notes about Langlands' work is based on six lectures of Weng at the Mathematics Department, University of Toronto, between October and November, 2005. Finally, the Program paper, of which the first version was circulated around the turn of the millennium, is revised significantly for this publication and is indeed the driving force for the whole project[1].

The Editors

[1]This project is partially supported by JSPS.

Contents

On Local γ-Factors

Dihua JIANG

Contents

1 Introduction

Let G be a reductive algebraic group defined over a p-adic local field F. We assume that F is a finite extension of $\mathbb{Q}_p$ for simplicity. Let $\mathcal{W}_F$ be the local Weil group of F and ${}^LG$ be the Langlands dual group of G, which is a semi-product of the complex dual group $G^\vee$ and the absolute Galois group $\Gamma_F = \mathrm{Gal}(\overline{F}/F)$. Consider continuous homomorphisms ϕ from the Weil-Deligne group $\mathcal{W}_F \times \mathrm{SL}_2(\mathbb{C})$ to the Langlands dual group ${}^LG$, which is admissible in the sense of [B79]. The $G^\vee$-conjugacy class of such a homomorphism ϕ is called a local Langlands parameter. The set of local Langlands parameters is denoted by $\Phi(G/F)$. Let $\Pi(G/F)$ be the set of equivalence classes of irreducible admissible complex representations of $G(F)$.

The local Langlands conjecture for G over F asserts that for each local Langlands parameter $\phi \in \Phi(G/F)$, there should be a finite subset $\Pi(\phi)$, which is called the local L-packet attached to ϕ such that the set $\{\Pi(\phi) \mid \phi \in \Phi(G/F)\}$ is a partition of $\Pi(G/F)$, among other required properties ([B79]). The map $\phi \mapsto \Pi(\phi)$ is called the local Langlands correspondence or the local Langlands reciprocity law for G over F.

The main problem is of course how to construct the local Langlands reciprocity map $\phi \mapsto \Pi(\phi)$. From the classification-theoretic point of view, the local Langlands conjecture provides a classification for irreducible admissible representations up to L-packet. It is interesting to characterize the local L-packets in general. The most well-known approach to characterize local L-packets is in terms of stability of distribution characters following from the idea of Arthur trace formula approach to the discrete spectrum of automorphic forms. We refer to [MW03], [KV05], [DR05], [R05] and [V93] for further discussions.

In this note, we discuss the roles of local factors attached to irreducible admissible representations of $G(F)$. They yield information about the classification theory and the functorial structures of irreducible admissible representations of $G(F)$.

First, we recall the local Langlands conjecture for GL_n over F, which is proved by Harris-Taylor [HT01] and by Henniart [H00].

Theorem 1.1 ([HT01], [H00], [H93]). *There is a unique collection of bijections*

$$\mathrm{rec}_F \ : \ \Pi(\mathrm{GL}_n/F) \to \Phi(\mathrm{GL}_n/F)$$

for every $n \geq 1$ such that

1. *for $\pi \in \Pi(\mathrm{GL}_1/F)$, $\mathrm{rec}_F(\pi) = \pi \circ \mathrm{Art}_F^{-1}$, where Art_F is the local Artin reciprocity map from $F^\times$ to $\mathcal{W}_F^{\mathrm{ab}}$;*

2. *for $\pi_1 \in \Pi(GL_{n_1}/F)$ and $\pi_2 \in \Pi(GL_{n_2}/F)$,*

$$L(s, \pi_1 \times \pi_2) = L(s, \mathrm{rec}_F(\pi_1) \otimes \mathrm{rec}_F(\pi_2))$$

and

$$\epsilon(s, \pi_1 \times \pi_2, \psi) = \epsilon(s, \mathrm{rec}_F(\pi_1) \otimes \mathrm{rec}_F(\pi_2), \psi)$$

where ψ is a given nontrivial character of F;

3. *for $\pi \in \Pi(GL_n/F)$ and $\chi \in \Pi(\mathrm{GL}_1/F)$,*

$$\mathrm{rec}_F(\pi \otimes (\chi \circ \det)) = \mathrm{rec}_F(\pi) \otimes \mathrm{rec}_F(\chi);$$

4. *for $\pi \in \Pi(GL_n/F)$ with central character $\omega_\pi = \chi$,*

$$\det \circ \mathrm{rec}_F(\pi) = \mathrm{rec}_F(\chi);$$

5. *for $\pi \in \Pi(GL_n/F)$, $\mathrm{rec}_F(\pi^\vee) = \mathrm{rec}_F(\pi)^\vee$, where $\vee$ denotes the contragredient.*

We note that the existence of the local Langlands correspondence (the reciprocity map satisfies conditions (1)–(5) is proved in [HT01] and [H00]. The uniqueness of the such maps is proved in [H93]. We refer to [HT01] for historical remarks on the proof of the local Langlands conjecture for $\mathrm{GL}_n(F)$. The local factors on the $\mathrm{GL}_n(F)$ side is given [JPSS83] and the local factors on the $\mathcal{W}_F \times \mathrm{SL}_2(\mathbb{C})$ side is given in [T79]. One can define as in [JPSS83] the local γ-factors by

$$\gamma(s, \pi_1 \times \pi_2, \psi) = \epsilon(s, \pi_1 \times \pi_2, \psi) \cdot \frac{L(1-s, \pi_1^\vee \times \pi_2^\vee)}{L(s, \pi_1 \times \pi_2)}. \tag{1.1}$$

On the $\mathcal{W}_F \times \mathrm{SL}_2(\mathbb{C})$ side, one defines the γ-factor in the same way [T79]. Note that for $\mathrm{GL}_n(F)$, the local L-packets always contains one member. This fact follows from [H93] and the Bernstein-Zelevinsky classification theory ([BZ77] and [Z80]).

For general reductive groups local factors have been defined for many cases. When irreducible admissible representations π of $G(F)$ are generic, i.e. have nonzero Whittaker models, the Shahidi's theory of local coefficients defines the local L-, ϵ-, and γ-factors. It is expected that the local factors defined by Shahidi should be essentially the same as the ones defined by the Rankin-Selberg method if they are available, although it has to be verified case by case. It should be mentioned that for nongeneric representations, there are cases where the local factors can be defined by the Rankin-Selberg method ([GPSR97] and [LR05]), and also that the work [FG99] has the potential to define the local factors for nongeneric representations, which can be viewed as the natural extension of Shahidi's work. Of course, one may define the local factors by means of the conjectured local Langlands conjecture for G over F, and this definition should be consistent with all other definitions.

We recall the local Langlands functoriality principle. Let G and H be reductive algebraic groups defined over F. For an admissible homomorphism ${}^L\rho$ ([B79]) from the Langlands dual group ${}^LH$ to the Langlands dual group ${}^LG$, there should be a functorial transfer ρ from $\Pi(H/F)$ to $\Pi(G/F)$, which takes L-packets of $H(F)$ to L-packets of $G(F)$, and satisfies the following conditions.

1. For any local Langlands parameter $\phi_H \in \Phi(H/F)$, ${}^L\rho \circ \phi_H$ is a local Langlands parameter in $\Phi(G/F)$, such that the functorial transfer ρ takes the local L-packet $\Pi(\phi_H)$ to the local L-packet $\Pi({}^L\rho \circ \phi_H)$.

2. For any finite-dimensional complex representation r of ${}^LG$ and $\sigma \in \Pi(\phi_H)$, one has

   $$L(s, \rho(\sigma), r) = L(s, \sigma, r \circ {}^L\rho),$$

   and

   $$\epsilon(s, \rho(\sigma), r, \psi) = \epsilon(s, \sigma, r \circ {}^L\rho, \psi).$$

It follows that $\gamma(s, \rho(\sigma), r, \psi) = \gamma(s, \sigma, r \circ {}^L\rho, \psi)$. From the formulation of the local Langlands conjecture for $\mathrm{GL}_n(F)$ (Theorem 1.1), the functorial transfer should be characterized by the conditions similar to conditions (1)–(5) in Theorem 1.1, in particular, by

$$L(s, \sigma \times \tau) = L(s, \rho(\sigma) \times \tau)$$

and

$$\epsilon(s, \sigma \times \tau, \psi) = \epsilon(s, \rho(\sigma) \times \tau, \psi),$$

for all irreducible supercuspidal representations τ of $\mathrm{GL}_n(F)$ for all integers $n \geq 1$. For the local γ-factors, one expects

$$\gamma(s, \sigma \times \tau, \psi) = \gamma(s, \rho(\sigma) \times \tau, \psi). \tag{1.2}$$

Of course, if one assumes the validity of the local Langlands functoriality from reductive groups to the general linear group, then one may use (1.2) to define the twisted local γ-factors in general. We refer [K95], [K00], [BK00] and [GK99] for some very interesting discussions in this aspect.

For a given $\pi \in \Pi(G/F)$, we have two collections of local γ-factors: one is

$$\{\gamma(s, \pi \times \tau, \psi) \mid \textit{for all } \tau \in \Pi(\mathrm{GL}_n/F), \textit{and for all } n = 1, 2, \ldots\}, \tag{1.3}$$

the twisted local γ-factors of π, and the other is

$$\{\gamma(s, \pi, r, \psi) \mid \textit{for all } r\}, \tag{1.4}$$

the local γ-factors attached to all finite-dimensional complex representations r of ${}^LG$. Although the exact definition of these collections of local γ-factors is still conjectural in general, it is clear that they are invariants attached to irreducible admissible representations π of $G(F)$ up to equivalence. The basic questions are the following.

1. How do the collections of local γ-factors classify the irreducible admissible representations? (the Local Converse Theorem)
2. How do the explicit analytic properties of the local γ-factors determine the functorial structures of the irreducible admissible representations? (a local version of Langlands problem)

We first recall some basic properties of local γ-factors, and then discuss these two basic problems in details, including some typical known examples in the following sections.

2 Basic Properties of Local γ-Factors

We recall briefly some basic properties of the local γ-factors. Among them are mainly the multiplicativity and stability of the local γ-factors.

2.1 Multiplicativity

For the twisted local γ-factors, one expects the multiplicativity holds. More precisely, it can be stated as follows. For an irreducible admissible representation π of $G(F)$, there is a supercuspidal datum $(M(F), \sigma)$ where $P = MN$ is a parabolic subgroup of G defined over F and σ is an irreducible supercuspidal representation of $M(F)$ such that π is isomorphic to an irreducible subrepresentation of the induced representation $\mathrm{Ind}_{P(F)}^{G(F)}(\sigma)$. Then one expects

$$\gamma(\pi \times \tau, s, \psi) = \gamma(\sigma \times \tau, s, \psi) \tag{2.1}$$

for all irreducible supercuspidal representations τ of $\mathrm{GL}_l(F)$ with $l = 1, 2, \ldots$. Further, if $M = \mathrm{GL}_r \times H$, then one may write $\sigma = \tau_r \otimes \sigma'$, and expect

$$\gamma(\sigma \times \tau, s, \psi) = \gamma(\tau_r \times \tau, s, \psi) \cdot \gamma(\sigma' \times \tau, s, \psi). \tag{2.2}$$

Properties (2.1) and (2.2) are called the *Multiplicativity* of the local γ-factors. F. Shahidi proved in [Sh90b] the multiplicativity for irreducible generic representations π of all F-quasisplit reductive algebraic groups $G(F)$ by using his theory on the local coefficients. For $G = \mathrm{GL}(n)$, it is proved by Soudry by the Rankin-Selberg method ([S00]). One may expect that the work ([GPSR97] and [FG99]) has implication in this aspect for irreducible admissible representations, which may not be generic.

2.2 Stability

Another significant property of twisted local γ-factors is the *Stability*, which can be stated as follows. For irreducible admissible representations π_1 and π_2 of $G(F)$, there exists a highly ramified character χ of $F^\times$ such that

$$\gamma(\pi_1 \times \chi, s, \psi) = \gamma(\pi_2 \times \chi, s, \psi).$$

It was proved by Jacquet-Shalika ([JS85]) for the group $\mathrm{GL}_m \times \mathrm{GL}_n$. For irreducible generic representations of classical groups, it was proved in [CPS98] and [CKPSS04]. For irreducible generic representations of general F-quasisplit groups, the approach is taken in [CPSS05]. For F-split classical groups of either symplectic or orthogonal type, the stability of local γ-factors has be proved for general irreducible admissible representations via the doubling method ([RS05]).

2.3 Remarks

It is important to mention that E. Lapid and S. Rallis ([LR05]) determines the sign of the local ϵ-factors via the doubling method. They introduce the ten properties, called the *Ten Commandments* of the local γ-factors, which determines the local γ-factors uniquely.

We would also like to mention the explicit calculations of the local γ-factors for irreducible supercuspidal representations via the local Rankin-Selberg method (see [jK00] for example).

3 Local Converse Theorems

The local converse theorem is to find the smallest subcollection of twisted local γ-factors $\gamma(s, \pi \times \tau, \psi)$ which classifies the irreducible admissible representation π up to equivalence. However, this is usually not the case in general. From the local Langlands conjecture, one may expect a certain subcollection of local γ-factors classifies the irreducible representation π up to L-packet. On the other hand, if the irreducible admissible representations under consideration have additional structures, then one may still expect that a certain subcollection of local γ-factors classifies the irreducible representation π up to equivalence.

3.1 The case of $\mathrm{GL}_n(F)$

Let π be an irreducible admissible representation of $\mathrm{GL}_n(F)$. Then there is a partition $n = \sum_{j=1}^{r} n_j$ $(n_j > 0)$ and an irreducible supercuspidal representation

$$\tau_1 \otimes \cdots \otimes \tau_r$$

of $\mathrm{GL}_{n_1}(F) \times \cdots \times \mathrm{GL}_{n_r}(F)$ such that the representation π can be realized as a subrepresentation of the (normalized) induced representation

$$I(\tau_1, \ldots, \tau_r) = \mathrm{Ind}_{P_{[n_1 \cdots n_r]}(F)}^{\mathrm{GL}_n(F)}(\tau_1 \otimes \cdots \otimes \tau_r). \tag{3.1}$$

By the multiplicativity of the local γ-factors ([Sh90b] and [S00]), we have

$$\gamma(s, \pi \times \tau, \psi) = \prod_{j=1}^{r} \gamma(s, \tau_j \times \tau, \psi) \tag{3.2}$$

for all irreducible admissible representations τ of $\mathrm{GL}_l(F)$ for all $l \geq 1$. It reduces the problem for the case of general irreducible admissible representations to the case when the irreducible admissible representations are supercuspidal. It should be remarked that even if the irreducible supercuspidal representations can be determined by the twisted local γ-factors up to equivalence, it is the best one can expect that in general the twisted local γ-factors determines the irreducible admissible representations up to the equivalence of supercuspidal data.

We first consider the case of irreducible supercuspidal representations of $\mathrm{GL}_n(F)$. The first local converse theorem (**LCT**) for $\mathrm{GL}_n(F)$ is proved by G. Henniart in [H93], which can be stated as follows.

Theorem 3.1 (LCT(n,n-1) [H93]). *Let π_1, π_2 be irreducible supercuspidal representations of $\mathrm{GL}_n(F)$ with the same central character. If the twisted local γ-factors are the same, i.e.*

$$\gamma(s, \pi_1 \times \tau, \psi) = \gamma(s, \pi_2 \times \tau, \psi)$$

for all irreducible supercuspidal representations τ of $\mathrm{GL}_l(F)$ with $l = 1, 2, \ldots, n-1$, then π_1 and π_2 are equivalent.

It follows that an irreducible supercuspidal representation π can be determined up to equivalence by the subcollection of twisted local γ-factors

$$\{\gamma(s, \pi \times \tau, \psi) \mid \textit{for all } \tau \textit{ as in the theorem}\}.$$

The remaining problem is to reduce the 'size' of the subcollection of twisted local γ-factors, that is, to prove **LCT(n,r)** for $r < n-1$. In this direction, we have

Theorem 3.2 (LCT(n,n-2) [C96], [CPS99]). *Let π_1, π_2 be irreducible supercuspidal representations of $\mathrm{GL}_n(F)$. If the twisted local γ-factors are the same, i.e.*

$$\gamma(s, \pi_1 \times \tau, \psi) = \gamma(s, \pi_2 \times \tau, \psi)$$

for all irreducible supercuspidal representations τ of $\mathrm{GL}_l(F)$ with $l = 1, 2, \ldots, n-2$, then π_1 and π_2 are equivalent.

This theorem is proved in [C96] by a purely local argument, and prove in [CPS99] as a consequence of the global converse theorem for automorphic forms. It is well known to expect

Conjecture 3.3 (Jacquet). *Let π_1, π_2 be irreducible supercuspidal representations of $\mathrm{GL}_n(F)$. If the twisted local γ-factors are the same, i.e.*

$$\gamma(s, \pi_1 \times \tau, \psi) = \gamma(s, \pi_2 \times \tau, \psi)$$

for all irreducible supercuspidal representations τ of $\mathrm{GL}_l(F)$ with $l = 1, 2, \ldots, [\frac{n}{2}]$, then π_1 and π_2 are equivalent.

There are not strong evidence to support this conjecture, which is known for $n = 2, 3, 4$ for example. On the other hand, one may expect an even stronger version of this conjecture from the conjectural global converse theorem in [CPS94]. In order to prove a better local converse theorem, it is expected to use the explicit construction of irreducible supercuspidal representations of $\mathrm{GL}_n(F)$ and reduce to the case over finite fields. On the other hand, it is also important to consider the local converse theorem for general reductive groups.

3.2 A conjectural LCT

For a general reductive algebraic group G defined over F, the collection of twisted local γ-factors $\gamma(s, \pi \times \tau, \psi)$ is expected to determine the irreducible supercuspidal representation π up to the local

L-packet. Note that all irreducible supercuspidal representations of $\mathrm{GL}_n(F)$ are generic, i.e. have nonzero Whittaker models. It is natural to consider the local converse theorem for irreducible generic supercuspidal representations of $G(F)$ in general.

We recall the notion of Whittaker models for F-quasisplit reductive algebraic group $G(F)$. Fix an F-Borel subgroup $B = TU$. Let $\Phi(G,T)$ be the root system with the positive roots Φ^+ determined by U and Δ be the set of the simple roots. Choose an F-split $\{X_\alpha\}$, where X_α is a basis vector in the one-dimensional F-root space of α. Then we have

$$U/[U,U] \cong \oplus_{\alpha\in\Delta} F \cdot X_\alpha. \tag{3.3}$$

Let ψ be a character of $U(F)$. Then ψ factorizes through the quotient $U(F)/[U(F),U(F)]$, which is isomorphic as abelian groups to $\oplus_{\alpha\in\Delta} F \cdot X_\alpha$. A character ψ of $U(F)$ is called *generic* if ψ is nontrivial at each of the simple root α, via the isomorphism above. By the Pontriagin duality, such characters of $U(F)$ is parametrized by r-tuples

$$\underline{a} = (a_1, \dots, a_r) \in (F^\times)^r,$$

where r is the F-rank of G, i.e. the number of simple roots in Δ. An irreducible admissible representation (π, V_π) of $G(F)$ is called generic or ψ-generic if the following space

$$\mathrm{Hom}_{U(F)}(V_\pi, \psi) \cong \mathrm{Hom}_{G(F)}(V_\pi, \mathrm{Ind}_{U(F)}^{G(F)}(\psi))$$

is nonzero. For any nonzero functional $\ell_\psi \in \mathrm{Hom}_{U(F)}(V_\pi, \psi)$, under the above isomorphism, there is $G(F)$-equivariant homomorphism

$$v \in V_\pi \mapsto W_v^\psi(g) = \ell_\psi(\pi(g)(v)).$$

The subspace $\{W_v^\psi(g) \mid v \in V_\pi\}$ is called the ψ-Whittaker model associated to π. By the uniqueness of local Whittaker models ([Shl74]), the functional ℓ_ψ is unique up to scalar multiple.

For $t \in T(F)$, we define $t \circ \psi(u) = \psi(t^{-1}ut)$. If ψ is generic, then $t \circ \psi$ is generic for all $t \in T(F)$. Also it is clear that

$$\mathrm{Ind}_{U(F)}^{G(F)}(\psi) \cong \mathrm{Ind}_{U(F)}^{G(F)}(t \circ \psi)$$

for all $t \in T(F)$. It follows that if an irreducible admissible representation π of $G(F)$ is ψ-generic, then π is also $t \circ \psi$-generic for all $t \in T(F)$. In other words, the genericity of irreducible admissible representations of $G(F)$ depends on the $T(F)$-orbit of the generic characters. It is an easy exercise to show ([K02]) that

Proposition 3.4. *The set of $T(F)$-orbits the generic characters of $U(F)$ is in one-to-one correspondence with $H^1(\Gamma_F, Z_G)$, where Γ_F is the absolute Galois group of F and Z_G is the center of G.*

For an irreducible admissible representation π of $G(F)$, we define

$$\mathcal{F}(\pi) = \{\psi \mid \pi \text{ is } \psi\text{-generic}\}, \tag{3.4}$$

and call it the set of generic characters attached to π. It is clear that the set $\mathcal{F}(\pi)$ is $T(F)$-stable, and the $T(F)$-orbits $\mathcal{F}(\pi)/T(F)$ determines the genericity of π. The following conjecture of Shahidi ([Sh90a]) is fundamental to the harmonic analysis of p-adic groups and representations.

Conjecture 3.5 (Shahidi [Sh90a]). *Every tempered local L-packet contains a generic member.*

In general some nontempered local L-packets may also contains generic members ([JS04]). We call a local L-packet with generic members a generic local L-packet. Shahidi's conjecture is known to be true for the case of $\mathrm{GL}_n(F)$ ([BZ77], [Z80]), for the case of $\mathrm{SL}_n(F)$ ([LS86]), for the case of $\mathrm{SO}_{2n+1}(F)$ ([JS04], [M98]), and for the case of $\mathrm{U}_{2,1}(F)$ ([GRS97]). More recently, it is proved to be true for cuspidal local L-packets of F-quasisplit groups ([DR05]). For reductive algebraic groups at archemidean local fields, it follows from [L89]. Some relevant discussions can be found in [MT02], and global applications can be found in [Ar89] and [KS99].

The author proposes the following refinement of the Shahidi conjecture.

Conjecture 3.6 (Refinement of Shahidi's Conjecture). *In a generic local L-packet $\Pi(\phi)$, for any generic members $\pi_1, \pi_2 \in \Pi(\phi)$, the sets $\mathcal{F}(\pi_1)$ and $\mathcal{F}(\pi_2)$ are disjoint and the union of $T(F)$-orbits* *of generic characters over the subset* $\Pi^g(\phi)$ *of* $\Pi(\phi)$ *consisting of all generic members in* $\Pi(\phi)$, *i.e.*

$$\cup_{\pi\in\Pi^g(\phi)}\mathcal{F}(\pi)/T(F)$$

is in one-to-one correspondence with $H^1(\Gamma_F, Z_G)$.

It is not difficult to check that Conjecture 3.6 holds if one knows the complete structure of local L-packets. This should be the case for the cases where Shahidi's conjecture holds. However, Conjecture 3.6 may be verified before one knows completely the structure of local l-packets. We will explain in the next section that Conjecture 3.6 holds for $G = \mathrm{SO}_{2n+1}$ by means of the local converse theorem.

Based on Conjecture 3.6, we formulate the following general version of the local converse theorem.

Conjecture 3.7 (LCT). *For any irreducible admissible generic representations* π_1 *and* π_2 *of* $G(F)$, *if the following two conditions hold*

1. *the intersection of* $\mathcal{F}(\pi_1)$ *and* $\mathcal{F}(\pi_2)$ *is not empty, and*
2. *the twisted local* γ*-factors are equal, i.e.*

$$\gamma(s, \pi_1 \times \tau, \psi) = \gamma(s, \pi_2 \times \tau, \psi)$$

for all irreducible supercuspidal representations τ *of* $\mathrm{GL}_l(F)$ *with* $l = 1, 2, \ldots, [\frac{r}{2}]$, *where* r *is* F*-rank of* G,

then $\pi_1 \cong \pi_2$.

We remark that any theorem of this nature should be called a local converse theorem. The number of twists up to the half of the F-rank of G is an imitation of Jacquet's conjecture for GL_n. We have no strong evidence about this claim. In the next section we discuss the author's joint work with David Soudry for SO_{2n+1}.

3.3 The case of $\mathrm{SO}_{2n+1}(F)$

We review briefly here the joint work with Soudry ([JS03]) on **LCT** for SO_{2n+1}.

Theorem 3.8 (LCT for SO_{2n+1} [JS03]). *Let π and π' be irreducible admissible generic representations of $\mathrm{SO}_{2n+1}(k)$. If the twisted local gamma factors $\gamma(\pi \times \tau, s, \psi)$ and $\gamma(\pi' \times \tau, s, \psi)$ are the same, i.e.*

$$\gamma(\pi \times \tau, s, \psi) = \gamma(\pi' \times \tau, s, \psi)$$

for all irreducible supercuspidal representations τ of $\mathrm{GL}_l(k)$ with $l = 1, 2, \ldots, 2n-1$, then the representations π and π' are equivalent.

We remark that this theorem was proved in [JS03] by using Theorem 2.1, from the work [H93]. It is clear that we can improve this theorem to **LCT(n,n-2)** by using the work of [C96] or [CPS99]. We did not do this in [JS03] because it was good enough for applications in that paper. We remark that the **LCT** for generic representations of $\mathrm{U}(2,1)$ and for $\mathrm{GSp}(4)$ was established by E. M. Baruch in [B95] and [B97].

In order to point out the essence of the local converse theorem for general reductive groups, we would like to recall some important applications of the local converse theorem for SO_{2n+1} to the theory of automorphic forms.

Let k be a number field and $\mathbb{A} = \mathbb{A}_k$ be the ring of adeles of k, First we obtain the injectivity of the weak Langlands functorial lifting established in [CKPSS01] (which is proved for example in [JS03] and [JS04] to strong Langlands functoriality).

Theorem 3.9 (Theorem 5.2 [JS03]). *Let $\Pi^{gca}(\mathrm{SO}_{2n+1}/\mathbb{A})$ be the set of all equivalence classes of irreducible generic cuspidal automorphic representations of $\mathrm{SO}_{2n+1}(\mathbb{A})$ and $\Pi^a(\mathrm{GL}_{2n}/\mathbb{A})$ be the set of all equivalence classes of irreducible automorphic representations of $\mathrm{GL}_{2n}(\mathbb{A})$. Then the Langlands functorial lifting from $\Pi^{gca}(\mathrm{SO}_{2n+1}/\mathbb{A})$ to $\Pi^a(\mathrm{GL}_{2n}/\mathbb{A})$ is an injective map.*

The second global application is to determine the generic cuspidal data for an irreducible cuspidal automorphic representations. It is a well-known theorem that every irreducible cuspidal automorphic representation of $\mathrm{GL}_n(\mathbb{A})$ is generic (i.e. having nonzero Whittaker-Fourier coefficients). This follows from the Whittaker-Fourier expansion of cuspidal automorphic forms of $\mathrm{GL}_n(\mathbb{A})$. In general, we consider a cuspidal datum (P, σ), where P is a parabolic subgroup of GL_n and σ is an irreducible cuspidal automorphic representation of $M(\mathbb{A})$ with $P = MN$ being the Levi decomposition. Jacquet and Shalika proved the following theorem.

Theorem 3.10 (Theorem 4.4 [JS81]). *Let $(P;\sigma)$ and $(Q;\tau)$ be two pairs of cuspidal data of $\mathrm{GL}(n)$. If the two induced representations $\mathrm{Ind}_{P(\mathbb{A})}^{\mathrm{GL}_n(\mathbb{A})}(\sigma)$ and $\mathrm{Ind}_{Q(\mathbb{A})}^{\mathrm{GL}_n(\mathbb{A})}(\tau)$ share the same irreducible unramified local constituents at almost all places, then the two pairs of cuspidal data are associate.*

By the Langlands functorial lifting from SO_{2n+1} to GL_{2n} and the local converse theorem for SO_{2n+1}, we prove in [JS05] an analogue of Jacquet-Shalika's Theorem for SO_{2n+1} with generic cuspidal data. For the trivial parabolic subgroup $P = \mathrm{SO}_{2n+1}$, this was proved in [JS03] (and also in [GRS01]).

Theorem 3.11 (Theorem 3.2 [JS05]). *Let $(P;\sigma)$ and $(Q;\tau)$ be two pairs of generic cuspidal data of $\mathrm{SO}_{2n+1}(\mathbb{A})$. If the two induced representations $\mathrm{Ind}_{P(\mathbb{A})}^{\mathrm{SO}_{2n+1}(\mathbb{A})}(\sigma)$ and $\mathrm{Ind}_{Q(\mathbb{A})}^{\mathrm{SO}_{2n+1}(\mathbb{A})}(\tau)$ share the same irreducible unramified local constituent at almost all places, then $(P;\sigma)$ and $(Q;\tau)$ are associate.*

It has the following consequences which are important to the understanding of structure of the discrete spectrum of $\mathrm{SO}_{2n+1}(\mathbb{A})$.

Theorem 3.12 (Corollary 3.3 [JS05]). *With notations as above, we have*

(1) Irreducible generic cuspidal automorphic representations π of the group $\mathrm{SO}_{2n+1}(\mathbb{A})$ cannot be a CAP with respect to a generic, proper, cuspidal datum (P, σ), i.e π cannot be nearly equivalent to any irreducible constituent of $\mathrm{Ind}_{P(\mathbb{A})}^{\mathrm{SO}_{2n+1}(\mathbb{A})}(\sigma)$.

(2) If two pairs of generic cuspidal data $(P;\sigma)$ and $(Q;\tau)$, are nearly associate, i.e. their local components are associate at almost all local places, then they are globally associate.

(3) The generic cuspidal datum $(P;\sigma)$ is an invariant for irreducible automorphic representations of $\mathrm{SO}_{2n+1}(\mathbb{A})$ up to near equivalence.

We refer to [JS05] for more detailed discussions on their relation with the Arthur conjecture.

4 Poles of Local γ-Factors

In his recent paper ([L04]), R. Langlands gave detailed discussion on a conjecture relating the order of the pole at $s = 1$ of automorphic L-functions to his functoriality principle. This conjecture can be stated as follows.

Let k be a number field and $\mathbb{A}$ be the ring of adeles of k. For any reductive algebraic group G defined over k, ${}^LG$ denotes its Langlands dual group ([B79]). Let π be an irreducible cuspidal automorphic representation of $G(\mathbb{A})$. Langlands defined automorphic L-functions $L(s, \pi, \rho)$ for all finite-dimensional complex representation ρ of ${}^LG$ ([B79]). It is a theorem of Langlands that the automorphic L-function $L(s, \pi, \rho)$ converges absolutely for the real part of s large. It is a basic conjecture of Langlands that every automorphic L-function $L(s, \pi, \rho)$ should have meromorphic continuation to the whole complex plane $\mathbb{C}$ and satisfy a functional equation relating s to $1-s$. This basic conjecture has in fact been verified in many cases through the spectral theory of automorphic forms. See [Bmp04] and [GS88] for some detailed account on this aspect.

Problem 4.1 (Langlands [L04]). *For a given irreducible cuspidal automorphic representation π of $G(\mathbb{A})$, there exists an algebraic subgroup $\mathcal{H}_\pi$ of ${}^LG$ such that for every finite-dimensional complex representation ρ of ${}^LG$, the order of the pole at $s = 1$, denoted by $m_\pi(\rho)$, of the automorphic L-function $L(s, \pi, \rho)$ is equal to the multiplicity, denoted by $m_{\mathcal{H}_\pi}(\rho)$, of the trivial representation of $\mathcal{H}_\pi$ occurring in the representation ρ of ${}^LG$ when restricted to the subgroup $\mathcal{H}_\pi$. That is, the following identity*

$$m_\pi(\rho) = m_{\mathcal{H}_\pi}(\rho) \tag{4.1}$$

holds for all finite-dimensional complex representations ρ of ${}^LG$.

In [L04], Langlands discussed in length some relations of this conjecture to many basic problems in arithmetic and number theory and suggested a trace formula approach to certain important cases. Some detailed discussions for a special case, along the main idea of [L04] has been carried out by A. Venkatesh in his thesis [V04]. For $G = \mathrm{SO}_{2n+1}$, the relation between the order of the pole at $s = 1$ of the automorphic L-functions attached to irreducible unitary generic cuspidal automorphic representation π and the fundamental representations of the complex dual group $\mathrm{Sp}_{2n}(\mathbb{C})$ and the endoscopy structure of π has been discussed in detail in [J05], along the line of the Langlands Problem.

The main point here is to develop the local theory for the Langlands problem and the local analogy of [J05]. For simplicity of discussions below, it is assumed that all reductive algebraic groups considered below are F-split, and the Langlands dual group ${}^LG$ is taken to be the complex dual group $G^\vee(\mathbb{C})$, without action of the absolute Galois group.

Let π be an irreducible supercuspidal representation of $G(F)$ and ρ be a finite-dimensional complex representation of the complex dual group $G^\vee(\mathbb{C})$. One may define the local γ-factor attached to (π, ρ, ψ) (for a fixed nontrivial additive character of F) to be

$$\gamma(s, \pi, \rho, \psi) := \epsilon(s, \pi, \rho, \psi) \times \frac{L(1-s, \pi^\vee, \rho^\vee)}{L(s, \pi, \rho)}, \tag{4.2}$$

where $\epsilon(s, \pi, \rho, \psi)$ is the local ϵ-factor attached to (π, ρ, ψ). This definition is based on the assumption of the local Langlands conjecture for $G(F)$.

Problem 4.2 (Local Version of the Langlands Problem). *Let G be a reductive algebraic group defined over F. For an irreducible unitary supercuspidal representation π of $G(F)$, there exists an algebraic subgroup $\mathcal{H}_\pi$ of $G^\vee(\mathbb{C})$ such that if ρ is a finite-dimensional complex representation of $G^\vee(\mathbb{C})$, then the multiplicity $m_{\mathcal{H}_\pi}(\rho)$ of the trivial representation of $\mathcal{H}_\pi$ occurring in the restriction of ρ to $\mathcal{H}_\pi$ is the order $m_\pi(\rho)$ of the pole of the local γ-factor $\gamma(s, \pi, \rho, \psi)$ at $s = 1$, i.e. the following identity*

$$m_\pi(\rho) = m_{\mathcal{H}_\pi}(\rho)$$

holds for all finite-dimensional complex representations of $G^\vee(\mathbb{C})$.

We remark that the key point here is to define the group $\mathcal{H}_\pi$ for a given π and to study the relation between the structure of $\mathcal{H}_\pi$ and the endoscopy structure of π. We refer to [J05] for discussion of $\mathcal{H}_\pi$ in terms of the *observable groups* in the classical invariant theory. In the following we discuss the local analogy of [J05].

4.1 The case of $G = \mathrm{SO}_{2n+1}$

Let $G = \mathrm{SO}_{2n+1}$ be the F-split odd special orthogonal group. Then its complex dual group is $\mathrm{Sp}_{2n}(\mathbb{C})$. The fundamental representations of $\mathrm{Sp}_{2n}(\mathbb{C})$ are the finite-dimensional irreducible complex representations associated to the fundamental weights. They can be constructed by the following split exact sequence

$$0 \to V_{\rho_a}^{(2n)} \to \Lambda^a(\mathbb{C}^{2n}) \to \Lambda^{a-2}(\mathbb{C}^{2n}) \to 0, \tag{4.3}$$

where $\Lambda^a(\mathbb{C}^{2n})$ denotes the a-th exterior power of $\mathbb{C}^{2n}$, the contraction map from $\Lambda^a(\mathbb{C}^{2n})$ onto $\Lambda^{a-2}(\mathbb{C}^{2n})$ is as defined in Page 236, [GW98], and its kernel is denoted by $V_{\rho_a}^{(2n)}$. By Theorem 5.1.8 in [GW98], $V_{\rho_a}^{(2n)}$ is the space of the irreducible representation ρ_a of $\mathrm{Sp}_{2n}(\mathbb{C})$ with the a-th fundamental weight. Let ι be the natural embedding of $\mathrm{Sp}_{2n}(\mathbb{C})$ into $\mathrm{GL}_{2n}(\mathbb{C})$. Let Λ^2 be the exterior square representation of $\mathrm{GL}_{2n}(\mathbb{C})$ on the vector space $\Lambda^2(\mathbb{C}^{2n})$, which has dimension $2n^2 - n$. The composition $\Lambda^2 \circ \iota$ of Λ^2 with ι is a complex representation of $\mathrm{Sp}_{2n}(\mathbb{C})$. By (2.1) and by complete reducibility of representations of $\mathrm{Sp}_{2n}(\mathbb{C})$, we obtain

$$\Lambda^2 \circ \iota = \rho_2 \oplus \mathbf{1}_{\mathrm{Sp}_{2n}} \tag{4.4}$$

where ρ_2 is the second fundamental complex representation of the group $\mathrm{Sp}_{2n}(\mathbb{C})$, which is irreducible and has dimension $2n^2 - n - 1$, and $\mathbf{1}_{\mathrm{Sp}_{2n}}$ is the trivial representation of $\mathrm{Sp}_{2n}(\mathbb{C})$.

Let $\tau = \tau(\pi)$ be the image of π under the Langlands functorial transfer from $\mathrm{SO}_{2n+1}(F)$ to $\mathrm{GL}_{2n}(F)$ for irreducible admissible generic representations, which was established in [JS03] and [JS04]. One may expert the following identities hold,

$$L(s, \pi, \rho_a) = \frac{L(s, \tau(\pi), \Lambda^a)}{L(s, \tau(\pi), \Lambda^{a-2})} \tag{4.5}$$

and

$$\gamma(s, \pi, \rho_a, \psi) = \frac{\gamma(s, \tau(\pi), \Lambda^a, \psi)}{\gamma(s, \tau(\pi), \Lambda^{a-2}, \psi)}. \tag{4.6}$$

For the second exterior power representation ρ_2, the above identities have been verified by G. Henniart ([H03]). The following theorem relates the endoscopy structure of σ to the order of pole at $s = 1$ of the second fundamental local γ-factors.

Theorem 4.3. *Let π be an irreducible generic supercuspidal representation of $\mathrm{SO}_{2n+1}(F)$ and ρ_2 be the second fundamental complex representation of $\mathrm{Sp}_{2n}(\mathbb{C})$. Then the second fundamental local γ-factors $\gamma(s, \pi, \rho_2, \psi)$ are meromorphic functions over $\mathbb{C}$ and have the following properties.*

1. *The second fundamental local γ-factor $\gamma(s, \pi, \rho_2, \psi)$ has a pole of order $r - 1$ at $s = 1$ if and only if there exists a partition $n = \sum_{j=1}^{r} n_j$ with $n_j > 0$ such that π is a Langlands functorial lifting from an irreducible generic supercuspidal representation $\pi_1 \otimes \cdots \otimes \pi_r$ of*

$$\mathrm{SO}_{2n_1+1}(F) \times \cdots \times \mathrm{SO}_{2n_r+1}(F).$$

2. *The partition $[n_1 \cdots n_r]$ is uniquely determined by the irreducible generic supercuspidal representation π. More precisely, the set of positive integers*

$$\{n_1, n_2, \ldots, n_r\}$$

consists of all positive integers m such that there exists an irreducible supercuspidal representation τ of $\mathrm{GL}_m(F)$ such that the tensor product local γ-factor $\gamma(s, \pi \times \tau, \psi)$ has a pole at $s = 1$.

3. *The set $\{\pi_1, \pi_2, \ldots, \pi_r\}$ of irreducible generic supercuspidal representations of $\mathrm{SO}_{2n_i+1}(F)$ is completely determined by the irreducible generic supercuspidal representation π, up to equivalence, namely, it is the set of irreducible generic supercuspidal*

representations π' (up to equivalence) of $\mathrm{SO}_{2l+1}(F)$ *such that the tensor product local γ-factor*

$$\gamma(s, \pi \times \tau(\pi'), \psi)$$

has a pole at $s = 1$*, where* $\tau(\pi')$ *is the local Langlands transfer of* π' *to* $\mathrm{GL}_{2l}(F)$ *and is irreducible and supercuspidal.*

4. *We have* $m_\pi(\rho_2) = m_{H^\vee_{[n_1 \cdots n_r]}}(\rho_2)$*, where*

$$H^\vee_{[n_1 \cdots n_r]} = \mathrm{Sp}_{2n_1}(\mathbb{C}) \times \cdots \times \mathrm{Sp}_{2n_r}(\mathbb{C}).$$

The proof of this theorem follows essentially from [JS03] and [JS04], and the arguments in [J05].

The relation between the theorem and the endoscopy can be briefly discussed as below. The theory of twisted endoscopy can be found in [KS99]. For simplicity, we first recall from [Ar04] and [Ar05] the basic structure of all standard elliptic endoscopy groups of SO_{2n+1}. Let $n = n_1 + n_2$ with $n_1, n_2 > 0$. Take a semisimple element

$$s_{n_1,n_2} = \begin{pmatrix} -I_{n_1} & & 0 \\ & I_{2n_2} & \\ 0 & & -I_{n_1} \end{pmatrix} \in \mathrm{Sp}_{2n}(\mathbb{C}).$$

Then the centralizer of s_{n_1,n_2} in $\mathrm{Sp}_{2n}(\mathbb{C})$ is given by

$$H^\vee_{[n_1,n_2]} = \mathrm{Cent}_{\mathrm{Sp}_{2n}(\mathbb{C})}(s_{n_1,n_2}) = \mathrm{Sp}_{2n_1}(\mathbb{C}) \times \mathrm{Sp}_{2n_2}(\mathbb{C}).$$

The standard elliptic endoscopy group associated to the partition $n = n_1 + n_2$ is

$$H_{[n_1,n_2]} = \mathrm{SO}_{2n_1+1} \times \mathrm{SO}_{2n_2+1},$$

and the groups $H_{[n_1,n_2]}$ exhaust all standard elliptic endoscopy groups of SO_{2n+1}, in the sense of [KS99].

In general, an endoscopy transfer of admissible representations from an endoscopy group H of G to G takes a local Arthur packet of admissible representations of $H(F)$ to a local Arthur packet of admissible representations of $G(F)$, which is characterized by the stability of certain distributions from the geometric side of the Arthur trace formula ([Ar05]). Since the admissible representations of $H(\mathbb{A})$ and $G(\mathbb{A})$ considered in this paper are generic and tempered, following the Arthur conjecture on the structure of the local Arthur packets, the admissible representations we are considering in this paper should be the distinguished representatives of the corresponding local Arthur packets. This must also take the Shahidi conjecture on the genericity of tempered local L-packets into account, which is the case in Arthur's formulation of his conjecture. Then by the relation between the local Arthur packets (A-packets) and the local Langlands packets (L-packets), the endoscopy transfer should take the distinguished member of a local Arthur packet to its image of the Langlands functorial lifting from H to G. In other words, the endoscopy transfer from H to G for the distinguished members of local Arthur packets should be the same as the Langlands functorial lifting from H to G. In the above theorem, the Langlands functorial lifting can be viewed as the local endoscopy transfer from $H_{[n_1,\dots,n_r]}$ to SO_{2n+1}.

As in the global case considered in [J05], we can discuss the order of the pole at $s = 1$ of the higher fundamental local γ-factors explicitly and its precise relation with the structure of the endoscopy group $H_{[n_1,\dots,n_r]}$ to SO_{2n+1}, i.e. complete determination of the set

$$\{n_1, n_2, \dots, n_r\}.$$

Since the argument for the local case is about the same as that for the global case, we omit the details here.

4.2 Other classical groups

We also remark that there exist analogy of the above discussions for other classical groups. The functoriality from the classical groups to the general linear groups for generic representations are known through more recent work of [AS], [CKPSS04], [KK05], and [S05]. On the other hand, the analogy of the local theory for SO_{2n+1} in [JS03] and [JS04] is still work in progress.

References

[Ar89] Arthur, J. *Unipotent automorphic representations: conjectures.* Astérisque, 171-172, 1989, 13–71.

[Ar04] Arthur, J., *Automorphic representations of* GSp(4). Contributions to automorphic forms, geometry, and number theory, 65–81, Johns Hopkins Univ. Press, 2004.

[Ar05] Arthur, J., *An introduction to the trace formula.* preprint (2005).

[AS] Asgari, M.; Shahidi, F., *Generic transfer for general spin groups.* preprint 2004

[B95] Baruch, E. M., *Local factors attached to representations of p-adic groups and strong multiplicity one.* Yale Thesis 1995.

[B97] Baruch, E. M., *On the gamma factors attached to representations of* $\mathrm{U}(2,1)$ *over a p-adic field.* Israel J. Math. 102 (1997), 317–345.

[BZ77] Bernstein, J.; Zelevinsky, A. *Induced representations of reductive p-adic groups. I.* Ann. Sci. École Norm. Sup. (4) 10 (1977), no. 4, 441–472.

[B79] Borel, A. *Automorphic L-functions.* Proc. Sympos. Pure Math., 33, Automorphic forms, representations and L-functions Part 2, pp. 27–61, Amer. Math. Soc., 1979.

[BK00] Braverman, A.; Kazhdan, D. *γ-functions of representations and lifting.* Visions in Mathematics, GAFA special volume, Vol. 1, 2000.

[Bmp04] Bump, D. *On the Rankin-Selberg Method.* preprint (2004)

[C96] Chen, J.-P. *Local factors, central characters, and representations of the general linear group over non-archimedean local fields.* PhD Thesis, Yale University May 1996.

[CPS94] Cogdell, J.; Piatetski-Shapiro, I. *Converse theorems for* GL_n. Inst. Hautes Études Sci. Publ. Math. No. 79 (1994), 157–214.

[CPS98] Cogdell, J.; Piatetski-Shapiro, I. *Stability of gamma factors for* $\mathrm{SO}(2n+1)$. Manuscripta Math. 95(1998), 437–461.

[CPS99] Cogdell, J.; Piatetski-Shapiro, I. *Converse theorems for* GL_n. *II.* J. Reine Angew. Math. 507 (1999), 165–188.

[CPS04] Cogdell, J.; Piatetski-Shapiro, I., *Remarks on Rankin-Selberg convolutions.* Contributions to automorphic forms, geometry, and number theory, 255–278, Johns Hopkins Univ. Press, 2004.

[CPSS05] Cogdell, J.; Piatetski-Shapiro, I.; Shahidi, F. *On stability of local* γ*-factors.* Automorphic Representations, L-Functions and Applications: Progress and Prospects. Ohio State University Mathematical Research Institute Publications 11. (OSU 11) 2005.

[CKPSS01] Cogdell, J.; Kim, H.; Piatetski-Shapiro, I.; Shahidi, F. *On lifting from classical groups to* $GL(n)$. Publ. Math. Inst. Hautes Études Sci. No. 93(2001), 5–30.

[CKPSS04] Cogdell, J.; Kim, H.; Piatetski-Shapiro, I.; Shahidi, F., *Functoriality for the classical groups.* Publ. Math. Inst. Hautes Études Sci. No. 99 (2004), 163–233.

[DR05] DeBacker, S.; Reeder, M. *Depth-zero supercuspidal L-packets and their stability.* preprint 2005.

[FG99] Friedberg, S.; Goldberg, D. *On local coefficients for non-generic representations of some classical groups.* Compositio Math. 116 (1999), no. 2, 133–166.

[GRS97] Gelbart, S.; Rogawski, J.; Soudry, D. *Endoscopy, theta liftings, and period integrals for unitary group in three variables.* Ann. of Math., 145(1997)(3), 419–476.

[GS88] Gelbart, S.; Shahidi, F., *Analytic properties of automorphic L-functions.* Perspectives in Mathematics, 6. Academic Press, Inc., Boston, MA, 1988.

[GK99] Gelfand, S.; Kazhdan, D. *Conjectural algebraic formulas for representations of* $\mathrm{GL}(n)$. Asian J. Math. 3(1999)(1), 17–48.

[GPSR97] Ginzburg, D.; Piatetski-Shapiro, I.; Rallis, S. *L functions for the orthogonal group.* Mem. Amer. Math. Soc. 128 (1997), no. 611.

[GRS01] Ginzburg, D.; Rallis, S.; Soudry, D. *Generic automorphic forms on* $\mathrm{SO}(2n+1)$*: functorial lift to* $\mathrm{GL}(2n)$*, endoscopy, and base change.* Internat. Math. Res. Notices 2001, no. 14, 729–764.

[GW98] Goodman, R.; Wallach, N. *Representations and Invariants of the Classical Groups.* Cambridge University Press, 1998.

[G97] Grosshans, F. *Algebraic homogeneous spaces and invariant theory.* Lecture Notes in Mathematics 1673, 1997, Springer.

[HT01] Harris, M.; Taylor, R. *The geometry and cohomology of some simple Shimura varieties.* Annals of Math. Studies, 151, 2001, Princeton Unversity Press.

[H93] Henniart, G. *Caractérisation de la correspondance de Langlands locale par les facteurs* ϵ *de paires.* (French) Invent. Math. 113 (1993), no. 2, 339–350.

[H00] Henniart, G. *Une preuve simple des conjectures de Langlands pour* $\mathrm{GL}(n)$ *sur un corps* p*-adique.* (French) Invent. Math. 139 (2000), no. 2, 439–455.

[H03] Henniart, G. *Correspondance de Langlands et fonctions L des carrés extérieur et symétrique.* preprint 2003.

[JPSS83] Jacquet, H.; Piatetskii-Shapiro, I.; Shalika, J. *Rankin-Selberg convolutions.* Amer. J. Math. 105 (1983), no. 2, 367–464.

[JS81] Jacquet, H.; Shalika, J. *On Euler products and the classification of automorphic forms. II.* Amer. J. Math. 103 (1981), no. 4, 777–815.

[JS85] Jacquet, H.; Shalika, J. *A lemma on highly ramified ϵ-factors.* Math. Ann. 271(1985), 319–332.

[J05] Jiang, D. *On the fundamental automorphic L-functions for* SO_{2n+1}. preprint (2005).

[JS03] Jiang, D.; Soudry, D., *The local converse theorem for* $\mathrm{SO}(2n+1)$ *and applications.* Ann. of Math. 157(2003), 743–806.

[JS04] Jiang, D.; Soudry, D., *Generic Representations and the Local Langlands Reciprocity Law for p-adic* $\mathrm{SO}(2n+1)$. Contributions to automorphic forms, geometry, and number theory, 457–519, Johns Hopkins Univ. Press, Baltimore, MD, 2004.

[JS05] Jiang, D.; Soudry, D., *On the genericity of cuspidal automorphic forms of* SO_{2n+1}. preprint 2005.

[K95] Kazhdan, D. *Forms of the principal series for* $\mathrm{GL}(n)$. Functional Analysis on the Eve of the 21st Century, Vol. 1, 153–171, Progr. Math. 131, Birkhauser, 1995.

[K00] Kazhdan, D. *An algebraic integration.* Mathemtaics: frontiers and perspectives, Amer. Math. Soc. 2000.

[KV05] Kazhdan, D.; Varshavsky, V. *On endoscopic decomposition of certain depth zero representations.* preprint (2005).

[hK00] Kim, H. *Langlands-Shahidi method and poles of automorphic L-functions. II.* Israel J. Math. 117 (2000), 261–284.

[KK05] Kim, H.; Krishnamurthy, M., *Stable base change lift from unitary groups to* GL_N. Intern. Math. Research Papers, No. 1, 2005.

[KS04] Kim, H.; Shahidi, F., *On simplicity of poles of automorphic L-functions.* J. Ramanujan Math. Soc. 19(2004) 267-280.

[jK00] Kim, J. *Gamma factors of certain supercuspidal representations.* Math. Ann., 317(2000)(4), 751–781.

[KS99] Kottwitz, R.; Shelstad, D., *Foundations of twisted endoscopy.* Astérisque 255, 1999.

[K02] Kuo, W. *Principal nilpotent orbits and reducible principal series.* Represent. Theory 6 (2002), 127–159 (electronic).

[LS86] Labesse, J.-P.; Schwermer, J., *On liftings and cusp cohomology of arithmetic groups.* Invent. Math. 83(1986), no. 2, 383–401.

[L89] Langlands, R. *On the classification of irreducible representations of real algebraic groups.* Representation Theory and Harmonic Analysis on Semisimple Lie Groups, Amer. Math. Soc., 1989, 101–170.

[L04] Langlands, R. *Beyond Endoscopy.* Contributions to automorphic forms, geometry, and number theory, 611–697, Johns Hopkins Univ. Press, 2004.

[LR05] Lapid, E.; Rallis, S. *On the local factors of representations of classical groups.* Automorphic Representations, L-Functions and Applications: Progress and Prospects. Ohio State University Mathematical Research Institute Publications 11. (OSU 11) 2005.

[LP90] Larsen, M.; Pink, R. *Determining representations from invariant dimensions.* Invent. Math. 102 (1990), no. 2, 377–398.

[MW89] Moeglin, C; Waldspurger, J.-L. *Le spectre résiduel de* GL(n). (French) Ann. Sci. École Norm. Sup. (4) 22 (1989), no. 4, 605–674.

[MW03] Moeglin, C; Waldspurger, J.-L. *Paquets stables de représentations tempérées et de réduction unipotente pour* SO($2n+1$). (French) Invent. Math. 152 (2003), no. 3, 461–623.

[MT02] Moeglin, C.; Tadic, M. *Construction of discrete series for classical p-adic groups.* J. of Amer. Math. Soc., 15(2002)(3), 715–786.

[M98] Muic, D. *On generic irreducible representations of* $\mathrm{Sp}(n, F)$ *and* $\mathrm{SO}(2n+1, F)$. Glasnik Math. 33(1998), 19–31.

[RS05] Rallis, S.; Soudry, D. *Stability of the local gamma factor arising from the doubling method.* preprint 2005.

[R05] Reeder, M. *Some supercuspidal L-packets of positive depth.* preprint 2005.

[Sh88] Shahidi, F. *On the Ramanujan conjecture and finiteness of poles for certain L-functions.* Ann. of Math. (2) 127 (1988), no. 3, 547–584.

[Sh90a] Shahidi, F. *A proof of Langlands' conjecture on Plancherel measures; complementary series for p-adic groups.* Ann. of Math. (2) 132 (1990), no. 2, 273–330.

[Sh90b] Shahidi, F. *On multiplicativity of local factors.* Israel Math. Conf. Proc., 3, Weizmann, Jerusalem, 1990, 279–289.

[Sh02] Shahidi, F. *Local coefficients as Mellin transforms of Bessel functions; towards a general stability.* IMRN. 39(2002), 2075–2119.

[Shl74] Shalika, J. *The multiplicity one theorem for* GL_n. Ann. of Math. (2) 100 (1974), 171–193.

[S00] Soudry, D. *Full multiplicativity of gamma factors for* $\mathrm{SO}_{2l+1} \times \mathrm{GL}_n$. Proceedings of the Conference on p-adic Aspects of the Theory of Automorphic Representations (Jerusalem, 1998). Israel J. Math. 120 (2000), part B, 511–561.

[S05] Soudry, D., *On Langlands functoriality from classical groups to* GL_n. Automorphic forms. I. Astrisque No. 298 (2005), 335–390.

[Td86] Tadic, M. *Classification of unitary representations in irreducible representations of general linear group (non-archimedead case).* Ann. Scient. Éc. Norm. Sup. 4(1986), 335–382.

[Tk97] Takano, K. *On standard L-functions for unitary groups.* Proc. Japan Acad. Ser. A, Math. Sci., 73(1997)(1), 5–9.

[T79] Tate, J., *Number theoretic background.* Proc. Sympos. Pure Math., 33, Automorphic forms, representations and L-functions Part 2, pp. 3–26, Amer. Math. Soc., 1979.

[V04] Venkatesh, A. *"Beyond endoscopy" and special forms on GL(2).* J. Reine Angew. Math. 577 (2004), 23–80.

[V93] Vogan, D. *The local Langlands conjecture.* Representation theory of groups and algebras, 305–379, Contemp. Math., 145, Amer. Math. Soc., Providence, RI, 1993.

[Z80] Zelevinsky, A. *Induced representations of reductive p-adic groups. II. On irreducible representations of* $\mathrm{GL}(n)$. Ann. Sci. École Norm. Sup. (4) 13 (1980), no. 2, 165–210.

Dihua JIANG
School of Mathematics,
University of Minnesota,
Minneapolis, MN 55455, USA
Email: dhjiang@math.umn.edu

Deligne Pairings over Moduli Spaces of Punctured Riemann Surfaces

K. OBITSU, W.-K. TO and L. WENG

1. WP Metrics and TZ Metrics

(1.1) Teichmüller Spaces and Moduli Spaces For $g \geq 0$ and $N > 0$, we denote by $T_{g,N}$ the Teichmüller space of Riemann surfaces of type (g, N). Each point of $T_{g,N}$ is a Riemann surface M^0 of type (g, N), i.e., $M^0 = M \backslash \{P_1, \ldots, P_N\}$, where M is a compact Riemann surface of genus g, and the punctures $P_1, \ldots, P_N$ of M^0 are N distinct points in M. We will always assume that $2g - 2 + N > 0$. The Teichmüller space $T_{g,N}$ is naturally a complex manifold of dimension $3g - 3 + N$.

The moduli space $\mathcal{M}_{g,N}$ of Riemann surfaces of type (g, N) is obtained as the quotient of $T_{g,N}$ by the Teichmüller modular group $\mathrm{Mod}_{g,N}$, i.e., $\mathcal{M}_{g,N} \simeq T_{g,N}/\mathrm{Mod}_{g,N} := \{(M; P_1, P_2, \ldots, P_N) : M$ cpt Riemann surface of genus g, $P_i \in M$, $P_i \neq P_j$, $\}/\sim_{\text{iso.}}$ So $\mathcal{M}_{g,N}$ is naturally endowed with the structure of a complex V-manifold. However, $\mathcal{M}_{g,N}$ is not compact. The so-called Deligne-Mumford compactification $\overline{\mathcal{M}_{g,N}}$ is obtained by adding the so-called stable curves M. Like $\mathcal{M}_{g,N}$, $\overline{\mathcal{M}_{g,N}}$ admits a V-manifold structure. It is well known that $\Delta_{\text{bdy}} := \overline{\mathcal{M}_{g,N}} - \mathcal{M}_{g,N}$ is a normal crossing divisor.

The Riemann surfaces on the boundary may be understood as follows. Denote by $\delta_{\gamma_1,\ldots,\gamma_m} T_{g,N}$ the boundary Teichmüller space of $T_{g,N}$ arising from pinching m distinct points. Take a point $M_0 \in \delta_{\gamma_1,\ldots,\gamma_m} T_{g,N}$. Then M_0 is a Riemann surface with N punctures $P_1, \ldots, P_N$ and m nodes $Q_1, \ldots, Q_m$, and $M_0^o := M_0 \backslash \{Q_1, \ldots, Q_m\}$ is a non-singular Riemann surface with $N+2m$ punctures. Each node Q_i corresponds to two punctures on M_0^o (other than $P_1, \ldots, P_N$). Denote the components of M_0^o by C_α, $\alpha = 1, 2, \ldots, r$. Each C_α is a Riemann surface of genus g_α and with n_α punctures, i.e., C_α is of type (g_α, n_α). As such, the stable condition is equivalent to that $2g_\alpha - 2 + n_\alpha > 0$ for each α, that is to say, each C_α also admits the complete hyperbolic metric of constant sectional curvature -1. It is easy to see that $\sum_{\alpha=1}^r (3g_\alpha - 3 + n_\alpha) + m = 3g - 3 + N$. With respect to the disjoint union $M_0^o = \cup_{\alpha=1}^r C_\alpha$, one easily sees that $\delta_{\gamma_1,\ldots,\gamma_m} T_{g,N}$ is a product of lower dimensional Teichmüller spaces given by

$$\delta_{\gamma_1,\ldots,\gamma_m} T_{g,N} = T_{g_1,n_1} \times T_{g_2,n_2} \times \cdots \times T_{g_r,n_r}$$

with each $C_\alpha \in T_{g_\alpha,n_\alpha}$, $\alpha = 1, 2, \ldots, r$.

(1.2) Weil-Petersson Metric and Takhtajan-Zograf Metric For any $M^0 \in T_{g,N}$, M^0 admits the complete hyperbolic metric of constant sectional curvature -1. By the uniformization theorem, M^0 can be represented as a quotient $\mathcal{H}\backslash\Gamma$ of the upper half plane $\mathcal{H} := \{z \in \mathbb{C} : \operatorname{Im} z > 0\}$ by the natural action of Fuchsian group $\Gamma \subset \mathrm{PSL}(2,\mathbb{R})$ of the first kind. Γ is generated by $2g$ hyperbolic transformations $A_1, B_1, \ldots, A_g, B_g$ and N parabolic transformations $S_1, \ldots, S_N$ satisfying the relation

$$A_1 B_1 A_1^{-1} B_1^{-1} \cdots A_g B_g a_g^{-1} B_g^{-1} \cdot S_1 S_2 \cdots S_N = \mathrm{Id}.$$

Let $z_1, \ldots, z_N \in \mathbb{R} \cup \{\infty\}$ be the fixed points of the parabolic transformations $S_1, \ldots, S_N$ respectively, which are also called cusps, such that they correspond to the punctures $P_1, \ldots, P_N$ of M under the projection $\mathcal{H}^* := \mathcal{H} \cup \{z_1, \ldots, z_N\} \to \mathcal{H}^*\backslash\Gamma \simeq M$ accordingly. For each $i = 1, 2, \ldots, N$, it is well known that S_i generates an infinite cyclic subgroup of Γ, and we can select $\sigma_i \in \mathrm{PSL}(2,\mathbb{R})$ so that $\sigma_i(\infty) = z_i$ and $\sigma_i^{-1} P_i \sigma_i$ is the transformation $z \mapsto z + 1$ on $\mathcal{H}$. For $s \in \mathbb{C}$ with $\operatorname{Re} s > 1$, the Eisenstein series $E_i(z,s)$ attached to the cusp z_i is given by

$$E_i(z,s) := \sum_{\gamma \in <S_i>\backslash\Gamma} \operatorname{Im}(\sigma_i^{-1}\gamma z)^s, \qquad z \in \mathcal{H}.$$

It is uniformly convergent on compact subsets of $\mathcal{H}$, and invariant under Γ. Thus $E_i(z,s)$ descends to a function on M.

To describe the tangent and cotangent spaces at a point M of $T_{g,N}$, we first denote by $Q(M)$ the space of holomorphic quadratic differentials $\phi = \phi(z)\,dz^2$ on M with finite L^1 norm, i.e., $\int_M |\phi| < \infty$. Also, we denote by $B(M)$ the space of L^∞ measurable Beltrami differentials $\mu = \mu(z)\,d\bar{z}/dz$ on M (i.e., $\|\mu\|_\infty := \text{ess.sup}_{z\in M}|\mu(z)| < \infty$). Let $HB(M)$ be the subspace of $B(M)$ which can be represented as $\rho\bar{\phi}$ for some $\phi \in Q(M)$. Here $\rho = \rho(z)\,dz\,d\bar{z}$ denotes the hyperbolic metric on M. Elements of $HB(M)$ are called harmonic Beltrami differentials. There is a natural Kodaira-Serre pairing $\langle\, ,\, \rangle : B(M) \times Q(M) \to \mathbb{C}$ given by

$$\langle \mu, \phi \rangle = \int_M \mu(z)\phi(z)\,dzd\bar{z}, \qquad \text{where } \mu \in B(M),\ \phi \in Q(M).$$

Let $Q(M)^\perp \subset B(M)$ be the annihilator of $Q(M)$ under this pairing. Then one has the decomposition $B(M) = HB(M) \oplus Q(M)^\perp$, and natural isomorphisms $T_M T_{g,N} \simeq B(M)/Q(M)^\perp \simeq HB(M)$ and $T_M^* T_{g,N} \simeq Q(M)$ with the duality between $T_M T_{g,N}$ and $T_M^* T_{g,N}$ given by the pairing $\langle\cdot,\cdot\rangle$ above.

The *Weil-Petersson metric* g^{WP} and the *Takhtajan-Zograf metric* g^{TZ} on $T_{g,N}$ (the latter being introduced in [TZ1,2]) are defined as follows: for $X \in T_{g,N}$ and μ, $\nu \in HB(M)$, one has

$$g^{\mathrm{WP}}(\mu,\nu) = \int_M \mu\bar{\nu}\rho, \qquad \text{and} \qquad g^{\mathrm{TZ}}(\mu,\nu) = \int_M \sum_{i=1}^N E_i(\cdot,2)\cdot\mu\bar{\nu}\rho.$$

In particular, $g^{\mathrm{TZ}}(\mu,\nu) = \sum_{i=1}^N g^{(i)}(\mu,\nu)$ with $g^{(i)}(\mu,\nu) = \int_M E_i(\cdot,2)\cdot \mu\bar{\nu}\rho$. We will call $g^{(i)}$ the Takhtajan-Zograf metric on $T_{g,N}$ associated to the cusp z_i (or the puncture p_i). It is well known that the Weil-Petersson metrics g^{WP} is Kählerian, non-complete ([Wo1]) and whose holomorphic sectional curvature is bounded from above by $-\frac{1}{\pi(2g-2+N)}$ ([Wo2] for $N = 0$ and [We1] for $N > 1$). Moreover, we know that the Takhtajan-Zograf metric is also Kählerian ([TZ1,2]) and non-complete ([Ob1,2]).

The metrics g^{WP} and g^{TZ} (but not each individual $g^{(i)}$ unless $N = 1$) are invariant under $\mathrm{Mod}_{g,N}$ and thus they descend to Kähler metrics on (the smooth points of) $\mathcal{M}_{g,N}$.

2. Line Bundles over Moduli Spaces

(2.1) Deligne Pairing Deligne pairing, a refined version of intersection, plays a key role in understanding the Weil-Petersson and Takhtajan-Zograf metrics. To start with, we use a simple example to explain the essential point of such pairings.

Example. Let C_1 and C_2 be two prime divisors, i.e., curves, on a surface S. Assume that they intersect transversally at three points P_1, P_2 and P_3. Then $C_1 \cdot C_2 = \# C_1 \cap C_2 = \#\{P_1, P_2, P_3\} = 3$. Moreover, for any meromorphic function f on S, $\Big(C_1 + \mathrm{div}(f)\Big) \cdot C_2 = 3 = C_1 \cdot C_2$, where $\mathrm{div}(f)$ is defined as the zeros minus the poles. Thus if set $\{C_1 + \mathrm{div}(f) : \forall f\} =: \mathcal{O}_S(C_1)$, then $\mathcal{O}_S(C_1) \cdot \mathcal{O}_S(C_2) = 3$ is well-defined.

One may try to refine the intersection with $C_1 \cdot C_2 :\overset{\text{refined}}{=} P_1 + P_2 + P_3$. But for our purpose, consider a relative picture $\pi : S \to B$ for a fibration π over the curve B. Assume then that C_1 and C_2 are horizontal, i.e., $\pi(C_i) = B$ and denote the images of P_1, P_2 and P_3 by Q_1, Q_2 and Q_3 respectively. (Assume that $Q_i \neq Q_j$.) Then, viewing from B, we get $(C_1 \cdot C_2)_B = Q_1 + Q_2 + Q_3$, or better

$$\begin{cases} \langle \mathcal{O}_S(C_1), \mathcal{O}_S(C_1) \rangle_B & = \mathcal{O}_B(Q_1 + Q_2 + Q_3), \\ \qquad \langle \mathbf{1}_{C_1}, \mathbf{1}_{C_2} \rangle_B & = \mathbf{1}_{Q_1+Q_2+Q_3}, \end{cases} \tag{$*$}$$

where $\mathbf{1}$ denotes the defining section.

In summary, if the relative dimension is 1, then for any two line bundles L_1 and L_2 (together with two sections s_1, s_2 whose divisors intersect transversally) over the total space, we get a line bundle $\langle L_1, L_2 \rangle$ (together with a section $\langle s_1, s_2 \rangle$) on the base B in a canonical way.

More generally, if $\pi : X \to Y$ is 'nice', of relative dimension n, then for any $(n+1)$ line bundles $L_0, L_1, \ldots, L_n$ on X, we get a unique line bundle $\langle L_0, L_1, \ldots, L_n \rangle$ on Y, the so-called *Deligne tuple* associated to $L_0, L_1, \ldots, L_n$ with respect to π ([De]).

(2.2) Universal Curves Pretend that we are using the V-manifold language (so that what follows can be justified). We have a universal curve $\pi : \overline{C_{g,N}} \to \overline{\mathcal{M}_{g,N}}$, together with N sections $\mathbb{P}_i$'s of π. Hence, for $x = [(M; P_1, \ldots, P_N)] \in \overline{\mathcal{M}_{g,N}}$, $\pi^{-1}(x) = M$ and $\mathbb{P}_i(x) = P_i$.

In fact, $\overline{C_{g,N}} = \overline{\mathcal{M}_{g,N+1}}$, and essentially π is the map of dropping the last puncture. In particular, the fiber of π at $[(M; P_1, \ldots, P_N)] \in \mathcal{M}_{g,N}$ is the compact Riemann surface M together with punctures $P_1, \ldots, P_N$. Hence, by gluing K_M and $P_1, \ldots, P_N$ along $\mathcal{M}_{g,N}$, and extend to $\overline{\mathcal{M}_{g,N}}$, then on $\overline{C_{g,N}}$, we get
(i) K_π, the relative canonical line bundles associated to π; and
(ii) $\mathbb{P}_1, \ldots, \mathbb{P}_N$; sections of π viewed by abuse of terminology as line bundles.

(**2.3**) **Primitive Line Bundles** Using Deligne pairing formalism, we now introduce some primitive line bundles on $\overline{\mathcal{M}_{g,N}}$ (Weng [We1,2]).

(i) The *Weil-Petersson line bundle*

$$\Delta_{\mathrm{WP}} := \big\langle K_\pi(\mathbb{P}_1 + \cdots + \mathbb{P}_N), K_\pi(\mathbb{P}_1 + \cdots + \mathbb{P}_N)\big\rangle;$$

(ii) The *Takhtajan-Zograf line bundle*

$$\Delta_{\mathrm{TZ}} := \big\langle K_\pi, \mathbb{P}_1 + \cdots + \mathbb{P}_N\big\rangle;$$

(iii) The *m-th Mumford (type) line bundle*, for $m \geq 1$,

$$\lambda_m := \lambda\Big(mK_\pi + (m-1)\big(\mathbb{P}_1 + \cdots + \mathbb{P}_N\big)\Big).$$

Essentially, $\lambda_m|_{[(M;P_1,\ldots,P_N)]} = \det\Big(H^0\Big(M, mK_\pi + (m-1)(P_1 + \cdots + P_N)\Big)\Big)$, the determinant of the space Γ_m of cusp forms of weight $2m$. Moreover, for $m \leq 0$, λ_m can be defined by using Grothendieck-Mumford determinant formalism and Serre duality. (See [We1,2].)

3. Fundamental Relations on $\overline{\mathcal{M}_{g,N}}$: Algebraic Story

Among line bundles λ_m, Δ_{WP}, Δ_{TZ} and Δ_{bdy}, there are the following fundamental relations.

(**FR I**) (Deligne [De], Mumford [Mu] N=0; Weng [We1,2] $N > 0$)

$$\lambda_m^{\otimes 12} \simeq \Delta_{\mathrm{WP}}^{\otimes(6m^2-6m+1)} \otimes \Delta_{\mathrm{TZ}}^{\otimes -1} \otimes \Delta_{\mathrm{bdy}};$$

(**FR II**) (Weng [We1,2])

$$\Delta_{\mathrm{WP}}^{\otimes N^2} \leq \Delta_{\mathrm{TZ}}^{\otimes (2g-2+N)^2};$$

(**FR III**) (i) (Xiao [X] & Cornalba-Harris [CH]) On $\overline{\mathcal{M}_g} = \overline{\mathcal{M}_{g,0}}$,

$$\left(8 + \frac{4}{g}\right)\lambda_1 \geq \Delta_{\mathrm{bdy}};$$

(**FR III**) (ii) (Weng [We1,2]) On $\overline{\mathcal{M}_{g,N}}$, $N \geq 1$,

$$\left(8 + \frac{2N}{g-1+N}\right)\lambda_1 + \Delta_{\mathrm{TZ}} \geq \Delta_{\mathrm{bdy}}.$$

Remarks. (1) $L/\overline{\mathcal{M}_{g,N}} \geq 0$, by definition, if $\forall B$ curves in $\mathcal{M}_{g,N}$, $\deg(L|_{\bar{B}}) \geq 0$;
(2) When $N = 0$, Moriwaki [Mo] has a sharp version for FR III(i); When $N = 1$, Harris-Morrison (resp. Hain) obtained a similar result as FR II (resp. FR III(ii)). For example,

Basic Inequality. (Harris-Morrison [HM]): *On* $\overline{\mathcal{M}_{g,1}}$,

$$4g(g-1)K \geq 12\lambda_1 - \Delta_{\mathrm{bdy}}.$$

Here K := *relative canonical line bundle of* $\overline{C_g} = \overline{C_{g,0}} \to \overline{\mathcal{M}_g}$.
We point out that this Basic Inequality is equivalent to $\Delta_{\mathrm{WP}}^{\otimes 1^2} \leq \Delta_{\mathrm{TZ}}^{\otimes (2g-2+1)^2}$, i.e., our FR II with $N = 1$. (For details, please refer to [We1,2].)
(3) Fundamental relations above in fact expose certain intrinsic relations between two different kinds of geometries for the moduli space: The discrete spectrum geometry represented by the Weil-Petersson line bundle and the continuous spectrum geometry represented via the Takhtajan-Zograf line bundle. (See below.)

4. Fundamental Relation on $\mathcal{M}_{g,N}$: Arithmetic Story

(4.1) Basic Relations For any $[(M; P_1, \ldots, P_N)] \in \mathcal{M}_{g,N}$, we may view the fiber of π at this point as a punctured Riemann surface

$M^0 := M\backslash\{P_1, \ldots, P_N\}$. Thus if $2g - 2 + N \geq 1$, by uniformization theorem, M^0 is covered by the upper half plane $\mathcal{H}$. So $M^0 \simeq \Gamma\backslash\mathcal{H}$ with a certain $\Gamma \subset \mathrm{PSL}_2(\mathbf{R})$.

The Poincaré metric on $\mathcal{H}$ given by $\frac{dx\wedge dy}{y^2}$ is $\mathrm{PSL}_2(\mathbf{R})$-invariant, and hence can be descended to $\Gamma\backslash\mathcal{H} \simeq M^0$. Denote this metric by τ_{hyp}.

On the other hand, from M point of view, the metric τ_{hyp} may be better understood in terms of the logarithmic tangent bundle $\Big(K_M(P_1 + \cdots + P_N)\Big)^\vee$. That is, we get then a natural *singular* metric on $K_M(P_1 + \cdots + P_N)$. Gluing them together along $\mathcal{M}_{g,N}$, we obtain the nice hyperbolic metric on $K_\pi(\mathbb{P}_1 + \cdots + \mathbb{P}_N)$. Denote this metrized line bundle by

$$\underline{K_\pi(\mathbb{P}_1 + \cdots + \mathbb{P}_N)}.$$

By developing an arithmetic intersection for *singular* metrics (as a part of our ω-admissible theory [We1,2]), we obtain a natural smooth metric on

$$\Delta_{\mathrm{WP}} = \big\langle K_\pi(\mathbb{P}_1 + \cdots + \mathbb{P}_N), K_\pi(\mathbb{P}_1 + \cdots + \mathbb{P}_N)\big\rangle.$$

Denote the resulting metrized line bundle by $\underline{\Delta_{\mathrm{WP}}}$.

(**FR IV**) (Wolpert [Wo2], Weng [We1,2]) *On* $\mathcal{M}_{g,N}$,

$$c_1(\underline{\Delta_{\mathrm{WP}}}) = \frac{\omega_{\mathrm{WP}}}{\pi^2}.$$

Here c_1 *denotes the first Chern form,* ω_{WP} *denotes the Weil-Petersson Kähler form.*

As for $\Delta_{\mathrm{TZ}} = \langle K_\pi, \mathbb{P}_1 + \cdots + \mathbb{P}_N\rangle$, we may also get a very nice metric by some really very very hard work. (See 4.2 below.) Denote the resulting metrized line bundle by $\underline{\Delta_{\mathrm{TZ}}}$.

(**FR V**) (Weng [We1,2]) *On* $\mathcal{M}_{g,N}$,

$$c_1(\underline{\Delta_{\mathrm{TZ}}}) = \frac{4}{3}\omega_{\mathrm{TZ}}.$$

Here ω_{WP} *denotes the Takhtajan-Zograf Kähler form.*

FR IV and FR V show that Weil-Petersson metrics and Takhtajan-Zograf metrics are all algebraic and also justify our terminology.

(4.2) Backgrounds (i) According to Kodaira-Spencer's deformation theory, tangent vectors on the base may be viewed as elements in $H^1\big(M, K_M^\vee(-P_1 - \cdots - P_N)\big)$, over which there is the Petersson metric. The Weil-Petersson metric on $\mathcal{M}_{g,N}$ is defined by

$$\langle \mu, \nu \rangle_{\mathrm{WP}}\Big([M; P_1, \ldots, P_N]\Big) := \int_{\mathcal{D}(\Gamma)} \mu\bar{\nu}\frac{dx \wedge dy}{y^2}.$$

Here $\mathcal{D}(\Gamma)$ denotes a fundamental domain of Γ in $\mathcal{H}$. (So $M \simeq \Gamma\backslash\mathcal{H}$.) Similarly, the Takhtajan-Zograf metric on $\mathcal{M}_{g,N}$ is defined to be

$$\langle \mu, \nu \rangle_{\mathrm{TZ}}\Big([M; P_1, \ldots, P_N]\Big) := \int_{\mathcal{D}(\Gamma)} \mu\bar{\nu} \cdot \sum_{i=1}^{N} E_i(z; 2) \cdot \frac{dx \wedge dy}{y^2}.$$

Here $E_i(z, s)$ denotes the i-th Eisenstein series associated to Γ.

(ii) In order to get a natural metric on $\Delta_{\mathrm{TZ}} = \langle K_\pi, \mathbb{P}_1 + \cdots + \mathbb{P}_N \rangle$, we need to construct nice metrics on K_M as well as on the line bundle corresponding to $P_1 + \cdots + P_N$, starting from the (singular) hyperbolic metric on $K_M(P_1 + \cdots + P_N)$.

Difficulty: $P_1, \ldots, P_N$ are black holes for the hyperbolic metric τ_{hyp}: the metric becomes singular at these points;

Solution: Use arithmetic construction;

Keys: (a) K_M has a natural metric, the so-called Arakelov metric ([Ar]), induced from $M \hookrightarrow J(M) = \mathbf{C}^g/\Lambda$. This metric is Euclidean in nature. However we can use a fundamental result in ω-admissible theory, called the **Mean Value Lemma**, as a bridge to get a τ_{hyp}-admissible metric, the so-called *arithmetic hyperbolic metric* on K_M, which is then hyperbolic in nature.

(b) Still, this new metric is not quite right: It is nice up to a certain constant factor, depending on punctured surfaces. To pin down this factor, we need to introduce a new invariant called *Arakelov-Poincaré volumes* for punctured Riemann surfaces. This new invariant is built from the so-called Faltings δ-function for M ([F]) and special values of Selberg zeta function for M^0, motivated by a result of D'Hoker-Phong [DP] and Sarnak [Sa].

(iii) Till now, we have shown how arithmetic intersection can be used to study moduli spaces. To go further, we still need to develop an arithmetic cohomology for singular metrics. The details can be found in [We1,2]. All in all, what we should say is that without using the analytic torsion approach ([Qu]), which is quite impossible for singular metrics, we successfully introduce a nice determinant metric and then prove the following

Arithmetic Riemann-Roch Isometry for Singular Metrics (Weng [We1,2]) *Associated to a fixed normalized volume ω with at most quasi-hyperbolic singularities, there is an ω-admissible theory from which we have the natural isometry*

$$\underline{\lambda(L)}^{\otimes 12} \simeq \langle \underline{L}, \underline{L-K} \rangle^{\otimes 6} \otimes \langle \underline{K}, \underline{K} \rangle$$

up to a constant factor depending only on the genus.

(4.3) Application On $\lambda_m = \det \Gamma_m$, the determinant of the associated Petersson metric is denoted by $\rho_{L^2}(m)$. Moreover, following Takhtajan-Zograf ([TZ1,2]), set

$$h_Q(m) := \rho_{L^2}(m) \cdot \begin{cases} \frac{1}{\sqrt{Z'_{M^0}(1)}}, & \text{if } m = 1, \\ \frac{1}{\sqrt{Z_{M^0}(m)}}, & \text{if } m > 1. \end{cases}$$

Here

$$Z_{M^0}(s) := \prod_{\gamma:\ \text{primitive geodesic}} \prod_{n=0}^{\infty} \Big(1 - e^{-(s+n)l(\gamma)}\Big), \qquad \mathrm{Re}\, s > 1$$

denotes the standard Selberg zeta function, with l the length. (See e.g., [He].) Denote the metrized line bundle $(\lambda_m, h_Q(m))$ simply by $\underline{\lambda_m}$.

(**FR VI**) (Deligne [De] $N = 0$; Weng [We1,2] $N \geq 1$) *On $\mathcal{M}_{g,N}$, up to a universal constant factor,*

$$\underline{\lambda_m}^{\otimes 12} \simeq \underline{\Delta_{\mathrm{WP}}}^{\otimes(6m^2-6m+1)} \otimes \underline{\Delta_{\mathrm{TZ}}}^{\otimes -1}.$$

Hence, we recover the following fundamental result of Takhtajan-Zograf, which is the staring point for our study.

Theorem. (Takhtajan-Zograf [TZ1,2])

$$c_1\Big(\lambda_m, h_Q(m)\Big) = \frac{6m^2 - 6m + 1}{12} \cdot \frac{\omega_{\mathrm{WP}}}{\pi^2} - \frac{1}{9} \cdot \omega_{\mathrm{TZ}}.$$

Remark. Recall that as line bundles λ_m, Δ_{WP} and Δ_{TZ} make sense even over the compactified moduli $\overline{\mathcal{M}_{g,N}}$. Hence FR IV – VI indeed give a good way to understand the factorization of Selberg zeta functions, Eisenstein series, Weil-Petersson metrics and Takhtajan-Zograf metrics. (See also 6 below.)

5. Deligne Tuple in General

For the universal curve $\pi_N : \overline{C_{g,N}} \xrightarrow{\pi} \overline{\mathcal{M}_{g,N}}$, define the *i-th Takhtajan-Zograf line bundle* by $\Delta_i := \Delta_i(N) := \langle K_\pi, \mathbb{P}_i \rangle(\pi)$, $i = 1, \ldots, N$. Then

(i) $\Delta_i(N) = \pi_{N-1}^*\Big(\Delta_i(N-1)\Big)(\mathbb{P}_i)$ for $i = 1, \ldots, N-1$;

(ii) $\Delta_N(N) = K_{\pi_{N-1}}\Big(\mathbb{P}_1 + \cdots + \mathbb{P}_{N-1}\Big)$.

Now identify $\overline{C_{g,N}}$ with $\overline{M_{g,N+1}}$ and set $\lambda_N : \overline{C_{g,N+1}} \to \overline{C_{g,N}}$ be the map of 'forgetting the second to the last puncture'. Then

(iii) $\lambda_N^*\Bigg(K_{\pi_N}\Big(\mathbb{P}_1 + \cdots + \mathbb{P}_N\Big)\Bigg) = K_{\pi_{N+1}}\Big(\mathbb{P}_1 + \cdots + \mathbb{P}_N\Big)$.

Let $\Delta = \Delta(N)$ be the Weil-Petersson line bundle on $\overline{M_{g,N}}$, then

(iv) $\Delta(N+1) = \pi_{N+1}^*\Big(\Delta(N)\Big) + \Delta_{N+1}(N+1)$.

More generally, for any $m \in \mathbb{Z}_{\geq 0}$, we have the natural map $\pi_{N+m,N} : \overline{M_{g,N+m}} \to \overline{M_{g,N}}$ of 'forgetting the last m punctures'. The relative dimension of $\pi_{N+m,N}$ is $m+1$, it makes sense to talk about

$$\Big\langle \Delta_1(N+m), \ldots, \Delta_1(N+m), \ldots, \Delta_{N+m}(N+m), \ldots,$$
$$\Delta_{N+m}(N+m), \Delta(N+m), \ldots, \Delta(N+m)\Big\rangle(\pi_{N+m,N}),$$

where $\Delta_i(N+m)$ appears a_i-times, $i = 1, \ldots, N+m$, and Δ appears b-times. (Clearly $a_1 + \cdots + a_{N+m} + b = m+1$.) For simplicity, denote this line bundle (on $\overline{M_{g,N}}$) by

$$\Big\langle \Delta_1^{\cdot a_1}, \ldots, \Delta_{N+m}^{\cdot a_{N+m}}, \Delta^{\cdot b}\Big\rangle(\pi_{N+m,N}).$$

Line bundles $\Big\langle \Delta_1^{\cdot a_1}, \ldots, \Delta_{N+m}^{\cdot a_{N+m}}, \Delta^{\cdot b}\Big\rangle(\pi_{N+m,N})$ satisfy many nice properties ([WZ]). We will not try to list all of them here, but merely point out that

(I) (**String Equation**) $\Big\langle \Delta_1^{\cdot d_1}, \ldots, \Delta_{N+m}^{\cdot d_{N+m}}\Big\rangle(\pi_{N+m+1,N})$

$$= \sum_{j=1}^{N+m} \Big\langle \Delta_1^{\cdot d_1}, \ldots, \Delta_j^{\cdot d_j - 1}, \ldots, \Delta_{N+m}^{\cdot d_{N+m}}\Big\rangle(\pi_{N+m,N});$$

(II) (**Dilaton Equation**) $\Big\langle \Delta_1^{\cdot d_1}, \ldots, \Delta_{N+m}^{\cdot d_{N+m}}, \Delta_{N+m+1}\Big\rangle(\pi_{N+m+1,N})$

$$= (2g - 2 + N + m)\Big\langle \Delta_1^{\cdot d_1}, \ldots, \Delta_{N+m}^{\cdot d_{N+m}}\Big\rangle(\pi_{N+m,N}).$$

6. Degeneration of TZ Metrics: Analytic Story

(6.1) Pseudo-Metrics on Boundary of $T_{g,N}$ $M_0 \in \overline{\mathcal{M}_{g,N}} \backslash \mathcal{M}_{g,N}$ gives a stable Riemann surface with N punctures $P_1, \ldots, P_N$ and m nodes $Q_1, \ldots, Q_m$. Thus we may regard M_0 as a point in $\delta_{\gamma_1,\ldots,\gamma_m} T_{g,N}$. Write $M_0 \backslash \{Q_1, \ldots, Q_m\} =: \cup_{1 \leq \alpha \leq r} C_\alpha$, $\delta_{\gamma_1,\ldots,\gamma_m} T_{g,N} =: \prod_{\alpha=1}^{r} T_{g_\alpha, n_\alpha}$ with each component $C_\alpha \in T_{g_\alpha, n_\alpha}$. As said above, since M_0 is stable, each C_α admits the complete hyperbolic metric of constant sectional curvature -1.

The punctures of C_α arise from either the punctures or the nodes of M_0, and for simplicity, they will be called old cusps and new cusps of S_α respectively. Denote the number of old cusps (resp. new cusps) of S_α by n'_α (resp. n''_α), so that $n_\alpha = n'_\alpha + n''_\alpha$. We index the punctures of C_α such that $\{P_{\alpha,i}\}_{1\le i\le n'_\alpha}$ denotes the set of old cusps, and $\{P_{\alpha,i}\}_{n'_\alpha+1\le i\le n_\alpha}$ denotes the set of new cusps. For each α and i, we denote by $g^{(\alpha,i)}$ the Takhtajan-Zograf metric on T_{g_α,n_α} with respect to the puncture $P_{\alpha,i}$. Now we define a pseudo-metric $\hat{g}^{\mathrm{TZ},\alpha}$ on T_{g_α,n_α} by summing the $g^{(\alpha,i)}$'s over the old cusps, i.e.,

$$\hat{g}^{\mathrm{TZ},\alpha} := \sum_{1\le i\le n'_\alpha} g^{(\alpha,i)}.$$

If none of the punctures of C_α are old cusps, then $\hat{g}^{\mathrm{TZ},\alpha}$ is simply defined to be zero identically. With this, the nodally depleted Takhtajan-Zograf pseudo-metric $\hat{g}^{\mathrm{TZ},(\gamma_1,\ldots,\gamma_n)}$ on $\delta_{\gamma_1,\ldots,\gamma_m}T_{g,N}$ is defined to be the product pseudo-metric of the $\hat{g}^{\mathrm{TZ},\alpha}$'s on the T_{g_α,n_α}'s, i.e.,

$$\left(\delta_{\gamma_1,\ldots,\gamma_m}T_{g,N}, \hat{g}^{\mathrm{TZ},(\gamma_1,\ldots,\gamma_n)}\right) = \prod_{i=1}^{r}\left(T_{g_\alpha,n_\alpha}, \hat{g}^{\mathrm{TZ},\alpha}\right).$$

(6.2) Degeneration Coordinates Each node Q_j in M_0 admits an open neighborhood $N_j = \{(z_j, w_j) \in \mathbb{C}^2 : |z_j|, |w_j| < 1,\ z_j \cdot w_j = 0\}$ so that $N_j = N_j^1 \cup N_j^2$. Here N_j^1 and N_j^2 are regarded as the coordinate discs in $\mathbb{C}^2$. For each α, we choose $3g_\alpha - 3 + n_\alpha$ linearly independent Beltrami differentials $\nu_i^{(\alpha)}, 1 \le i \le 3g_\alpha - 3 + n_\alpha$, which are supported on $C_\alpha \setminus \cup_{j=1}^{n} N_j$, so that they form a basis of $T_{C_\alpha}T_{g_\alpha,n_\alpha}$. For simplicity, we rewrite $\{v_i^{(\alpha)}\}_{1\le\alpha\le r, 1\le i\le 3g_\alpha-3+n_\alpha}$ as $\{v_i\}_{1\le\alpha\le r,1\le i\le 3g-3+N-m}$. Then one has an associated local coordinate neighborhood V of M_0 in $\delta_{\gamma_1,\ldots,\gamma_m}T_{g,N}$ with coordinates $\tau = (\tau_1, \ldots, \tau_{3g-3+N-m})$ such that M_0 corresponds to 0. Shrinking V if necessary, we may assume $V \simeq \Delta^{3g-3+N-m}$. For a point $\tau \in V$, one has the associated Beltrami differential $\mu(\tau) = \sum_{i=1}^{3g-3+N-m} \tau_i v_i$ and a quasi-conformal homeomorphism $w^{\mu(\tau)} : M_0 \to M_\tau$ onto a Riemann surface M_τ satisfying $\frac{\partial w^{\mu(\tau)}}{\partial \bar{z}} = \mu(z)\frac{\partial w^{\mu(\tau)}}{\partial z}$. The map $w^{\mu(\tau)}$ is conformal on each $N_j, j = 1, \ldots, m$, so that we may regard $N_j \subset M_\tau$ for each j. Then for each $t = (t_1, \ldots, t_m)$ with each $|t_j| < 1$, we obtain a new Riemann surface $M_{t,\tau}$ for M_τ by removing the disks $\{z_j \in N_j^1 : |z_j| < |t_j|\}$ and $\{w_j \in N_j^2 : |w_j| < |t_j|\}$ and identifying $z_j \in N_j^1$ with $w_j = t_j/z_j \in N_j^2$, $j = 1, \ldots, m$. Then one obtains a holomorphic family of Riemann surfaces $\{M_{t,\tau}\}$ parametrized by the coordinates $(t,\tau)=(t_1,\ldots,t_m,\tau_1,\ldots,\tau_{3g-3+N-m})$ of $\Delta^m\times V\simeq\Delta^m\times\Delta^{3g-3+N-m}$. Moreover, the Riemann surfaces $M_{t,\tau}$ with $(t,\tau)\in(\Delta^*)^m\times V$ are of type (g,N), where $\Delta^*=\Delta\backslash\{0\}$. The coordinates $t=(t_1,\ldots,t_m)$ will be called pinching coordinates, and $\tau=(t_1,\ldots,t_{3g-3+N-m})$ will be called boundary coordinates. For $1\leq j\leq m$, let α_j denote the simple closed curve $|z_j|=|w_j|=|t_j|^{\frac{1}{2}}$ on $X_{t,\tau}$. Shrinking Δ^m and V if necessary, it is known that the universal cover of $(\Delta^*)^m\times V$ is naturally a domain in $T_{g,N}$ and the corresponding covering transformations are generated by Dehn twist about the α_j's. Since Dehn twists are elements of $\mathrm{Mod}_{g,N}$, the $\mathrm{Mod}_{g,N}$-invariant metrics g^{WP} and g^{TZ} descend to metrics on $(\Delta^*)^m\times V$. It is well known that $M_0\in\overline{\mathcal{M}_{g,N}}\backslash\mathcal{M}_{g,N}$ always admits an open neighborhood $\hat{U}$ in $\overline{\mathcal{M}_{g,N}}$ together with a local uniformizing chart $\chi:U\simeq\Delta^m\times V\to\hat{U}$ for some $\Delta^m\times V$ as described above, where χ is a finite ramified cover. Obviously the metrics g^{WP} and g^{TZ} on $(\Delta^*)^m\times V\subset U$ may also be regarded as extensions of the pull-back of the corresponding metrics on the smooth points of $\hat{U}\cap\mathcal{M}_{g,N}$ via the map χ.

(6.3) Degeneration Let M_0 be a Riemann surface with N punctures $P_1,\ldots,P_N$ and m nodes $Q_1,\ldots,Q_m$. A node Q_i is said to be adjacent to punctures if the component of punctured Riemann surface $M_0\backslash\{Q_1,\ldots,Q_{i-1},Q_{i+1},\ldots,Q_m\}$ containing Q_i also contains at least one of the P_j's. Otherwise, it is said to be non-adjacent to punctures.

As an analog of Masur's well-known result [Ma] on degenerations of Weil-Petersson metrics, we have the following

Theorem. (Obitsu-To-Weng [OTW]) *For $g\geq 0$ and $N>0$, let $M_0\in\overline{\mathcal{M}_{g,N}}\backslash\mathcal{M}_{g,N}$ be a stable Riemann surface with N punctures $P_1,\ldots,P_N$ and m nodes $Q_1,\ldots,Q_m$ arranged in such a way that $Q_i, 1\leq i\leq m'$ are adjacent to punctures and $Q_i, m'+1\leq i\leq m$ are non-adjacent to punctures. Let U be an open neighborhood of M_0 in $\overline{\mathcal{M}_{g,N}}$, together with a local uniformizing chart $\psi:U\simeq\Delta^m\times V\to\hat{U}$, where $V\simeq\Delta^{3g-3+N-m}$ is a domain in an appropriate boundary Teichmüller space $\delta_{\gamma_1,\ldots,\gamma_m}T_{g,N}$ corresponding to M_0, and* *let* $(s_1,\ldots,s_{3g-3+N})=(t_1,\ldots,t_m,\tau_1,\ldots,\tau_{3g-3+N-m})=(t,\tau)$ *be the pinching and boundary coordinates of* $U\simeq\Delta^m\times V$ *with each* t_i *and* δ_i *corresponding to* Q_i. *Let the components of the Takhtajan-Zograf metric* g^{TZ} *be given by*

$$g_{i\bar{j}}^{\mathrm{TZ}}=g^{\mathrm{TZ}}\left(\frac{\partial}{\partial s_i},\frac{\partial}{\partial s_j}\right),\qquad 1\le i,j\le 3g-3+N,$$

on $(\Delta^*)^m\times V\subset\tilde{U}$. *Denote by* $\hat{g}_{j\bar{k}}^{\mathrm{TZ},(\gamma_1,\ldots,\gamma_m)}$ *the nodally depleted Takhtajan-Zograf pseudo-metric on the boundary* $\delta_{\gamma_1,\ldots,\gamma_m}T_{g,N}$. *Then*

(i) *For each* $1\le j\le m$ *and any* $\varepsilon>0$,

$$\limsup_{(t,\tau)\to(0,0),(t,\tau)\in(\Delta^*)^m\times V}\Big(|t_j|^2(-\log|t_j|)^{4-\varepsilon}\Big)\cdot g_{j\bar{j}}^{\mathrm{TZ}}(t,\tau)=0;$$

(ii) *For each* $1\le j\le m'$ *and any* $\varepsilon>0$,

$$\liminf_{(t,\tau)\to(0,0),(t,\tau)\in(\Delta^*)^m\times V}\Big(|t_j|^2(-\log|t_j|)^{4+\varepsilon}\Big)\cdot g_{j\bar{j}}^{\mathrm{TZ}}(t,\tau)=+\infty;$$

(iii) *For* $1\le j,\,k\le m$ *and* $j\ne k$, *as* $(t,\tau)\in(\Delta^*)^m\times V\to(0,0)$,

$$g_{j\bar{k}}^{\mathrm{TZ}}(t,\tau)=O\left(\frac{1}{|t_j|\,|t_k|\,(\log|t_j|)^3(\log|t_k|)^3}\right);$$

(iv) *For each* $j,\,k\ge m+1$,

$$\liminf_{(t,\tau)\to(0,0),(t,\tau)\in(\Delta^*)^m\times V}\Big(|t_j|^2(-\log|t_j|)^{4+\varepsilon}\Big)\cdot g_{j\bar{k}}^{\mathrm{TZ}}(t,\tau)=\hat{g}_{j\bar{k}}^{\mathrm{TZ},(\gamma_1,\ldots,\gamma_m)}(0,0);$$

(v) *For* $j\le m$ *and* $k\ge m+1$, *as* $(t,\tau)\in(\Delta^*)^m\times V\to(0,0)$,

$$g_{j\bar{k}}^{\mathrm{TZ}}(t,\tau)=O\left(-\frac{1}{|t_j|(-\log|t_j|)^3}\right).$$

Remarks. (1) (i) (resp. (ii)) is equivalent to the following statement: For each $1 \leq j \leq m$ and any $\varepsilon > 0$, there exists a constant $C_{1,\varepsilon} > 0$ (resp. $C_{2,\varepsilon} > 0$) such that for all $(t,\tau) \in (\Delta^*)^m \times V$,

$$g_{j\bar{j}}^{\mathrm{TZ}}(t,\tau) \leq \frac{C_{1,\varepsilon}}{|t_j|^2(-\log|t_j|)^{4-\varepsilon}} \qquad \left(\text{resp.} \geq \frac{C_{2,\varepsilon}}{|t_j|^2(-\log|t_j|)^{4+\varepsilon}}\right).$$

Accordingly, it is natural to ask whether the stronger estimate

$$g_{j\bar{j}}^{\mathrm{TZ}}(t,\tau) \sim \frac{C_{1,\varepsilon}}{|t_j|^2(-\log|t_j|)^4}$$

actually holds for $1 \leq j \leq m'$ and $(t,\tau) \in (\Delta^*)^m \times V$;
(2) There are also discussions on degenerations of Green's functions (intersection), Selberg zeta functions, Faltings δ-invariants and Arakelov-Poincare volumes (cohomology). For more details, please refer to [Jo], [Wen], [TW1,2,3] and [We1,2].

References

[Ar] S. Arakelov, Intersection theory of divisors on an arithmetic surface, Izv Akad. Nauk SSSR ser Mat., 38 (1974) no. 6, 1179–1192

[CH] M. Cornalba, J. Harris, Divisor classes associated to families of stable varieties, Ann Sci. Ec. Norm. Sup 4 serie, 21 (1988), 455-475

[De] P. Deligne, Le déterminant de la cohomologie, *Contemporary Math.* Vol. **67**, (1987), 93-178

[D'HP] E. D'Hoker and D.H. Phong, On determinants of Laplacians on Riemann surfaces, *Comm. Math. Physics,* **104**, (1986) 537-545

[F] G. Faltings, Calculus on arithmetic surfaces. Ann. of Math. (2) 119 (1984), no. 2, 387–424.

[HM] J. Harris, I. Morrison, Moduli of curves. GTM 187. Springer-Verlag, 1998.

[He] D.A. Hejhal, *The Selberg trace formula for* $PSL_2(\mathbf{R})$, vol. 2 Springer LNM **1001**, 1983

[Jo] J. Jorgenson, Asymptotic behavior of Faltings's delta function. Duke Math. J. 61 (1990), no. 1, 221–254.

[Ma] H. Masur, The extension of the Weil-Petersson metric to the boundary of Teichmüller space, *Duke Math. J.*, **43** (1976) 623–635

[Mo] A. Moriwaki, Relative Bogomolov's inequality and the cone of positive divisors on the moduli space of stable curves, JAMS **11**, (1998) no. 3, 569-600

[Mu] D. Mumford, Stability of projective varieties, *L'Ens. Math.*, **24** (1977) 39-110

[Ob1] K. Obitsu, Non-completeness of Takhtajan-Zograf Kähler metric for Teichmüller space of punctured Riemann surfaces, Comm. Math. Phys., 205 (1999) no. 2, 415-420

[Ob2] K. Obitsu, The asymptotic behavior of Eisenstein series and a comparison of Weil-Petersson and the Takhtajan-Zograf metrics, Publ. RIMS, Kyoto Univ., Vol. 37 (2001) 459–478

[OTW] K. Obitsu, W. To & L. Weng, The Asymptotic Behavior of the Takhtajan-Zograf Metrics, in preperation

[Qu] D. Quillen, Determinants of Cauchy-Riemann operators over a Riemann surface, *Funk. Anal. i Prilozhen* **19** (1985) 31-34

[Sa] P. Sarnak, Determinants of Laplacians, *Commun. Math. Phys.* **110**, (1987) no. 1, 113–120

[TW1] W.-K. To & L. Weng: The asymptotic behavior of Green's functions for quasi-hyperbolic metrics on degenerating Riemann surfaces, *Manuscripta Math.* **93** (1997) 465-480

[TW2] W.-K. To & L. Weng: Green's functions for quasi-hyperbolic metrics on degenerating Riemann surfaces with a separating node, *Ann. of Global Analysis and Geometry* (1999) no. 3, 239–265

[Tw3] W.-K. To & L. Weng, Admissible Hermitian metrics on families of line bundles over degenerating Riemann surfaces, Pacific J. Math. 197 (2001), no. 2, 441–489

[TZ1] L. Takhtajan & P. Zograf, The Selberg zeta function and a new Kähler metric on the moduli space of punctured Riemann surfaces, *J. Geo. Phys.* **5**, (1988) 551-570

[TZ2] L. Takhtajan & P. Zograf, A local index theorem for families of $\bar{\partial}$-operators on punctured Riemann surfaces and a new Kähler metric on their moduli spaces, *Commun. Math. Phys.* **137**, (1991) 399-426

[We1] L. Weng, Ω-admissible theory II. Deligne pairings over moduli spaces of punctured Riemann surfaces. Math. Ann. 320 (2001), no. 2, 239–283.

[We2] L. Weng, *Hyperbolic Metrics, Selberg Zeta Functions and Ara- kelov Theory for Punctured Riemann Surfaces*, Lecture Note Series in Mathematics, Vol. **6**, Osaka, 1998

[WZ] L. Weng & D. Zagier, Deligne Products of Line Bundles over Moduli Spaces of Marked Curves, in preparation

[Wen] R. Wentworth, The asymptotics of the Arakelov-Green's function and Faltings' delta invariant. Comm. Math. Phys. 137 (1991), no. 3, 427–459

[Wo1] S. Wolpert, Noncompleteness of the Weil-Petersson metric for Teichmüler space. Pacific J. Math. 61 (1975), no. 2, 573–577

[Wo2] S. Wolpert, Chern forms and the Riemann tensor for the moduli space of curves. Invent. Math. 85 (1986), no. 1, 119–145

[Wo3] S. Wolpert, The hyperbolic metric and the geometry of universal curve. *J. Differ. Gemo.* **31** (1990) no. 2, 417–472

[Wo4] S. Wolpert, Cusps and the family hyperbolic metric, preprint, 2005

[X] G. Xiao, Fibered algebraic surfaces with low slope, Math. Ann. 276 (1987), 449-466

Kunio OBITSU
Department of Mathematics and Computer Science,
Kagoshima University
Kagoshima 890-0065, Japan
Email: obitsu@sci.kagoshima-u.ac.jp

Wing-Keung TO
Department of Mathematics
National University of Singapore
Singapore 119260
Email: mattowk@nus.edu.sg

Lin WENG
Graduate School of Mathematics
Kyushu University
Fukuoka 812-8581, Japan
Email: weng@math.kyushu-u.ac.jp

Vector Bundles on Curves over $\mathbb{C}_p$

Annette WERNER

1. Introduction

This paper is a report on joint work with Christopher Deninger published in [De-We1], [De-We2] and [De-We3]. We define a certain class of vector bundles on p-adic curves which can be endowed with parallel transport along étale paths. In particular, all those bundles induce representations of the algebraic fundamental group.

This article is intended as a survey, explaining some results more leisurely than in the original papers. We focus mainly on the definitions and constructions in [De-We2], outlining the ideas rather than reproducing the formal proofs.

The last section deals with Mumford curves. We explain results of Herz in [He] which relate our construction to the paper [Fa1].

Note that in [Fa2], Faltings develops a p-adic non-abelian Hodge theory, which in some aspects is more general than the present theory.

2. Complex Vector Bundles

Let X be a compact Riemann surface with base point $x \in X$. Then every complex representation $\rho : \pi_1(X, x) \to \mathrm{GL}_r(\mathbb{C})$ of the fundamental group gives rise to a flat vector bundle E_ρ on X, i.e. a vector bundle with locally constant transition function. Namely, let $\pi : \tilde{X} \to X$ be the universal covering of X. Then E_ρ is defined as the quotient of the trivial bundle $\tilde{X} \times \mathbb{C}^r$ by the $\pi_1(X, x)$-action given by combining the natural action of $\pi_1(X, x)$ on the first factor with the action induced by ρ on the second factor. Conversely, if E is a flat vector bundle on X, its pullback $\pi^* E$ is trivial as a flat bundle on $\tilde{X}$, i.e. $\pi^* E \simeq \mathbb{C}^r \times \tilde{X}$. Since $\pi_1(X, x)$ acts in a natural way on the pullback bundle $\pi^* E$, it also acts on the right hand side, which gives a representation ρ of $\pi_1(X,x)$ on $\mathbb{C}^r$. This representation satisfies $E \simeq E_\rho$.

A similar construction gives for every flat bundle E and every continuous path γ from x to x' in X an isomorphism of parallel transport along γ from the fiber E_x of E in x to the fiber $E_{x'}$. Namely, choose a point y in the universal covering $\tilde{X}$ over x. Then γ can be lifted to a continuous path in $\tilde{X}$ starting in y. The endpoint of the lifted path is a point y' in $\tilde{X}$ lying over x'. Since π^*E is trivial, there is a trivial parallel transport $(\pi^*E)_y \xrightarrow{\sim} (\pi^*E)_{y'}$. Since $\pi(y) = x$ and $\pi(y') = x'$, there are natural isomorphisms $(\pi^*E)_y \xrightarrow{\sim} E_x$ and $(\pi^*E)_{y'} \xrightarrow{\sim} E_{x'}$. Putting all these isomorphisms together, parallel transport along γ is given by

$$E_x \xrightarrow{\sim} (\pi^*E)_y \xrightarrow{\sim} (\pi^*E)_{y'} \xrightarrow{\sim} E_{x'}.$$

Regarding E_ρ as a holomorphic bundle on X, a theorem of Weil [Weil] says that a holomorphic bundle E on X is isomorphic to some E_ρ (i.e. E comes from a representation of $\pi_1(X,x)$) if and only if $E = \bigoplus_i E_i$ with indecomposable subbundles E_i of degree zero. A famous result by Narasimhan and Seshadri [Na-Se] says that a holomorphic vector bundle E of degree zero on X is stable if and only if E is isomorphic to E_ρ for some irreducible unitary representation ρ. Hence a holomorphic vector bundle comes from a unitary representation ρ if and only if it is of the form $E = \bigoplus_i E_i$ for stable (and hence indecomposable) subbundles of degree zero.

3. Fundamental Groups of p-Adic Curves

It is a natural question to look for a p-adic analogue of the results described in the previous section. We denote by $\mathbb{C}_p$ the completion of an algebraic closure $\overline{\mathbb{Q}}_p$ of $\mathbb{Q}_p$. Besides, $\overline{\mathbb{Z}}_p$ and $\mathfrak{o}$ denote the rings of integers in $\overline{\mathbb{Q}}_p$ and $\mathbb{C}_p$, respectively. By $k = \overline{\mathbb{F}}_p$ we denote the residue field of $\overline{\mathbb{Z}}_p$ and $\mathfrak{o}$.

We call any purely one-dimensional separated scheme of finite type over a field a curve. Let X be a smooth, projective and connected curve over $\overline{\mathbb{Q}}_p$ and $X_{\mathbb{C}_p}$ its base change to $\mathbb{C}_p$. We are looking for a relation between vector bundles on $X_{\mathbb{C}_p}$ and representations of the fundamental group of X.

First we have to clarify what we mean by fundamental group. Of course, in the algebraic setting there is no topological fundamental groups defined with closed paths. However, there is an algebraic fundamental group $\pi_1(X, x)$ for a base point $x \in X(\mathbb{C}_p)$. It is defined as the group of automorphisms of the fiber functor F_x. This fiber functor F_x maps a finite étale covering Y of X to the set of $\mathbb{C}_p$-valued points of Y lying over x. Hence an automorphism of F_x induces in a functorial way for every finite étale covering $Y \to X$ a permutation of the points in the fiber over x. Note that on a Riemann surface X, a closed topological path γ on X induces such an automorphism $\gamma : F_x \xrightarrow{\sim} F_x$ of the fiber functor. Namely, for every finite étale cover $Y \to X$ and every point $y \in Y$ over x we can lift γ to a continuous path in Y starting in y. Its endpoint y' also lies over x, and one can define γ by mapping $y \in F_x(Y)$ to $y' \in F_x(Y)$.

Since the algebraic fundamental group involves only the finite étale coverings, it is in fact analogous to the profinite completion of the topological fundamental group on Riemann surfaces.

There is also an analogue of non-closed paths. For two points x and x' in $X(\mathbb{C}_p)$ we call any isomorphism $F_x \xrightarrow{\sim} F_{x'}$ of the corresponding fiber functors an étale path from x to x. Such an étale path associates for all finite, étale $Y \to X$ to every point in the fiber of Y over x a point in the fiber of Y over x'.

The étale fundamental groupoid of $\Pi_1(X)$ of X is defined as the category such that the points in $X(C_p)$ are the objects and such the set of morphisms from $x \in X(\mathbb{C}_p)$ to $x' \in X(\mathbb{C}_p)$ is the set of étale paths from x to x', i.e. the set of isomorphisms of fiber functors $F_x \xrightarrow{\sim} F_{x'}$.

4. Finite Vector Bundles

The algebraic fundamental group only involves finite étale coverings, hence in the algebraic setting there is no universal covering. Therefore we can imitate the constructions on Riemann surfaces in the p-adic situation only for finite vector bundles, i.e. for vector bundles E on $X_{\mathbb{C}_p}$ such that there is a finite étale covering $\pi : Y_{\mathbb{C}_p} \to X_{\mathbb{C}_p}$ for which $\pi^* E$ is trivial.

Namely, for every finite vector bundle E on $X_{\mathbb{C}_p}$ we choose a finite, étale and Galois covering $\pi : Y_{\mathbb{C}_p} \to X_{\mathbb{C}_p}$ trivializing E and a point y in $Y_{\mathbb{C}_p}(\mathbb{C}_p)$ lying over x. Then there is a short exact sequence

$$0 \to \pi_1(Y_{\mathbb{C}_p}, y) \to \pi_1(X_{\mathbb{C}_p}, x) \to \mathrm{Gal}(Y_{\mathbb{C}_p}/X_{\mathbb{C}_p}) \to 0.$$

Besides, since π^*E is trivial, the fiber map $\Gamma(Y_{\mathbb{C}_p}, \pi^*E) \to (\pi^*E)_y$ is an isomorphism. As a pullback bundle, π^*E and also its set of global sections $\Gamma(Y_{\mathbb{C}_p}, \pi^*E)$ carries a natural $\mathrm{Gal}(Y_{\mathbb{C}_p}/X_{\mathbb{C}_p})$-action. Via the fiber isomorphism, this action induces a $\mathrm{Gal}(Y_{\mathbb{C}_p}/X_{\mathbb{C}_p})$-action on $(\pi^*E)_y$, which can be identified with the fiber E_x of E in x. Therefore we have defined a representation

$$\pi_1(X, x) = \pi_1(X_{\mathbb{C}_p}, x) \to \mathrm{Gal}(Y_{\mathbb{C}_p}/X_{\mathbb{C}_p}) \to \mathrm{Aut}(E_x).$$

In fact, this construction works for curves over arbitrary fields. In [La-Stu], Lange and Stuhler investigate finite vector bundles on a smooth, projective curve C over a field of characteristic p. Let us denote by F the absolute Frobenius on C, defined by the p-power-map on the structure sheaf. By [La-Stu], 1.4, a vector bundle E on C is finite if and only if for a suitable power F^n we have $F^{n*}E \xrightarrow{\sim} E$.

5. A Bigger Category of Vector Bundles

Let us again consider a smooth, projective and connected curve X over $\overline{\mathbb{Q}}_p$. We are interested in vector bundles on $X_{\mathbb{C}_p}$. Here we denote by $X_{\mathbb{C}_p}$ the base change of X with $\mathbb{C}_p$. This kind of notation for base changes will be used throughout this paper.

We have seen that finite vector bundles on $X_{\mathbb{C}_p}$ give rise to representations of the fundamental group. However, in this way we only get representations factoring over a finite quotient of $\pi_1(X, x)$.

The main idea for the construction of representations for a more general class of vector bundles is the following: We consider vector bundles with integral models (in a sense to be made precise below) which are "finite modulo p^n" for all n. A similar construction as the one for finite bundles then gives for all n representations of $\pi_1(X, x)$ modulo p^n, i.e. over $\mathfrak{o}/p^n\mathfrak{o}$. In the limit we get a representation of $\pi_1(X, x)$ over $\mathfrak{o}$ which we can tensor with $\mathbb{C}_p$. In fact, we will more generally define parallel transport along étale paths.

To be more precise, let us call any finitely presented, flat and proper $\overline{\mathbb{Z}}_p$-scheme (respectively $\mathfrak{o}$-scheme) with generic fiber X (respectively $X_{\mathbb{C}_p}$) a model of X (respectively $X_{\mathbb{C}_p}$).

Note that any model of X descends to a finite extension of $\mathbb{Z}_p$ and is irreducible and reduced by [Liu], 4.3.8.

We assume that the curves and their models are finitely presented over $\overline{\mathbb{Q}}_p$, repectively $\overline{\mathbb{Z}}_p$, so that they can be descended to a finite extension of $\mathbb{Q}_p$, respectively $\mathbb{Z}_p$. We need this descent to a noetherian situation in some arguments. Our vector bundles and their models however live over $\mathbb{C}_p$, respectively $\mathfrak{o}$.

Definition 1. *We denote by $\mathfrak{B}_{X_{\mathbb{C}_p}}$ the full subcategory of all vector bundles on $X_{\mathbb{C}_p}$ for which there is a model $\mathcal{X}$ of X over $\overline{\mathbb{Z}}_p$ and a vector bundle $\mathcal{E}$ on $\mathcal{X}_{\mathfrak{o}}$ with generic fiber E such that for all natural numbers $n \geq 1$ there is a finitely presented proper $\overline{\mathbb{Z}}_p$-morphism $\pi : \mathcal{Y} \to \mathcal{X}$ with the following two properties:*

i) The generic fiber $\pi_{\overline{\mathbb{Q}}_p} : Y = \mathcal{Y} \otimes \overline{\mathbb{Q}}_p \to X$ is finite and étale.

ii) The pullback bundle $\pi_{\mathfrak{o}}^* \mathcal{E}$ becomes trivial on $\mathcal{Y}_{\mathfrak{o}}$ after base change with $\mathfrak{o}/p^n\mathfrak{o}$.

Hence $\mathfrak{B}_{X_{\mathbb{C}_p}}$ can be viewed as the category of all vector bundles on $X_{\mathbb{C}_p}$ which are in some sense finite modulo all p^n. Note however that the special fiber of the coverings $\pi : \mathcal{Y} \to \mathcal{X}$ will in general be neither finite nor étale. Only the generic fiber has these properties.

Note that every vector bundle E on $X_{\mathbb{C}_p}$ can be extended to a bundle on a suitable model of $X_{\mathbb{C}_p}$, see e.g. [De-We2], theorem 5. Hence the important point in definition 1 is the existence of the coverings π. Although we omit it in our notation, π depends of course on n.

This definition of the category $\mathfrak{B}_{X_{\mathbb{C}_p}}$ is useful for the construction of parallel transport, as we will see in the next section. However, it is difficult to check whether a bundle fulfils the conditions in definition 1. In sections 10 and 11 we give a more intrinsic characterizations of $\mathfrak{B}_{X_{\mathbb{C}_p}}$ (and also of a bigger category of bundles to be defined below).

6. Parallel Transport on Bundles in $\mathfrak{B}_{X_{\mathbb{C}_p}}$

Let us fix a bundle E in $\mathfrak{B}_{X_{\mathbb{C}_p}}$. For all n, the morphism $\pi : \mathcal{Y} \to \mathcal{X}$ descends to a morphism $\pi_R : \mathcal{Y}_R \to \mathcal{X}_R$ over a discrete valuation ring R finite over $\mathbb{Z}_p$. If R is chosen big enough, there is a semistable R-curve $\mathcal{Y}'$ over $\mathcal{Y}$ with geometrically connected generic fiber such that the induced map $\pi' : \mathcal{Y}' \otimes \overline{\mathbb{Z}}_p \to \mathcal{X}$ also has the properties i) and ii) in definition 1, see e.g. [De-We2], theorem 1. Hence we can assume that all the $\mathcal{Y}$ in definition 1 are semistable with connected generic fibers. It follows that the structure maps $\lambda : \mathcal{Y} \to \mathrm{Spec}\overline{\mathbb{Z}}_p$ are cohomologically flat in dimension zero. This means that the formation of $\lambda_* \mathcal{O}_{\mathcal{Y}}$ commutes with arbitrary base changes. As a consequence, we have $(\lambda \otimes R)_* \mathcal{O}_{\mathcal{Y} \otimes R} = \mathcal{O}_R$ for every $\overline{\mathbb{Z}}_p$-algebra R.

Now let us fix two points x and x' in $X(\mathbb{C}_p)$ and an étale path γ from x to x', i.e. an isomorphism of fiber functors $F_x \xrightarrow{\sim} F_{x'}$. Besides, we fix some $n \geq 1$. Then there is a morphism $\pi : \mathcal{Y} \to \mathcal{X}$ as in definition 1 with a semistable $\mathcal{Y}$. We fix some y in $\mathcal{Y}(\mathbb{C}_p)$ with $\pi(y) = x$. The fiber functor γ maps y to a point $y' \in \mathcal{Y}(\mathbb{C}_p)$ with $\pi(y') = x'$. The points y and y' can be extended to $\mathfrak{o}$-rational points on the proper model $\mathcal{Y}$, and x and x' can be extended to $\mathfrak{o}$-rational point $x_{\mathfrak{o}}$ and $x'_{\mathfrak{o}}$ on $\mathcal{X}$.

By x_n, x'_n, y_n respectively y'_n we denote the induced $\mathfrak{o}_n$-rational points, where we put $\mathfrak{o}_n = \mathfrak{o}/p^n\mathfrak{o} = \overline{\mathbb{Z}}_p/p^n\overline{\mathbb{Z}}_p$. Besides, we write $\pi_n : \mathcal{Y}_n \to \mathcal{X}_n$ for the base change of π with $\mathfrak{o}_n$. In particular, $\mathcal{X}_n = \mathcal{X} \otimes_{\overline{\mathbb{Z}}_p} \mathfrak{o}_n$ and $\mathcal{Y}_n = \mathcal{Y} \otimes_{\overline{\mathbb{Z}}_p} \mathfrak{o}_n$. Let $\mathcal{E}_n = \mathcal{E} \otimes_{\mathfrak{o}} \mathfrak{o}_n$ be the induced vector bundle on $\mathcal{X}_n$.

For the $\overline{\mathbb{Z}}_p$-algebra $R = \mathfrak{o}_n$ the equality above gives $(\lambda \otimes \mathfrak{o}_n)_* \mathcal{O}_{\mathcal{Y}_n} = \mathfrak{o}_n$. This implies that the $\mathfrak{o}_n$-rational point $y_n : \mathrm{Spec}(\mathfrak{o}_n) \to \mathcal{Y}_n$ induces by pullback an isomorphism

$$y_n^* : \Gamma(\mathcal{Y}_n, \mathcal{O}_{\mathcal{Y}_n}) \xrightarrow{\sim} \Gamma(\mathrm{Spec}(\mathfrak{o}_n), y_n^* \mathcal{O}_{\mathcal{Y}_n}) = \mathfrak{o}_n.$$

Since $\pi_n^* \mathcal{E}_n$ is trivial on $\mathcal{Y}_n$, pullback by y_n also induces an isomorphism

$$\begin{aligned} y_n^* : \Gamma(\mathcal{Y}_n, \pi_n^* \mathcal{E}_n) &\xrightarrow{\sim} \Gamma(\mathrm{Spec}(\mathfrak{o}_n), y_n^* \pi_n^* \mathcal{E}_n) \\ &= \Gamma(\mathrm{Spec}(\mathfrak{o}_n), x_n^* \mathcal{E}_n) =: \mathcal{E}_{x_n}. \end{aligned}$$

Similarly, pullback by y'_n induces an isomorphism

$$y_n'^* : \Gamma(\mathcal{Y}_n, \pi_n^* \mathcal{E}_n) \xrightarrow{\sim} \Gamma(\mathrm{Spec}(\mathfrak{o}_n), y_n'^* \pi_n^* \mathcal{E}_n) = \Gamma(\mathrm{Spec}(\mathfrak{o}_n), x_n'^* \mathcal{E}_n) =: \mathcal{E}_{x'_n}.$$

Now we define $\rho_{\mathcal{E},n}(\gamma)$ to be the composition

$$\rho_{\mathcal{E},n}(\gamma) = y_n'^* \circ (y_n^*)\mathrm{inv} : \mathcal{E}_{x_n} \xleftarrow{\sim} \Gamma(\mathcal{Y}_n, \pi_n^* \mathcal{E}_n) \xrightarrow{\sim} \mathcal{E}_{x'_n}.$$

Then the maps $\rho_{\mathcal{E},n}(\gamma)$ form a projective system. We denote by

$$\mathcal{E}_{x_\mathfrak{o}} = \varprojlim_n \mathcal{E}_{x_n}$$

the fiber of $\mathcal{E}$ over $x_\mathfrak{o}$ and by $\mathcal{E}_{x'_\mathfrak{o}}$ the fiber of $\mathcal{E}$ over $x'_\mathfrak{o}$. In the limit we get a map

$$\rho_{\mathcal{E}}(\gamma) : \mathcal{E}_{x_\mathfrak{o}} \to \mathcal{E}_{x'_\mathfrak{o}}.$$

Tensoring with $\mathbb{C}_p$ we finally get a "parallel transport" map

$$\rho_E(\gamma) : E_x \to E_{x'},$$

where E_x is the fiber of E in x. By construction, this isomorphism is continuous in the p-adic topology. In [De-We2], section 3, we show that the construction is independent of all choices.

7. Working Outside a Divisor on $X_{\mathbb{C}_p}$

The definition of parallel transport in the last section can be extended to open subcurves of X in the following way:

Let D be a divisor on X, and put $U = X \backslash D$. In the following, only the support of D plays a role. By $\mathfrak{B}_{X_{\mathbb{C}_p},D}$ we denote the category of all vector bundles E on $X_{\mathbb{C}_p}$ such that there exists a model $\mathcal{X}$ of X and a vector bundle $\mathcal{E}$ on $\mathcal{X}_\mathfrak{o}$ with generic fiber E such that for all natural numbers $n \geq 1$ there is a finitely presented proper $\overline{\mathbb{Z}}_p$-morphism $\pi : \mathcal{Y} \to \mathcal{X}$ with the following two properties:

i) The generic fiber $\pi_{\overline{\mathbb{Q}}_p} : Y = \mathcal{Y} \otimes \overline{\mathbb{Q}}_p \to X$ is finite and its restriction to $\pi_{\overline{\mathbb{Q}}_p}\mathrm{inv}(U)$ is étale.

ii) The pullback bundle $\pi_\mathfrak{o}^* \mathcal{E}$ becomes trivial on $\mathcal{Y}_\mathfrak{o}$ after base change with $\mathfrak{o}/p^n\mathfrak{o}$.

Thus $\mathfrak{B}_{X_{\mathbb{C}_p}} = \mathfrak{B}_{X_{\mathbb{C}_p},\emptyset}$. Let γ be an étale path from $x \in U(\mathbb{C}_p)$ to $x' \in U(\mathbb{C}_p)$ in U, i.e. an isomorphism of fiber functors defined on the open curve U. The same construction as above gives for every bundle E in $\mathfrak{B}_{X_{\mathbb{C}_p},D}$ an isomorphism "of parallel transport"

$$\rho_E(\gamma) : E_x \to E_{x'}.$$

8. Properties of Parallel Transport

By [De-We2], theorem 22, the association $\rho : \gamma \mapsto \rho_E(\gamma)$ is functorial in γ, i.e. for $E \in \mathfrak{B}_{X_{\mathbb{C}_p},D}$ and étale paths γ from x to x' and γ' from x' to x'' we have $\rho_E(\gamma' \circ \gamma) = \rho_E(\gamma') \circ \rho_E(\gamma)$ as isomorphisms from E_x to $E_{x''}$.

Recall that for $U = X \backslash D$ the fundamental groupoid $\Pi_1(U)$ is the category with object set $U(\mathbb{C}_p)$ and étale paths as morphisms. In fact, for x and x' in $U(\mathbb{C}_p)$ the morphism set $\mathrm{Mor}(x, x') = \{$étale paths from x to $x'\} = \mathrm{Iso}(F_x, F_{x'})$ carries a natural topology, since it is profinite. A functor from $\Pi_1(U)$ to the category of finite-dimensional $\mathbb{C}_p$-vector spaces which is continuous on the morphism spaces is called a representation of $\Pi_1(U)$ on finite-dimensional $\mathbb{C}_p$-vector spaces.

With this terminology, for every vector bundle $E \in \mathfrak{B}_{X_{\mathbb{C}_p},D}$ the functor

$$\rho_E : \Pi_1(U) \to \{\text{finite dimensional } \mathbb{C}_p - \text{vector spaces}\}$$

given by $x \mapsto E_x$ on objects and $\gamma \mapsto \rho_E(\gamma)$ on morphisms is a representation of $\Pi_1(U)$.

By [De-We2], theorem 28, we have

Theorem 2. *i) The association $E \mapsto \rho_E$ is functorial for morphisms in $\mathfrak{B}_{X_{\mathbb{C}_p},D}$, exact and commutes with tensor products, duals, internal homs and exterior powers.*
ii) If $f : X \to X'$ is a morphism between smooth, projective, connected curves over $\overline{\mathbb{Q}}_p$, and D' is a divisor on X', then pullback of vector bundles induces a functor $f^* : \mathfrak{B}_{X'_{\mathbb{C}_p},D'} \to \mathfrak{B}_{X_{\mathbb{C}_p},f^*D'}$ which commutes with tensor products, duals, internal homs and exterior powers.

Besides, for every bundle E *in* $\mathfrak{B}_{X'_{\mathbb{C}_p},D'}$ *we have* $\rho_{f^*E} = \rho_E \circ f_*$ *where* f_* *is the induced functor* $\Pi_1(X \backslash f^*D') \to \Pi_1(X' \backslash D')$ *on fundamental groupoids. Here we identify* $(f^*E)_x$ *with* $E_{f(x)}$.
iii) If $X = X_K \otimes_K \overline{\mathbb{Q}}_p$ *for some field* K *between* $\mathbb{Q}_p$ *and* $\overline{\mathbb{Q}}_p$*, then every element* σ *in* $\mathrm{Gal}(\overline{\mathbb{Q}}_p/K)$ *acts in a natural way on* $\mathfrak{B}_{X_{\mathbb{C}_p},D}$*. Besides,* σ *acts on* $\Pi_1(X \backslash D)$ *and on the category of finite-dimensional* $\mathbb{C}_p$*-vector spaces and hence on the category of representations of* $\Pi_1(X \backslash D)$*. The functor* ρ *commutes with these actions.*

9. Semistable Bundles

Recall that for a vector bundle E on a smooth, projective and connected curve over a field k the slope is defined by $\mu(E) = \deg(E)/\mathrm{rk}(E)$. The bundle E is called semistable (respectively stable), if for all proper non-zero subbundles F of E the inequality $\mu(F) \leq \mu(E)$ (respectively $\mu(F) < \mu(E)$) holds.

Let E be a bundle in $\mathfrak{B}_{X_{\mathbb{C}_p},D}$ for some divisor D. By definition, there exists a model $\mathcal{X}$ of X, a vector bundle $\mathcal{E}$ on $\mathcal{X}_{\mathfrak{o}}$ extending E and a finitely presented, flat morphism $\pi : \mathcal{Y} \to \mathcal{X}$ which is generically finite and étale over $X \backslash D$ such that the special fiber of $\pi_{\mathfrak{o}}^*\mathcal{E}$ is trivial. Here we only use the condition for $n = 1$ in the definition of $\mathfrak{B}_{X_{\mathbb{C}_p},D}$.

Using descent to a suitable discrete valuation ring and an argument due to Raynaud (see [De-We2], theorem 13) one can show that triviality of the special fiber of $\pi_{\mathfrak{o}}^*\mathcal{E}$ implies that the generic fiber $\pi_{\mathbb{C}_p}^* E$ is semistable of degree zero on $\mathcal{Y} \otimes \mathbb{C}_p$. Since $\pi_{\mathbb{C}_p}$ is finite, E is also semistable of degree zero on $X_{\mathbb{C}_p}$.

10. A Simpler Description of $\mathfrak{B}_{X_{\mathbb{C}_p},D}$

It turns out that the existence of some $\pi : \mathcal{Y} \to \mathcal{X}$ as above such that $\pi_{\mathfrak{o}}^*\mathcal{E}$ has a trivial special fiber is also sufficient for a bundle to lie in $\mathfrak{B}_{X_{\mathbb{C}_p},D}$. To be precise, by [De-We2], theorem 16 we have

Theorem 3. *A vector bundle* E *lies in* $\mathfrak{B}_{X_{\mathbb{C}_p},D}$ *if and only if there is a model* $\mathcal{X}$ *of* X*, a vector bundle* $\mathcal{E}$ *on* $\mathcal{X}_{\mathfrak{o}}$ *extending* E *and a finitely presented, proper morphism* $\pi : \mathcal{Y} \to \mathcal{X}$ *which is generically finite and étale over* $X \backslash D$ *such that the special fiber of* $\pi_{\mathfrak{o}}^*\mathcal{E}$ *is trivial.*

Let E be a bundle in $\mathfrak{B}_{X_{\mathbb{C}_p},D}$ with a model $\mathcal{E}$ on $\mathcal{X}_{\mathfrak{o}}$, and let $\pi : \mathcal{Y} \to \mathcal{X}$ be as in theorem 3. The image $\pi(\mathcal{Y})$ is closed in $\mathcal{X}$ and contains the generic fiber, since π is generically finite. Since the model $\mathcal{X}$ is irreducible, π must be surjective. Let $\pi_k : \mathcal{Y}_k \to \mathcal{X}_k$ be its special fiber (recall that $k = \overline{\mathbb{F}}_p$ is the residue field of $\overline{\mathbb{Z}}_p$ and $\mathfrak{o}$). Let $C_1, \ldots, C_r$ be the irreducible components of $\mathcal{X}_k$ with their reduced structure. Since π is surjective, every C_ν is finitely dominated by an irreducible component D_ν of $\mathcal{Y}_k$. Hence the restriction of $\mathcal{E}_k$ to C_ν is trivialized by a finite covering (which of course in general is not étale). Let $\tilde{C}_\nu$ be the normalization of C_ν. Then also the pullback of $\mathcal{E}_k$ to the smooth projective k-curve $\tilde{C}_\nu$ is trivialized by a finite covering, namely the normalization of D_ν.

11. Strongly Semistable Reduction

Let us denote by F the absolute Frobenius in characteristic p.

Definition 4. *Let E be a vector bundle on a smooth, projective, connected curve C over a field of characteristic p. Then E is called strongly semistable if and only if for all $n \geq 0$ the pullback $F^{n*}E$ is semistable on C.*

Let E be a vector bundle in the category $\mathfrak{B}_{X_{\mathbb{C}_p},D}$ with an extension $\mathcal{E}$ to a model $\mathcal{X}_{\mathfrak{o}}$ such that the special fiber of $\mathcal{E}$ becomes trivial after pullback to some $\mathcal{Y} \to \mathcal{X}$ as above. We have seen in the preceding section that for all irreducible components C_ν of the special fiber $\mathcal{X}_k$ the pullback of $\mathcal{E}_k$ to the normalization $\tilde{C}_\nu$ of C_ν becomes trivial on a finite covering. As the trivial vector bundle is strongly semistable of degree zero, we deduce that the pullback of $\mathcal{E}_k$ to any $\tilde{C}_\nu$ is also strongly semistable of degree zero.

One of the main results in [De-We2] shows that this property is also sufficient for a bundle to lie in $\mathfrak{B}_{X_{\mathbb{C}_p},D}$.

To be precise, we say that a vector bundle E on $X_{\mathbb{C}_p}$ has strongly semistable reduction of degree zero, if there is a model $\mathcal{X}$ of X and a vector bundle $\mathcal{E}$ on $\mathcal{X}_{\mathfrak{o}}$ extending E such that the pullback of the special fiber $\mathcal{E}_k$ of $\mathcal{E}$ to all normalized irreducible components of $\mathcal{X}_k$ is strongly semistable of degree zero. We denote the (full) category of all strongly semistable bundles on $X_{\mathbb{C}_p}$ by $\mathfrak{B}^s_{X_{\mathbb{C}_p}}$.

Theorem 5. *For every vector bundle E on $X_{\mathbb{C}_p}$ with strongly semistable reduction of degree zero there is a divisor D such that E is contained in $\mathfrak{B}_{X_{\mathbb{C}_p},D}$. In other words, we have $\mathfrak{B}^s_{X_{\mathbb{C}_p}} = \bigcup_D \mathfrak{B}_{X_{\mathbb{C}_p},D}$.*

For the **proof** see [De-We2], theorem 36. The idea is the following. We have just seen that all categories $\mathfrak{B}_{X_{\mathbb{C}_p},D}$ are contained in $\mathfrak{B}^s_{X_{\mathbb{C}_p}}$. Hence it remains to show that for every bundle E on $X_{\mathbb{C}_p}$ with strongly semistable reduction $\mathcal{E}_k$ of degree zero there is a divisor D on X such that E is contained in $\mathfrak{B}_{X_{\mathbb{C}_p},D}$. There is a finite extension K of $\mathbb{Q}_p$ such that the model $\mathcal{X}$ descends to a model $\mathcal{X}_{\mathfrak{o}_K}$ over the ring of integers in K and such that the special fiber $\mathcal{E}_k$ of $\mathcal{E}$ descends to a vector bundle $\mathcal{E}_{\mathbb{F}_q}$ on the special fiber $\mathcal{X}_{\mathbb{F}_q}$ of $\mathcal{X}_{\mathfrak{o}_K}$. By enlarging K if necessary, we can assume that the irreducible components $D_1, \ldots, D_r$ of $\mathcal{X}_{\mathbb{F}_q}$ are geometrically irreducible, hence they give the components $C_1, \ldots, C_r$ of $\mathcal{X}_k$ by base change. Then the pullback of $\mathcal{E}_{\mathbb{F}_q}$ to all normalized components $\tilde{D}_\nu$ is strongly semistable of degree zero.

Now on the smooth, projective curve $\tilde{D}_\nu$ over the finite field $\mathbb{F}_q$ there are only finitely many isomorphism classes of semistable bundles of degree zero. This implies that on $\mathcal{X}_{\mathbb{F}_q}$ there are only finitely many isomorphism classes of bundles whose pullbacks to all normalized components $\tilde{D}_\nu$ are semistable of degree zero (see the proof of [De-We2], theorem 18). For $q = p^r$ we put $F_q = F^r$, i.e. F_q is the Frobenius fixing $\mathbb{F}_q$. We regard F_q also as a $\mathbb{F}_q$-linear automorphism of $\mathcal{X}_{\mathbb{F}_q}$ and consider all pullbacks $F_q^{k*}\mathcal{E}_{\mathbb{F}_q}$. This gives an infinite collection of bundles whose pullback to all $\tilde{D}_\nu$ is semistable of degree zero. Since there are only finitely many isomorphism classes available, we find two bundles in this collection which are isomorphic. Hence there are natural numbers $r > s$ satisfying $F_q^{r*}\mathcal{E}_{\mathbb{F}_q} \simeq F_q^{s*}\mathcal{E}_{\mathbb{F}_q}$, i.e. for $n = r - s$ we find $F_q^{n*}(F_q^{s*}\mathcal{E}_{\mathbb{F}_q}) \simeq F_q^{s*}\mathcal{E}_{\mathbb{F}_q}$. By the theorem of Lange and Stuhler cited in section 4, this implies that $F_q^{s*}\mathcal{E}_{\mathbb{F}_q}$ is a finite bundle, i.e. it becomes trivial on a finite étale covering $\omega : Y_0 \to \mathcal{X}_{\mathbb{F}_q}$. Hence $\mathcal{E}_{\mathbb{F}_q}$ becomes trivial after pullback via $\omega \circ F_q^s$.

Now we have to lift this covering of the special fiber $\mathcal{X}_{\mathbb{F}_q}$ to a covering of the whole model $\mathcal{X}_{\mathfrak{o}_K}$. This is no problem for the étale part ω, which lifts to a finite, étale morphism $\omega_{\mathfrak{o}_K} : \tilde{\mathcal{Y}}_{\mathfrak{o}_K} \to \mathcal{X}_{\mathfrak{o}_K}$, but a non-trivial task for the Frobenius. As a first step we find a semistable, regular and projective $\mathfrak{o}_K$-scheme $\mathcal{Y}_{\mathfrak{o}_K}$ together with a morphism $\mathcal{Y}_{\mathfrak{o}_K} \to \tilde{\mathcal{Y}}_{\mathfrak{o}_K}$ (we might have to enlarge K here). The map $\theta : \mathcal{Y}_{\mathfrak{o}_K} \to \tilde{\mathcal{Y}}_{\mathfrak{o}_K} \to \mathcal{X}_{\mathfrak{o}_K}$ also has the property that $\theta_{\mathbb{F}_q} \circ F_q^s$ trivializes the special fiber of $\mathcal{E}_{\mathbb{F}_q}$ by pullback.

We embed $\mathcal{Y}_{\mathfrak{o}_K}$ in some projective space $\mathbb{P}^N_{\mathfrak{o}_K}$. Now we define a Frobenius lift f_{q^s} on $\mathbb{P}^N_{\mathfrak{o}_K}$ by $[x_0, \ldots, x_N] \mapsto [x_0^{q^s}, \ldots, x_N^{q^s}]$ on projective coordinates. The special fiber of f_{q^s} is the Frobenius F_q^s. Let $\mathcal{Y}'_{\mathfrak{o}_K}$ be the base change of $\mathcal{Y}_{\mathfrak{o}_K}$ by f_{q^s}, i.e. $\mathcal{Y}'_{\mathfrak{o}_K}$ sits in a cartesian diagram

$$\begin{array}{ccc} \mathcal{Y}'_{\mathfrak{o}_K} & \to & \mathcal{Y}_{\mathfrak{o}_K} \\ \downarrow & & \downarrow \tau \\ \mathbb{P}^N_{\mathfrak{o}_K} & \stackrel{f_{q^s}}{\to} & \mathbb{P}^N_{\mathfrak{o}_K}. \end{array}$$

Of course, the generic fiber of this Frobenius lift f_{q^s} is étale only outside the union of the coordinate hyperplanes. But we can twist the projective embedding τ of $\mathcal{Y}_{\mathfrak{o}_K}$ into $\mathbb{P}^N_{\mathfrak{o}_K}$ by an automorphism in PGL_N so that the morphism $\mathcal{Y}'_{\mathfrak{o}_K} \to \mathcal{Y}_{\mathfrak{o}_K} \to \mathcal{X}_{\mathfrak{o}_K}$ is generically étale outside a divisor D on the generic fiber X_K of $\mathcal{X}_{\mathfrak{o}_K}$.

Now we look at the special fiber of this morphism $\mathcal{Y}'_{\mathfrak{o}_K} \to \mathcal{Y}_{\mathfrak{o}_K} \to \mathcal{X}_{\mathfrak{o}_K}$. There is a natural map $i = (\tau_{\mathbb{F}_q}, F_q^s) : \mathcal{Y}_{\mathbb{F}_q} \to \mathcal{Y}'_{\mathbb{F}_q}$. By definition, it gives the Frobenius F_q^s after composition with the projection to $\mathcal{Y}_{\mathbb{F}_q}$. Since $\mathcal{Y}_{\mathbb{F}_q}$ is semistable, it is reduced. Besides, i induces an isomorphism with the reduced induced structure of $\mathcal{Y}'_{\mathbb{F}_q}$, see [De-We2], lemma 19. Now we dominate $\mathcal{Y}'_{\mathfrak{o}_K}$ by a semistable $\mathfrak{o}_K$-scheme $\mathcal{Z}_{\mathfrak{o}_K}$ (possibly after enlarging K). Thus we get a chain of morphisms $\mathcal{Z}_{\mathfrak{o}_K} \to \mathcal{Y}'_{\mathfrak{o}_K} \to \mathcal{Y}_{\mathfrak{o}_K} \to \mathcal{X}_{\mathfrak{o}_K}$. Let us look at the special fibers

$$\mathcal{Z}_{\mathbb{F}_q} \to \mathcal{Y}'_{\mathbb{F}_q} \to \mathcal{Y}_{\mathbb{F}_q} \to \mathcal{X}_{\mathbb{F}_q}.$$

Since the semistable curve $\mathcal{Z}_{\mathbb{F}_q}$ is reduced, the first map factors through the reduced induced structure of $\mathcal{Y}'_{\mathbb{F}_q}$, hence through i. But i composed with the projection $\mathcal{Y}'_{\mathbb{F}_q} \to \mathcal{Y}_{\mathbb{F}_q}$ is the Frobenius F_q^s. Hence $\mathcal{Z}_{\mathbb{F}_q} \to \mathcal{X}_{\mathbb{F}_q}$ factors through $\theta_{\mathbb{F}_q} \circ F_q^s$ which is equal to $F_q^s \circ \theta_{\mathbb{F}_q}$. Hence the pullback of $\mathcal{E}_{\mathbb{F}_q}$ is trivial.

Going up to $\mathfrak{o}$, we get a finitely presented, proper morphism $\pi : \mathcal{Z} \to \mathcal{X}$ which generically is finite and étale over $X \backslash D$ such that the special fiber of $\pi_{\mathfrak{o}}^* \mathcal{E}$ is trivial. Hence E lies in $\mathfrak{B}_{X_{\mathbb{C}_p},D}$ by theorem 3, which proves our claim. $\square$

In this argument, the freedom of changing the embedding of $\mathcal{Y}_{\mathfrak{o}_K}$ into projective space by an automorphism in PGL_N can be further exploited to find a second divisor D' on X which is disjoint from D such that E also lies in $\mathfrak{B}_{X_{\mathbb{C}_p},D'}$. Hence by the construction in section 2, we can define parallel transport on E along étale paths in $X \backslash D$ and along étale paths in $X \backslash D'$. It can be shown that these two constructions fit together on the intersection of the open subcurves $X \backslash D$ and $X \backslash D'$ and hence give rise to parallel transport along étale paths in the whole of X (see [De-We2], proposition 34). Applying theorem 2 we then deduce the following corollary (see [De-We2], theorem 36).

Corollary 6. *There is a functor ρ from $\mathfrak{B}^s_{X_{\mathbb{C}_p}}$ to the category of representations of $\Pi_1(X)$ on finite-dimensional $\mathbb{C}_p$-vector spaces, which is exact and commutes with tensor products, duals, internal homs and exterior powers. Besides, it behaves functorially with respect to pullbacks along morphisms of curves over $\overline{\mathbb{Q}}_p$ and is compatible with Galois-conjugation.*

Let $\mathfrak{B}^{ps}_{X_{\mathbb{C}_p}}$ be the (full) category of all vector bundles on $X_{\mathbb{C}_p}$ for which there exists a finite étale covering $\alpha_{\overline{\mathbb{Q}}_p} : Y \to X$ over $\overline{\mathbb{Q}}_p$ such that $\alpha^* E$ has strongly semistable reduction of degree zero on $Y_{\mathbb{C}_p}$, where $\alpha : Y_{\mathbb{C}_p} \to X_{\mathbb{C}_p}$ is the base change to $\mathbb{C}_p$. If E lies in $\mathfrak{B}^{ps}_{X_{\mathbb{C}_p}}$, we say that E has potentially strongly semistable reduction of degree zero.

Let E be a bundle on $X_{\mathbb{C}_p}$ with potentially strongly semistable reduction of degree zero, and let $\alpha_{\overline{\mathbb{Q}}_p} : Y \to X$ be a finite étale covering as above. We can assume that $Y \to X$ is a Galois covering. By corollary 6, $\alpha^* E$ defines a representation of the fundamental groupoid $\Pi_1(Y)$. Let γ be an étale path from x to x' on X. Once we choose a point y in Y lying over x, the path γ can be lifted to an étale path δ on Y from y to some point y' over x'. Then we define

$$\rho_E(\gamma) = \rho_{\alpha^* E}(\delta) : E_x = (\alpha^* E)_y \to (\alpha^* E)_{y'} = E_{x'}.$$

Choosing another lift δ amounts to choosing another point $\tilde{y}$ over x as starting point. Hence there is a Galois automorphism $\sigma \in \mathrm{Gal}(Y/X)$

with $\sigma(y) = \tilde{y}$. If $\sigma_* : \Pi_1(Y) \to \Pi_1(Y)$ denotes the natural functor given by σ, then the path $\sigma_*\delta$ is the lift of γ with starting point $\sigma(y) = \tilde{y}$. Since $\sigma^*\alpha^*E = \alpha^*E$, it follows from corollary 6 that $\rho_{\alpha^*E}(\delta) = \rho_{\alpha^*E}(\sigma_*\delta)$. Hence ρ_E is well-defined.

Therefore the construction of representations of the fundamental groupoid extends to bundles in $\mathfrak{B}^{ps}_{X_{\mathbb{C}_p}}$. All the properties stated in corollary 6 also hold for bundles in $\mathfrak{B}^{ps}_{X_{\mathbb{C}_p}}$.

12. How Big are our Categories of Bundles?

It is easy to see that all the categories $\mathfrak{B}_{X_{\mathbb{C}_p},D}$, $\mathfrak{B}^{s}_{X_{\mathbb{C}_p}}$ and $\mathfrak{B}^{ps}_{X_{\mathbb{C}_p}}$ are closed under direct sums, tensor products, internal homs and exterior powers. The following theorem collects more information about these categories.

Theorem 7. *i) The category $\mathfrak{B}^{ps}_{X_{\mathbb{C}_p}}$ contains all line bundles of degree zero.*

ii) For every divisor D, the categories $\mathfrak{B}_{X_{\mathbb{C}_p},D}$, $\mathfrak{B}^{s}_{X_{\mathbb{C}_p}}$ and $\mathfrak{B}^{ps}_{X_{\mathbb{C}_p}}$ are stable under extensions, i.e. if $0 \to E' \to E \to E'' \to 0$ is an exact sequence of vector bundles such that E' and E'' are in the respective category, the same holds for E.

iii) If E is contained in $\mathfrak{B}^{s}_{X_{\mathbb{C}_p}}$, respectively $\mathfrak{B}^{ps}_{X_{\mathbb{C}_p}}$, then every subbundle of degree zero and every quotient bundle of degree zero is also contained in $\mathfrak{B}^{s}_{X_{\mathbb{C}_p}}$, respectively $\mathfrak{B}^{ps}_{X_{\mathbb{C}_p}}$.

A **proof** for i) can be found in [De-We2], theorem 12, ii) is proven in [De-We2], theorem 11, and iii) is shown in [De-We3], theorem 9.

Theorem 7,iii) implies that $\mathfrak{B}^{ps}_{X_{\mathbb{C}_p}}$ together with its natural fiber functor $E \mapsto E_x$ for some fixed $x \in X(\mathbb{C}_p)$ is a neutral Tannaka category, see [De-We3], theorem 12.

If X is an elliptic curve, then by Atiyah's classification [At] of vector bundles on $X_{\mathbb{C}_p}$ the category $\mathfrak{B}^{ps}_{X_{\mathbb{C}_p}}$ contains all semistable bundles of degree zero, cf. [De-We2], corollary 15. Hence for elliptic curves $\mathfrak{B}^{ps}_{X_{\mathbb{C}_p}}$ coincide with the category of all semistable bundles of degree zero.

For curves of higher genus it is an open question, if the category $\mathfrak{B}^{ps}_{X_{\mathbb{C}_p}}$ coincides with the category of all semistable bundles of degree zero on $X_{\mathbb{C}_p}$.

In [De-We3], theorem 12, it is shown that this is the case if and only if the corresponding subcategories of polystable bundles of degree zero coincide. Here a vector bundle on $X_{\mathbb{C}_p}$ is called polystable of degree zero if it is isomorphic to a direct sum of stable bundles of degree zero. Denote by $\mathfrak{T}^{\mathrm{ss}}_{\mathrm{red}}$ (respectively $\mathfrak{B}^{\mathrm{ps}}_{\mathrm{red}}$) the category of all polystable bundles of degree zero on $X_{\mathbb{C}_p}$ (repectively of all polystable bundles of degree zero contained in $\mathfrak{B}^{ps}_{X_{\mathbb{C}_p}}$). In [De-We3], theorem 16, it is shown that the Tannaka groups of $\mathfrak{T}^{\mathrm{ss}}_{\mathrm{red}}$ and $\mathfrak{B}^{\mathrm{ps}}_{\mathrm{red}}$ have the same group of connected components.

13. Representations of the Fundamental Group

If we fix a base point $x \in X(\mathbb{C}_p)$ and restrict the functor ρ to closed étale paths in x, then every vector bundle E in $\mathfrak{B}^{ps}_{X_{\mathbb{C}_p}}$ gives rise to a continuous representation

$$\pi_1(X, x) \to \mathrm{Aut}(E_x)$$

of the fundamental group. In [De-We1], we define an analogue of $\mathfrak{B}_{X_{\mathbb{C}_p}}$ and ρ on an Abelian variety A over $\overline{\mathbb{Q}}_p$ with good reduction. We also show how our construction generalizes a map in [Ta] and how it is related to the Hodge-Tate decomposition of $H^1_{et}(A, \mathbb{Q}_p) \otimes \mathbb{C}_p$.

14. Mumford Curves

Finally, we consider the special case that X is a Mumford curve (see [Mu]), i.e. X descends to a curve X_K over a finite extension K of $\mathbb{Q}_p$ which has a rigid analytic uniformization as Ω/Γ. Here Γ is a Schottky group in $\mathrm{PGL}(2, K)$ and Ω is the open subset of the rigid analytic $\mathbb{P}^1_K$ where Γ acts discontinously. Let $\mathcal{X}_{\mathfrak{o}_K}$ be the minimal regular model of X_K, see e.g. [Liu], section 9.3. Then all normalized irreducible components in the special fiber of $\mathcal{X}_{\mathfrak{o}_K}$ are $\mathbb{P}^1$'s.

In [Fa2], Faltings shows that there is an equivalence of categories between semistable vector bundles E of degree zero on X_K and K-linear representations of the Schottky group Γ which satisfy a certain boundedness condition. In [Re-Pu], his results were generalized to non-discrete p-adic base fields, e.g. $\mathbb{C}_p$. In [He], Herz shows how Faltings' construction is related to the representations of the fundamental group given by bundles in $\mathfrak{B}_{X_{\mathbb{C}_p}}$.

Namely, let E be a vector bundle on $X_{\mathbb{C}_p}$ which can be extended to a vector bundle $\mathcal{E}$ on the minimal model $\mathcal{X}_\mathfrak{o}$. By the generalization [Re-Pu] of Faltings' results, E gives rise to a representation π of the Schottky group Γ. Herz shows that this representation comes in fact from a representation (also called π) of Γ on a free $\mathfrak{o}$-module. Moreover, he shows that E lies in $\mathfrak{B}_{X_{\mathbb{C}_p}}$. Let x be a base point in $X(\mathbb{C}_p)$ and $x_\mathfrak{o}$ in $\mathcal{X}_\mathfrak{o}(\mathfrak{o})$ its extension. By construction, the representation $\rho_E : \pi_1(X,x) \to \mathrm{Aut}(E_x)$ is induced from a representation

$$\rho_\mathcal{E} : \pi_1(X,x) \to \mathrm{Aut}(\mathcal{E}_{x_\mathfrak{o}})$$

on the $\mathfrak{o}$-lattice $\mathcal{E}_{x_\mathfrak{o}}$ in E_x, where $\rho_\mathcal{E}$ is the limit of representations $\rho_{\mathcal{E},n} : \pi_1(X,x) \to \mathrm{Aut}(\mathcal{E}_{x_n})$. It is shown in [He] that $\rho_{\mathcal{E},n}$ factors through a finite quotient of Γ for a suitably chosen model $\mathcal{E}$, and that the induced representation on this quotient is isomorphic to the reduction of π modulo p^n. Hence $\rho_\mathcal{E}$ factors through the profinite completion $\widehat{\Gamma}$ of the Schottky group Γ and is isomorphic to the profinite completion of Faltings' representation π.

References

[At] M. Atiyah: Vector bundles over an elliptic curve. Proc. London Math. Soc. **7** (1957), 414–452

[De-We1] C. Deninger, A. Werner: Line bundles and p-adic characters. In: G van der Geer, B. Moonen, R. Schoof (eds): Number Fields and Function Fields - Two Parallel Worlds. Birkhäuser Progress in Mathematics, Vol 239, 2005, 101-131.

[De-We2] C. Deninger, A. Werner: Vector bundles on p-adic curves and parallel transport. Ann. Scient. Éc. Norm. Sup. **38** (2005) 553-597.

[De-We3] C. Deninger, A. Werner: On Tannaka duality for vector bundles on p-adic curves. Preprint unter http://arxiv.org/abs/math/0506263.

[Fa1] G. Faltings: Semistable vector bundles on Mumford curves. Invent. Math. **74** (1983), 199–212

[Fa2] G. Faltings: A p-adic Simpson correspondence. Adv. Math. **198** (2005) 847-862.

[He] G. Herz: On representations attached to semistable vector bundles on Mumford curves. Thesis Münster 2005.

[La-Stu] H. Lange, U. Stuhler: Vektorbündel auf Kurven und Darstellungen der algebraischen Fundamentalgruppe. Math. Z. **156** (1977), 73–83

[Liu] Q. Liu: Algebraic Geometry and Arithmetic Curves. Oxford University Press 2002

[Mu] D. Mumford: An analytic construction of degenerating curves over complete local rings. Compositio Math. **24** (1972) 129-174.

[Na-Se] M.S. Narasimhan, C.S. Seshadri: Stable and unitary vector bundles on a compact Riemann surface. Ann. Math. **82** (1965), 540–567

[Re-Pu] M. Reversat, M. van der Put: Fibrés vectoriels semistables sur une courbe de Mumfort. Math. Ann. **273** (1986) 573-600.

[Ta] J. Tate: p-divisible groups. In: Proceedings of a Conference on local fields. Driebergen 1966, 158–183

[Weil] A. Weil: Généralisation des fonctions abéliennes. J. de Math. P. et App. (IX) 17 (1938) 47-87

Annette WERNER
Department of Mathematics
Universität Stuttgart
D - 70569 Stuttgart, Germany
Email: werner@mathematik.uni-stuttgart.de

Absolute CM-Periods — Complex and p-Adic

Hiroyuki YOSHIDA

1. Introduction

This paper is written under the request of the organizers of the Karatsu conference who asked the author to give an exposition of absolute CM-periods in an easily accessible fashion. In fact, it reproduces the three given talks at the conference rather faithfully. The paper is divided into two parts. In part I, we will give an overview of the results in the author's book [Y4]. In part II, we will review a p-adic analogue of part I. The research was done in collaboration with Tomokazu Kashio. A paper which will give full perspectives of p-adic theory is in preparation [KY2], but the reader can see a preliminary version [KY1].

Now let us begin with the simplest case. Let K be an imaginary quadratic field. Let E be an elliptic curve defined over $\overline{\mathbf{Q}}$ with complex multiplication by K and ϖ be a period of E. The Chowla–Selberg formula states

$$\varpi \sim \sqrt{\pi} \prod_{a=1}^{d-1} \Gamma\left(\frac{a}{d}\right)^{w\chi(a)/4h}. \tag{1}$$

Here $-d$, h and w denote the discriminant, the class number and the number of roots of unity of K respectively; χ denotes the Dirichlet character corresponding to K. For a, $b \in \mathbf{C}$, we write $a \sim b$ if $b \neq 0$ and $a/b \in \overline{\mathbf{Q}}$.

The absolute CM-periods give a conjectural generalization of (1). Let F be a totally real algebraic number field and K be an abelian extension of F. We assume that K is a CM-field. For $\tau \in \mathrm{Gal}(K/F)$, we define the absolute CM-period symbol $g_K(\mathrm{id}, \tau)$ using the multiple gamma function. We conjecture that $p_K(\mathrm{id},\tau) \sim g_K(\mathrm{id},\tau)$ where p_K is Shimura's period symbol. The symbol g_K is defined by

$$(2) \quad g_K(\mathrm{id},\tau) = \pi^{-\mu(\tau)/2} \exp\left(\frac{1}{|G|}\sum_{\chi\in\hat{G}_-}\frac{\chi(\tau)}{L(0,\chi)}\sum_{c\in C_{\mathfrak{f}(\chi)}}\chi(c)X(c)\right).$$

We refer the reader to §4 for the explanation of notation; we only note that $X(c)$ is a certain ray class invariant attached to an ideal class c and has the following form.

$$(3) \qquad X(c) = \sum_{j\in J}\sum_{z\in R(C_j,c)} \log\frac{\Gamma_{r(j)}(z,v_j)}{\rho_{r(j)}(v_j)} + \text{correction terms.}$$

Here $\Gamma_{r(j)}(z,v_j)$ is the $r(j)$-ple gamma function, J and $R(C_j,c)$ are finite sets. A remarkable feature is that $X(c)$ contains information on Stark–Shintani units. In this way we obtain a coherent picture unifying periods and units. We refer the reader to §6 for more details.

Now let us explain the contents of Part II briefly. The p-adic counter part of the Chowla–Selberg formula is the Gross–Koblitz formula. Let p be a prime number which splits completely in an imaginary quadratic field K. Let $(p) = \mathfrak{P}\mathfrak{P}^\rho$ be the prime ideal decomposition where ρ denotes the complex conjugation. We put $\mathfrak{P}^h = (\alpha)$. We choose an embedding $K \subset \mathbf{Q}_p$ so that it induces $\mathfrak{P}$. Then

$$(4) \qquad \log_p(\alpha^\rho/\alpha) = \frac{w}{2}\sum_{a=1}^{d-1}\chi(a)\log_p\left(\Gamma_p\left(\frac{a}{d}\right)\right).$$

Here $\log_p$ is the p-adic logarithmic function and Γ_p is the p-adic gamma function. Fix an embedding $F \subset \overline{\mathbf{Q}_p}$ and let $\mathfrak{p}$ be the prime ideal of F induced from this embedding. We define the p-adic (logarithmic) absolute CM-period symbol by

$$(5) \quad lg_{p,K}(\mathrm{id},\tau) = -\frac{\mu(\tau)}{2}\log\mathfrak{p} + \frac{1}{|G|}\sum_{\chi\in\hat{G}_-}\frac{\chi(\tau)}{L(0,\chi)}\sum_{c\in C_{\mathfrak{f}(\chi)\mathfrak{p}}}\chi(c)X_p(c),$$

$$(6) \qquad X_p(c) = \sum_{j\in J}\sum_{z\in R(C_j),c)} L\Gamma_{p,r(j)}(z,v_j) + \text{correction terms.}$$

Here $L\Gamma_{p,r(j)}(z, v_j)$ denotes the p-adic logarithmic $r(j)$-ple gamma function. We can make a very precise conjecture on $lg_{p,K}$ when $\mathfrak{p}$ splits completely in K (see §8). The p-adic analogue of the Stark–Shintani conjecture is a conjecture of Gross. It predicts, if $\mathfrak{p}$ decomposes completely in K, the value of the derivative of a p-adic L-function at $s = 0$, $L_p'(0, \omega\chi)$, where $\chi \in \widehat{G}_-$ and ω is the Teichmüller character. We can show that our conjecture implies the Gross conjecture, but not conversely. In §10, we will discuss the meaning of the symbol $lg_{p,K}$ in the general case in terms of p-adic periods.

2. Notation

For a, $b \in \mathbf{C}$, we write $a \sim b$ if $b \neq 0$ and a/b is an algebraic number. We fix an algebraic closure $\overline{\mathbf{Q}}$ of $\mathbf{Q}$ in $\mathbf{C}$. By an algebraic number field, we understand an algebraic extension of $\mathbf{Q}$ of finite degree contained in $\overline{\mathbf{Q}}$. We denote by ρ the complex conjugation. Let F be an algebraic number field. The ring of integers and the group of units of F are denoted by $\mathcal{O}_F$ and E_F respectively. We denote by E_F^+ the group of all totally positive units of F. The class number of F is denoted by h_F. We denote by J_F the set of all isomorphisms of F into $\mathbf{C}$. If F is a totally real algebraic number field of degree n, $\infty_1, \ldots, \infty_n$ denote all archimedean primes of F. For an integral ideal $\mathfrak{f}$ of F, $C_\mathfrak{f}$ denotes the ideal class group modulo $\mathfrak{f}\infty_1 \cdots \infty_n$. By a CM-field, we understand a totally imaginary quadratic extension of a totally real algebraic number field. For a prime number p, $\overline{\mathbf{Q}_p}$ denotes an algebraic closure of $\mathbf{Q}_p$ and $\mathbf{C}_p$ denotes the completion of $\overline{\mathbf{Q}_p}$.

Part I. Complex Theory

§1. *Gamma function*

For $x > 0$, we define the Hurwitz–Lerch zeta function by

$$(1.1) \qquad \zeta(s, x) = \sum_{m=0}^{\infty} (x + m)^{-s}, \qquad \mathrm{Re}(s) > 1.$$

This function can be continued meromorphically to the whole s-plane. It is holomorphic except for a simple pole at $s = 1$. If $x = 1$, we have $\zeta(s, 1) = \zeta(s)$. It is well known that

$$\zeta(0, x) = \frac{1}{2} - x, \tag{1.2}$$

$$\left.\frac{\partial}{\partial s}\zeta(s, x)\right|_{s=0} = \log\Gamma(x) - \frac{1}{2}\log 2\pi. \tag{1.3}$$

In particular, we have

$$\zeta(0) = -\frac{1}{2}, \qquad \zeta'(0) = -\frac{1}{2}\log 2\pi.$$

Let χ be a Dirichlet character to the modulus f. Then we have

$$L(s, \chi) = \sum_{n=1}^{n} \chi(n)n^{-s} = \sum_{a=1}^{f-1}\sum_{m=0}^{\infty} \chi(a)(mf + a)^{-s}.$$

Hence

$$L(s, \chi) = \sum_{a=1}^{f-1} \chi(a) f^{-s} \zeta\left(s, \frac{a}{f}\right). \tag{1.4}$$

Assume that the associated primitive character to χ is nontrivial. By (1.2) and (1.3), we obtain

$$L(0, \chi) = -\frac{1}{f}\sum_{a=1}^{f-1} \chi(a)a, \tag{1.5}$$

$$L'(0, \chi) = \sum_{a=1}^{f-1} \chi(a) \log\Gamma\left(\frac{a}{f}\right) - \log f \cdot L(0, \chi). \tag{1.6}$$

Let K be an algebraic number field. Let

$$\zeta_K(s) = \sum_{\mathfrak{a}} N(\mathfrak{a})^{-s}, \qquad \mathrm{Re}(s) > 1$$

be the Dedekind zeta function. The analytic class number formula of Dirichlet–Dedekind states that

$$\mathrm{Res}_{s=1}\,\zeta_K(s) = \frac{2^{r_1}(2\pi)^{r_2} h_K R_K}{w_K |D_K|^{1/2}}.$$

Here r_1 and r_2 are the numbers of real and imaginary infinite places of K respectively; h_K, R_K, D_K and w_K denote the class number, the regulator, the discriminant and the number of roots of unity of K respectively. Using the functional equation, we see that this formula is equivalent to

$$\zeta_K(s) \sim -\frac{h_K R_K}{w_K} s^{r_1+r_2-1}, \qquad s \to 0. \tag{1.7}$$

Now let us consider the case where K is a quadratic extension of $\mathbf{Q}$. Let d be the discriminant of K. Define a Dirichlet character χ by

$$\chi(n) = \left(\frac{d}{n}\right)$$

using the Kronecker symbol. Then χ is the primitive Dirichlet character to the modulus $|d|$. We have

$$\zeta_K(s) = \zeta(s) L(s,\chi).$$

We distinguish two cases.

(I) Assume $d > 0$. Then

$$\zeta_K(s) \sim -\frac{h_K R_K}{2} s, \qquad s \to 0$$

by (1.7). We have

$$L(0,\chi) = 0, \qquad L'(0,\chi) = h_K \log \epsilon_0$$

since $\zeta(0) = -1/2$, $R_K = \log \epsilon_0$. Here ϵ_0 denotes the fundamental unit of K. By (1.6), we have

$$\prod_{a=1}^{d-1} \Gamma\left(\frac{a}{d}\right)^{\chi(a)} = \epsilon_0^{h_K}. \tag{1.8}$$

We can transform the formula (1.8) to traditional Dirichlet's class number formula for a real quadratic field easily, using $\Gamma(s)\Gamma(1-s) = \frac{\pi}{\sin \pi s}$.

(II) Assume $d < 0$. By (1.7), we have $\zeta_K(0) = -\frac{h_K}{w_K}$. Hence

$$L(0,\chi) = \frac{2h_K}{w_K}.$$

By (1.5), we obtain

$$h_K = -\frac{w_K}{2|d|} \sum_{a=1}^{|d|-1} \chi(a)a. \tag{1.9}$$

This is Dirichlet's class number formula for an imaginary quadratic field.

Now it is natural to ask whether a similar expression of the left hand side of (1.8) has a meaning in the imaginary quadratic case. By (1.6), we have

$$\exp\left(\frac{L'(0,\chi)}{L(0,\chi)}\right) = \frac{1}{|d|} \prod_{a=1}^{|d|-1} \Gamma\left(\frac{a}{|d|}\right)^{w_K \chi(a)/2h_K}. \tag{1.10}$$

Let E be an elliptic curve defined over $\overline{\mathbf{Q}}$ with complex multiplication by K. Let ϖ be a period of E. Then the Chowla–Selberg formula states that

$$\prod_{a=1}^{|d|-1} \Gamma\left(\frac{a}{|d|}\right)^{w_K \chi(a)/2h_K} \sim \pi^{-1}\varpi^2 \sim \pi p_K(\mathrm{id}, \mathrm{id})^2. \tag{1.11}$$

Here $a \sim b$ for a, $b \in \mathbf{C}$ means $b \neq 0$, $a/b \in \overline{\mathbf{Q}}$; p_K is Shimura's period symbol which will be reviewed later.

Now let us explain the multiple gamma function. Let r be a positive integer. Let $\omega = (\omega_1, \omega_2, \ldots, \omega_r)$, $\omega_i > 0$ for $1 \leq i \leq r$ and $x > 0$. We define the r-ple zeta function by

$$\zeta_r(s, \omega, x) = \sum_{\Omega = m_1\omega_1 + m_2\omega_2 + \cdots + m_r\omega_r} (x + \Omega)^{-s}. \tag{1.12}$$

Here $(m_1, m_2, \ldots, m_r)$ extends over all r-tuples of nonnegative integers. It is known that $\zeta_r(s, \omega, x)$ can be continued meromorphically to the whole s-plane and is holomorphic at $s = 0$. For $\zeta_r(0, \omega, x)$, we have an elementary formula: if $x_k \in \mathbf{R}$ and $\sum_{k=1}^{r} \omega_k x_k > 0$, then (1.13)

$$\zeta_r\left(0, (\omega_1, \omega_2, \ldots, \omega_r), \sum_{k=1}^{r} \omega_k x_k\right) = (-1)^r$$

$$\times \sum_{l_1,\ldots,l_r \geq 0,\, l_1+\cdots+l_r = r} \omega_1^{l_1-1} \omega_2^{l_2-1} \cdots \omega_r^{l_r-1} \frac{B_{l_1}(x_1)}{l_1!} \frac{B_{l_2}(x_2)}{l_2!} \cdots \frac{B_{l_r}(x_r)}{l_r!}.$$

Here $B_m(x)$ denotes the Bernoulli polynomial defined by

$$\frac{e^{xt}}{e^t - 1} = \sum_{m=0}^{\infty} \frac{B_m(x)}{m!} t^{m-1}.$$

We put

$$-\log \rho_r(\omega) = \lim_{x \to +0} \left\{ \frac{\partial}{\partial s} \zeta_r(s, \omega, x) \bigg|_{s=0} + \log x \right\},$$

$$\frac{\partial}{\partial s} \zeta_r(s, \omega, x) \bigg|_{s=0} = \log \frac{\Gamma_r(x, \omega)}{\rho_r(\omega)}.$$

The function $\Gamma_r(x, \omega)$ is the r-ple gamma function introduced by Barnes ([Ba]). If $r = 1$, $\omega = 1$, we have

$$\Gamma_1(x, 1) = \Gamma(x), \qquad \rho_1(1) = \sqrt{2\pi}.$$

§2. *CM-periods*

We begin with an example due to A. Weil [W]. Let l be an odd prime. Take an integer a such that $1 \leq a \leq (l-1)/2$ and consider a curve

$$C : y^l = x^a (1 - x).$$

The genus g of C is $g = (l - 1)/2$. Put $\zeta = e^{2\pi i/l}$. The curve C has the automorphism

$$\alpha : (x, y) \longrightarrow (x, \zeta y).$$

For $x \in \mathbf{R}$, let $\langle x \rangle$ denote the fractional part of x. Namely $\langle x \rangle$ satisfies

$$0 \leq \langle x \rangle < 1, \qquad x - \langle x \rangle \in \mathbf{Z}.$$

For $t \in (\mathbf{Z}/l\mathbf{Z})^\times$, we consider a differential form

$$\omega_t = x^{\langle at/l \rangle - 1}(1-x)^{\langle t/l \rangle - 1} dx.$$

Then we can check that ω_t is a holomorphic 1-form[1] if and only if

$$\langle at/l \rangle + \langle t/l \rangle < 1.$$

Under the action of α, ω_t transforms as

$$\alpha^* \omega_t = \zeta^t \omega_t. \tag{2.1}$$

For the periods of ω_t, A. Weil showed that

$$\int_c \omega_t \sim B(\langle at/l \rangle, \langle t/l \rangle) \quad \text{for every} \quad c \in H_1(R, \mathbf{Z}). \tag{2.2}$$

Here B denotes the beta function:

$$B(p,q) = \frac{\Gamma(p)\Gamma(q)}{\Gamma(p+q)}.$$

Let J be the Jacobian variety of C; J is defined over $\mathbf{Q}$. From the existence of $\alpha \in \mathrm{Aut}(C)$, we see that $\mathrm{End}(J) \supset \mathbf{Z}[\zeta]$, $\mathrm{End}(J) \otimes \mathbf{Q} \supset \mathbf{Q}(\zeta)$. For $t \in (\mathbf{Z}/l\mathbf{Z})^\times$, define $\sigma(t) \in \mathrm{Gal}(\mathbf{Q}(\zeta)/\mathbf{Q})$ by $\zeta^{\sigma(t)} = \zeta^t$. We put

$$T_a = \{t \mid 1 \leq t \leq l-1, \langle at/l \rangle + \langle t/l \rangle < 1\},$$
$$\Phi_a = \{\sigma(t) \mid t \in T_a\}.$$

We have

$$|\Phi_a| = |T_a| = \frac{l-1}{2} = g = \dim J.$$

Furthermore, by (2.1), we see that the representation of $\mathbf{Q}(\zeta)$ on the space of holomorphic 1-forms on J is equivalent to Φ_a. (We identified Φ_a with $\sum_{\sigma \in \Phi_a} \sigma$.)

---

[1] The curve C has singularities except for the case $l = 3$, $a = 1$. We identify C with the corresponding compact Riemann surface R. By a holomorphic form, we mean a holomorphic form on R.

Next let us explain Shimura's period symbol p_K. Recall that an algebraic number field K is called *a CM-field* if it is a totally imaginary quadratic extension of a totally real algebraic number field F. Let K be a CM-field of degree $2n$ over $\mathbf{Q}$. Let J_K be the set of all isomorphisms of K into $\mathbf{C}$ and I_K be the free abelian group generated by J_K. A set Φ consisting of elements of J_K is called *a CM-type* if the restrictions of elements of Φ to F give all the distinct isomorphisms of F into $\mathbf{R}$. We have necessarily $|\Phi| = n$. We often identify Φ with $\sum_{\sigma\in\Phi}\sigma \in I_K$ and with the representation of K into $M(n, \mathbf{C})$, which it naturally defines.

Let A be an abelian variety of dimension n defined over $\mathbf{C}$. We say that A has complex multiplication by K if $\mathrm{End}(A) \otimes \mathbf{Q} \supset K$. Then we obtain a representation Φ of K on the space of holomorphic 1-forms. It can be shown that Φ is equivalent to the representation which comes from a CM-type of K. In this case, we say that A is of type (K, Φ). A typical example is the complex torus $\mathbf{C}^n/\Phi(\mathcal{O}_K)$. If A and A' have complex multiplication by K, they are isogenous if and only if they belong to the same CM-type.

Now let Φ be a CM-type of K. Then there exists an abelian variety of type (K, Φ) defined over $\overline{\mathbf{Q}}$. For every $\sigma \in \Phi$, we have a nonzero holomorphic 1-form rational over $\overline{\mathbf{Q}}$ such that

$$a^*\omega_\sigma = a^\sigma\omega_\sigma, \qquad a \in K.$$

Here we assume that a is in $\mathrm{End}(A)$ and a^* denotes its action. Then there exists a CM-period $p_K(\sigma, \Phi) \in \mathbf{C}^\times$ such that

$$(2.3) \qquad \int_c \omega_\sigma \sim \pi p_K(\sigma, \Phi) \qquad \text{for every} \quad c \in H_1(A, \mathbf{Z}).$$

We can show that $p_K(\sigma, \Phi) \mod \overline{\mathbf{Q}}^\times$ does not depend on the choice of A and ω_σ. Now the properties of the period symbol p_K can be summarized by the following theorem of Shimura. We denote by ρ the complex conjugation. For an extension L of K, $\mathrm{Res}_{L/K} : I_L \longrightarrow I_K$ denotes the restriction map. $\mathrm{Inf}_{L/K} : I_K \longrightarrow I_L$ denotes the inflation map: it is the homomorphism such that for every $\sigma \in J_K$, $\mathrm{Inf}_{L/K}(\sigma)$ is the sum of all elements in the inverse image of σ under the restriction map.

Theorem 2.1 (Shimura [Sh3], Theorem 32.5). *For every CM-field K, there exists a map $p_K : I_K \times I_K \longrightarrow \mathbf{C}^\times$ with the following properties.*

(1) $p_K(\sigma, \Phi)$ is defined by (2.3) if Φ is a CM-type of K and $\sigma \in \Phi$.

(2) $p_K(\xi_1 + \xi_2, \eta) \sim p_K(\xi_1, \eta) \cdot p_K(\xi_2, \eta)$, $p_K(\xi, \eta_1 + \eta_2) \sim p_K(\xi, \eta_1) \cdot p_K(\xi, \eta_2)$ for every ξ, ξ_1, ξ_2, η, η_1, $\eta_2 \in I_K$.

(3) $p_K(\xi\rho, \eta) \sim p_K(\xi, \eta\rho) \sim p_K(\xi, \eta)^{-1}$ for every ξ, $\eta \in I_K$.

(4) $p_K(\xi, \mathrm{Res}_{L/K}(\zeta)) \sim p_L(\mathrm{Inf}_{L/K}(\xi), \zeta)$ if $\xi \in I_K$, $\zeta \in I_L$ and $K \subset L$, L is a CM-field.

(5) $p_K(\mathrm{Res}_{L/K}(\zeta), \xi) \sim p_L(\zeta, \mathrm{Inf}_{L/K}(\xi))$ if $\xi \in I_K$, $\zeta \in I_L$ and $K \subset L$, L is a CM-field.

(6) $p_{K'}(\gamma\xi, \gamma\eta) \sim p_K(\xi, \eta)$ if γ is an isomorphism of K' onto K.

We will need one more theorem of Shimura in §4, §5. Let K be a CM-field and $\mathfrak{q}$ be an integral ideal of K. Let $I_\mathfrak{q}(K)$ be the ideal group modulo $\mathfrak{q}$, i.e., the group of all fractional ideals which are relatively prime to $\mathfrak{q}$. Take a CM-type Φ of K. Let λ be a homomorphism of $I_\mathfrak{q}(K)$ into $\mathbf{C}^\times$ such that

$$\lambda((\alpha)) = \prod_{\sigma \in \Phi} (\alpha^{\sigma\rho}/|\alpha^\sigma|)^{t_\sigma}, \qquad \alpha \equiv 1 \mod {}^\times\mathfrak{q}.$$

Here t_σ, $\sigma \in \Phi$ are nonnegative integers. In Weil's old terminology, λ is a Grössencharacter of type A_0.

Theorem 2.2 (Shimura [Sh3], Theorem 32.12). *Assume that t_σ mod 2 is independent of $\sigma \in \Phi$. Let m be an integer such that*

$$m \equiv t_\sigma \mod 2, \quad -t_\sigma < m \leq t_\sigma \qquad \text{for every } \sigma \in \Phi.$$

Then we have

$$L(m/2, \lambda) \sim \pi^{e/2} p_K\Big(\sum_{\sigma \in \Phi} t_\sigma \cdot \sigma, \Phi\Big),$$

where $e = m[K : \mathbf{Q}]/2 + \sum_{\sigma \in \Phi} t_\sigma$.

Now let us consider Weil's example again. Let J be the Jacobian variety of $C : y^l = x^a(1-x)$. Put $K = \mathbf{Q}(\zeta)$, $\zeta = e^{2\pi i/l}$. By (2.1), we see that J is of type (K, Φ_a). From the formula (2.2), we have
(2.4)
$$\pi p_K(\sigma(t), \Phi_a) \sim B(\langle at/l \rangle, \langle t/l \rangle) \quad \text{for} \quad 1 \le a \le (l-1)/2,\ \sigma(t) \in \Phi_a.$$
As an example, let $l = 7$, $a = 2$. Then we have
$$T_2 = \{1, 2, 4\}.$$
Let $K_0 = \mathbf{Q}(\sqrt{-7}) \subset K$. Taking $a = 2$, $t = 1$ in (2.4), we obtain
$$p_K(\mathrm{id}, \mathrm{Inf}_{K/K_0}(\mathrm{id})) = p_K(\mathrm{id}, \sigma(1)+\sigma(2)+\sigma(4)) \sim \pi^{-1}\frac{\Gamma(1/7)\Gamma(2/7)}{\Gamma(3/7)}.$$
By Theorem 2.1, (5), we get
$$\pi p_{K_0}(\mathrm{id}, \mathrm{id}) \sim \frac{\Gamma(1/7)\Gamma(2/7)}{\Gamma(3/7)}.$$
Using $\Gamma(s)\Gamma(1-s) = \frac{\pi}{\sin \pi s}$, we obtain
$$\Gamma(1/7)\Gamma(2/7)\Gamma(3/7)^{-1}\Gamma(4/7)\Gamma(5/7)^{-1}\Gamma(6/7)^{-1} \sim \pi p_{K_0}(\mathrm{id}, \mathrm{id})^2,$$
which is the Chowla–Selberg formula for the field K_0. In general, we can calculate $p_K(\mathrm{id}, \sigma(t))$, $1 \le t \le l-1$ in terms of the gamma function using (2.4). The proof is not very difficult. It is also possible to calculate the period symbol for $K = \mathbf{Q}(e^{2\pi i/n})$. In this case, we use factors of the Jacobian variety of the Fermat curve $x^n + y^n = 1$. We refer the reader to [Y4], Chapter III, §2. The result is

Theorem 2.3 (Anderson [A]). *Let K be a CM-field abelian over $\mathbf{Q}$. Put $G = \mathrm{Gal}(K/\mathbf{Q})$. Then we have*
$$p_K(\mathrm{id}, \tau) \sim \pi^{-\mu(\tau)/2} \prod_{\omega \in \hat{G}_-} \exp\left(\frac{\omega(\tau)}{|G|}\frac{L'(0,\omega)}{L(0,\omega)}\right), \qquad \tau \in G.$$
Here $\widehat{G}_-$ denotes the set of all odd characters of G and
$$\mu(\tau) = \begin{cases} 1 & \text{if } \tau = 1, \\ -1 & \text{if } \tau = \rho, \\ 0 & \text{otherwise.} \end{cases}$$

In view of this theorem, for non-abelian case, we naturally conjecture:

Conjecture A (Colmez [C], Yoshida [Y2]). *Let K be a CM-field normal over $\mathbf{Q}$. Put $G = \mathrm{Gal}(K/\mathbf{Q})$. Let c be a conjugacy class of G. Then we have*

$$\prod_{\tau \in c} p_K(\mathrm{id}, \tau) \sim \pi^{-\mu(c)/2} \prod_{\omega \in \hat{G}_-} \exp\left(\frac{|c|\chi_\omega(c)}{|G|} \frac{L'(0,\omega)}{L(0,\omega)}\right), \qquad \tau \in G.$$

Here $\widehat{G}_-$ denotes the set of equivalence classes of all irreducible odd representations of G, χ_ω denotes the character of ω and

$$\mu(c) = \begin{cases} 1 & \text{if } c = \{1\}, \\ -1 & \text{if } c = \{\rho\}, \\ 0 & \text{otherwise.} \end{cases}$$

Remark. We remind the reader that ρ denotes the complex conjugation; ρ belongs to the center of G. A representation ω of G is called odd if $\omega(\rho) = -\mathrm{id}$.

§3. *Shintani's formula*

Let $A = (a_{ij})$ be an $n \times r$-matrix whose entries a_{ij} are positive real numbers. Let $x = {}^t(x_1, x_2, \ldots, x_r)$ be a column vector such that $x_i \geq 0$, $1 \leq i \leq r$, $x \neq 0$. Shintani introduced the zeta function

$$\zeta(s, A, x) = \sum_{m_1,\ldots,m_r=0}^{\infty} \prod_{i=1}^{n} \left\{ \sum_{j=1}^{r} a_{ij}(x_j + m_j) \right\}^{-s}. \tag{3.1}$$

The series converges when $\mathrm{Re}(s) > n/r$ and can be continued meromorphically to the whole complex plane. It is holomorphic at $s = 0$. Let $A^{(i)} = (a_{i1}, a_{i2}, \ldots, a_{ir})$ denote the i-th row of A. Then comparing with (1.12), we see

$$\zeta(s, A^{(i)}, x) = \zeta_r\left(s, A^{(i)}, \sum_{j=1}^{r} a_{ij} x_j\right).$$

Here, of course, the left hand side is defined by (3.1) regarding $A^{(i)}$ as an $n \times 1$-matrix. Shintani proved that

$$\zeta(0, A, x) = \frac{1}{n} \sum_{i=1}^{n} \zeta(0, A^{(i)}, x), \tag{3.2}$$

$$
\begin{aligned}
\frac{\partial}{\partial s}\zeta(s,A,x)\bigg|_{s=0} &= \sum_{i=1}^{n} \log \frac{\Gamma_r(\sum_{j=1}^{r} a_{ij}x_j, (a_{i1}, a_{i2}, \ldots, a_{ir}))}{\rho_r(a_{i1}, a_{i2}, \ldots, a_{ir})} \\
&\quad + \frac{(-1)^r}{n} \sum_{l} C_l(A) \prod_{j=1}^{r} \frac{B_{l_j}(x_j)}{l_j!}.
\end{aligned} \tag{3.3}
$$

Here $l = (l_1, l_2, \ldots, l_r)$ extends over all r-tuples of nonnegative integers such that $l_1 + l_2 + \cdots + l_r = r$ and

$$C_l(A) = \sum_{1 \le j,k \le n} C_{l,j,k}(A),$$

$$C_{l,j,k}(A) = \int_0^1 \left\{ \prod_{m=1}^{r} (a_{jm} + a_{km}u)^{l_m - 1} - \prod_{m=1}^{r} a_{jm}^{l_m - 1} \right\} \frac{du}{u}.$$

The formula (3.3) can be written as

$$
\begin{aligned}
&\frac{\partial}{\partial s}\zeta(s,A,x)\bigg|_{s=0} \\
&= \sum_{i=0}^{n} \frac{\partial}{\partial s}\zeta(s,A^{(i)},x)\bigg|_{s=0} + \frac{(-1)^r}{n} \sum_{l} C_l(A) \prod_{j=1}^{r} \frac{B_{l_j}(x_j)}{l_j!}.
\end{aligned}
$$

The second term on the right hand side can be regarded as the correction term. The derivative $\frac{\partial}{\partial s}\zeta(s,A,x)\Big|_{s=0}$ is decomposed as the sum of the contribution from every row. (The same observation is true for (3.2), except for the proportionality factor $1/n$.) From our point of view, this is the most remarkable feature of Shintani's formula. The author generalized it to the second derivative of $\zeta(s,A,x)$ at $s = 0$ with the proportionality factor n ([Y4], Chapter I, §3).

Now let F be a totally real algebraic number field of degree n. Let $J_F = \{\sigma_1, \sigma_2, \ldots, \sigma_n\}$ denote the set of all isomorphisms of F into $\mathbf{R}$. We denote by ∞_1, ∞_2, …, ∞_n the corresponding archimedean primes of F. Let E_F^+ be the group of all totally positive units of F. We embed F into $\mathbf{R}^n$ by $x \to (x^{\sigma_1}, x^{\sigma_2}, \ldots, x^{\sigma_n})$. Then E_F^+ acts on $\mathbf{R}_+^n$ by componentwise multiplication, where $\mathbf{R}_+$ is the set of all positive real numbers. For linearly independent vectors v_1, v_2, ..., v_r, we set

$$C(v_1, v_2, \ldots, v_r) = \left\{ \sum_{i=1}^{r} t_i v_i \,\middle|\, t_1, t_2, \ldots, t_r > 0 \right\}$$

and call it *an r-dimensional open simplicial cone.*

Theorem 3.1 (Shintani [Sh1]). *There exist finitely many open simplicial cones*

$$C_j = C(v_{j1}, v_{j2}, \ldots, v_{jr(j)}), \qquad v_{ji} \in \mathcal{O}_F$$

such that

$$\mathbf{R}_+^n = \sqcup_{\epsilon \in E_F^+} \epsilon(\sqcup_{j \in J} C_j).$$

Here J is a finite set of indices.

We call this decomposition *a cone decomposition* of $\mathbf{R}_+^n$.

Example. Let F be a real quadratic field. Then $E_F^+ \cong \mathbf{Z}$. The generator $\epsilon > 1$ of E_F^+ is called *the totally positive fundamental unit* of F. A cone decomposition of $\mathbf{R}_+^2$ is given by taking

$$C_1 = C(1, \epsilon), \qquad C_2 = C(1), \quad J = \{1, 2\}.$$

Let $\mathfrak{f}$ be an integral ideal of F. We denote by $C_\mathfrak{f}$ the ideal class group modulo $\mathfrak{f}\infty_1\infty_2\cdots\infty_n$ of F. The conductor of this ideal class group is not necessarily equal to $\mathfrak{f}\infty_1\infty_2\cdots\infty_n$. The following formula for the order of $C_\mathfrak{f}$ is well known in class field theory.

$$|C_\mathfrak{f}| = \frac{2^n h_F \varphi(\mathfrak{f})}{[E_F : E_{F,\mathfrak{f}\infty_1\cdots\infty_n}]}.$$

Here $\varphi(\mathfrak{f}) = N(\mathfrak{f}) \prod_{\mathfrak{p}|\mathfrak{f}} (1 - \frac{1}{N(\mathfrak{p})})$ is the Euler function and

$$E_{F,\mathfrak{f}\infty_1\cdots\infty_n} = \{\epsilon \in E_F^+ \mid \epsilon \equiv 1 \mod \mathfrak{f}\}.$$

For $c \in C_\mathfrak{f}$, let

$$\zeta_F(s, c) = \sum_{\mathfrak{a} \in c} N(\mathfrak{a})^{-s}, \qquad \mathrm{Re}(s) > 1$$

be the partial zeta function of the class c. Here $\mathfrak{a}$ extends over integral ideals. This partial zeta function has meromorphic continuation to the whole complex plane, holomorphic except for a simple pole at $s = 1$. Let $\mathfrak{a}_1, \mathfrak{a}_2, \ldots, \mathfrak{a}_{h_0}$ be integral ideals which represent the narrow ideal classes of F. For a fractional ideal $\mathfrak{a}$ of F, we set

$$R(C_j, \mathfrak{a}) = \left\{ z \in \mathfrak{a} \cap C_j \,\middle|\, z = \sum_{i=1}^{r(j)} x_i(z) v_{ji}, \right.$$
$$\left. x_i(z) \in \mathbf{Q},\ 0 < x_i(z) \leq 1,\ 1 \leq i \leq r(j) \right\}.$$

This is a finite set, being contained in the intersection of a compact subset of C_j with the discrete set $\mathfrak{a}$. For $z \in R(C_j, \mathfrak{a})$, define the column vector $x(z)$ by $x(z) = {}^t(x_1(z), x_2(z), \ldots, x_{r(j)}(z))$.

Take a class $c \in C_{\mathfrak{f}}$. We take $\mathfrak{a}_\mu$ so that c and $\mathfrak{a}_\mu \mathfrak{f}$ belong to the same narrow ideal class. Put

$$R(C_j, c) = \{ z \in R(C_j, (\mathfrak{a}_\mu \mathfrak{f})^{-1}) \mid (z)\mathfrak{a}_\mu \mathfrak{f} \equiv c \text{ in } C_{\mathfrak{f}} \}.$$

For C_j, we attach an $n \times r(j)$-matrix $A_j = (v_{ji}^{\sigma_l})$; the (l, i)-component of A_j is $v_{ji}^{\sigma_l}$. Then we have

$$\zeta_F(s, c) = N(\mathfrak{a}_\mu \mathfrak{f})^{-s} \sum_{j \in J} \sum_{z \in R(C_j, c)} \zeta(s, A_j, x(z)). \tag{3.4}$$

Using (3.2) and (1.13), we have a closed formula of $\zeta_F(0, c)$. Using (3.3), we have an expression of $\zeta_F'(0, c)$ in terms of the multiple gamma functions.

§4. *Absolute CM-period symbol*

We understand that an algebraic number field is *a subfield* of $\mathbf{C}$, which is of finite degree over $\mathbf{Q}$. Let F be a totally real algebraic number field of degree n. We set $J_F = \{\sigma_1, \sigma_2, \ldots, \sigma_n\}$.

Proposition 4.1. *Let K be a CM-field which is abelian over F. Put $G = \mathrm{Gal}(K/F)$. Let $\widehat{G}$ denote the set of all characters and*

$\widehat{G}_-$ the set of all odd characters of G. Assume Conjecture A. Then, for $\tau \in G$, we have

$$\prod_{i=1}^{n} p_K(\sigma_i, \tau\sigma_i) \sim \pi^{-n\mu(\tau)/2} \prod_{\chi\in\hat{G}_-} \exp\left(\frac{\chi(\tau)}{|G|}\frac{L'(0,\chi)}{L(0,\chi)}\right). \tag{4.1}$$

Here $\mu(\tau)$ is defined by the same formula as in Theorem 2.3; we use the same letter σ_i for its extension to an element of J_K. By Theorem 2.1, (6), we see that $p_K(\sigma_i, \tau\sigma_i) \mod \overline{\mathbf{Q}}^\times$ does not depend on the choice of this extension. We also note that

$$p_K(\sigma_i, \tau\sigma_i) \sim p_{K^{\sigma_i}}(\mathrm{id}, \sigma_i^{-1}\tau\sigma_i)$$

which follows from Theorem 2.1, (6).

Now we are going to factorize the right hand side of (4.1). Let $\mathfrak{f}$ be an arbitrary integral ideal of F. We choose $\sigma_1 = \mathrm{id}$. For $c \in C_\mathfrak{f}$, we put

$$G(c) = \sum_{j\in J}\sum_{z\in R(C_j,c)} \log\frac{\Gamma_{r(j)}(z, A_j^{(1)})}{\rho_{r(j)}(A_j^{(1)})}. \tag{4.2}$$

Note that $z \in R(C_j, c)$ is a positive real number and that $A_j^{(1)} = (v_{j1}, v_{j2}, \ldots, v_{jr(j)})$ is the first row of A_j.

$$\begin{aligned} V(c) = \sum_{j\in J}(-1)^{r(j)}\sum_{l}\Bigg[\frac{1}{n}\sum_{k=2}^{n}(C_{l,1,k}(A_j) + C_{l,k,1}(A_j)) \\ -\frac{1}{n^2}\sum_{1\le i,k\le n, i\ne k} C_{l,i,k}(A_j)\Bigg] \times \sum_{z\in R(C_j,c)}\prod_{m=1}^{r(j)}\frac{B_{l_m}(x_m(z))}{l_m!}, \end{aligned} \tag{4.3}$$

$$W(c) = -\frac{1}{n}\log N(\mathfrak{a}_\mu\mathfrak{f})\zeta_F(0,c), \tag{4.4}$$

$$X(c) = G(c) + V(c) + W(c). \tag{4.5}$$

In (4.3), $l = (l_1, l_2, \ldots, l_{r(j)})$ extends over all $r(j)$-tuples of nonnegative integers such that $l_1 + l_2 + \cdots + l_{r(j)} = r(j)$. These quantities depend upon the choice of a cone decomposition $\{C_j\}_{j\in J}$ and the choice of the representatives $\{\mathfrak{a}_\mu\}$. We write them as $X(c; \{C_j\}_{j\in J}, \mathfrak{a}_\mu)$, etc., when we have to be precise. Now we have

Theorem 4.2.

$$\zeta_F'(0,c) = \sum_{\sigma \in J_F} X(c^\sigma; \{C_j^\sigma\}_{j\in J}, \mathfrak{a}_\mu^\sigma).$$

From our point of view, this theorem expresses the most remarkable feature of Shintani's formula reviewed in §3. Concerning the V-term, we have

Theorem 4.3. *We can write* $V(c) = \sum_{i=1}^m a_i \log \epsilon_i$ *with* $a_i \in F$, $\epsilon_i \in E_F^+$.

In [Y4], Chapter IV, §6, we proved the assertion with $\widetilde{F}$ in place of F, where $\widetilde{F}$ is the normal closure of F over $\mathbf{Q}$. Theorem 4.3 was left as Conjecture 6.5. The author proved it in collaboration with T. Kashio.

Let K be as in Proposition 4.1 and let $\mathfrak{f}\infty_1\infty_2\cdots\infty_n$ be the conductor of K over F. By class field theory, we can regard $\chi \in \widehat{G}$ as a character of $C_\mathfrak{f}$. For $\chi \in \widehat{G}$, let $\mathfrak{f}(\chi)$ denote the finite part of the conductor of χ. We have

$$\sum_{\chi\in\hat{G}_-} \frac{\chi(\tau)}{L(0,\chi)} L'(0,\chi) = \sum_{\chi\in\hat{G}_-} \frac{\chi(\tau)}{L(0,\chi)} \sum_{c\in C_{\mathfrak{f}(\chi)}} \chi(c)\zeta_F'(0,c).$$

For $\tau \in G$, we define

$$\text{(4.6)}\quad g_K(\mathrm{id},\tau) = \pi^{-\mu(\tau)/2} \exp\left(\frac{1}{|G|} \sum_{\chi\in\hat{G}_-} \frac{\chi(\tau)}{L(0,\chi)} \sum_{c\in C_{\mathfrak{f}(\chi)}} \chi(c)X(c)\right).$$

When we have to be precise, we write it as $g_K(\mathrm{id},\tau;\{C_j\}_{j\in J},\{\mathfrak{a}_\mu\})$. The formula (4.1), which is derived assuming Conjecture A, can be written as

$$\text{(4.7)}\quad \prod_{i=1}^n p_{K^{\sigma_i}}(\mathrm{id},\sigma_i^{-1}\tau\sigma_i) \sim \prod_{i=1}^n g_{K^{\sigma_i}}(\mathrm{id},\sigma_i^{-1}\tau\sigma_i;\{C_j^{\sigma_i}\}_{j\in J},\{\mathfrak{a}_\mu^{\sigma_i}\}).$$

Now it is natural to conjecture a stronger version of this formula.

Conjecture B. *Let K be an abelian extension of a totally real algebraic number field F. Put $G = \mathrm{Gal}(K/F)$. We assume that K is a CM-field. For $\tau \in G$, define $g_K(\mathrm{id}, \tau)$ by (4.6). Then*

$$p_K(\mathrm{id}, \tau) \sim g_K(\mathrm{id}, \tau).$$

We stress the importance of the factorization process described above; not only that Conjecture B is stronger than Conjecture A, it suggests possibilities of further generalizations.

We can show that the validity of Conjecture B does not depend on the choices of $\{C_j\}_{j\in J}$ and $\{\mathfrak{a}_\mu\}$. We call g_K *the absolute CM-period symbol.* Strictly speaking, $g_K(\mathrm{id}, \tau)$ depends also on the choice of the field F and should be written as $g_{K/F}(\mathrm{id}, \tau; \{C_j\}_{j\in J}, \{\mathfrak{a}_\mu\})$ in the full notation. For example, if K is abelian over F_1 and F_2 and $\tau \in \mathrm{Gal}(K/F_1) \cap \mathrm{Gal}(K/F_2)$, it is a highly nontrivial problem to show that $g_{K/F_1}(\mathrm{id}, \tau) \sim g_{K/F_2}(\mathrm{id}, \tau)$. For partial results, see [Y4], Chapter III, §3.8.

Conjecture B, if true, is powerful enough to give values of p_K symbol of Shimura modulo algebraic numbers. In fact, let us consider $p_K(\sigma, \tau)$, $\sigma, \tau \in J_K$ for an arbitrary CM-field K. Let L be the normal closure of K over $\mathbf{Q}$ and choose $\widetilde{\sigma} \in J_L$ so that $\mathrm{Res}_{L/K}(\widetilde{\sigma}) = \sigma$. By Theorem 2.1, (5), we have $p_K(\sigma, \tau) \sim p_L(\widetilde{\sigma}, \mathrm{Inf}_{L/K}(\tau))$. Put $\mathrm{Inf}_{L/K}(\tau) = \sum_i \tau_i$, $\tau_i \in J_L$. By Theorem 2.1, (6), we have

$$p_L(\widetilde{\sigma}, \mathrm{Inf}_{L/K}(\tau)) \sim \prod_i p_L(\widetilde{\sigma}, \tau_i) \sim \prod_i p_L(\mathrm{id}, \widetilde{\sigma}^{-1}\tau_i).$$

Let F_i be the fixed field by the group generated by τ_i and ρ. Then F_i is totally real and L is abelian over F_i. Therefore Conjecture B can predict the value of $p_L(\mathrm{id}, \widetilde{\sigma}^{-1}\tau_i)$.

Now we are going to state a refinement of Conjecture B which is amenable to numerical tests. In view of Theorem 2.2, we can find an integral ideal $\mathfrak{q}$ and homomorphisms $\lambda_1, \lambda_2, \ldots, \lambda_q$ of $I_{\mathfrak{q}}(K)$ into $\mathbf{C}^\times$ and integers $\epsilon_1, \epsilon_2, \ldots, \epsilon_q$ so that

$$\prod_{i=1}^{q} L(m/2, \lambda_i)^{\epsilon_i} \sim \pi^A \prod_{i=1}^{r} p_{K^{\sigma_i}}(\mathrm{id}, \sigma_i^{-1}\tau_i\sigma_i)^{a_i}$$

holds with $a_i \in \mathbf{Z}$, $\tau_i \in G$, $A \in 2^{-1}\mathbf{Z}$.

Conjecture C. *For $\sigma \in \mathrm{Gal}(\overline{\mathbf{Q}}/\mathbf{Q})$, we have*

$$\left(\frac{\prod_{i=1}^{q} L(m/2, \lambda_i)^{\epsilon_i}}{\pi^A \prod_{i=1}^{r} g_{K^{\sigma_i}}(\mathrm{id}, \sigma_i^{-1}\tau_i\sigma_i; \{C_j^{\sigma_i}\}, \{\mathfrak{a}_\mu^{\sigma_i}\})^{a_i}}\right)^{\sigma}$$
$$= \zeta(\sigma) \cdot \frac{\prod_{i=1}^{q} L(m/2, \lambda_i^{\sigma})^{\epsilon_i}}{\pi^A \prod_{i=1}^{r} g_{K^{\sigma_i}}(\mathrm{id}, \sigma^{-1}\sigma_i^{-1}\tau_i\sigma_i\sigma; \{C_j^{\sigma_i\sigma}\}, \{\mathfrak{a}_\mu^{\sigma_i\sigma}\})^{a_i}},$$

where $\zeta(\sigma)$ is a root of unity.

It seems that we can control the denominator of the quotient in this Conjecture if $\epsilon_i \geq 0$, $1 \leq i \leq r$. See [Y4], p. 108, Remark 4.5. The determination of ζ is a very interesting problem, the solution of which may be called a reciprocity law. For partial results, see [Y2].

§5. *Numerical examples*

Let $F = \mathbf{Q}(\sqrt{d})$, $0 < d \in \mathbf{Q}$ be a real quadratic field. We take a totally positive element $x + y\sqrt{d} \in F$, $x, y \in \mathbf{Q}$, $y \neq 0$ and put $\xi = \sqrt{x + y\sqrt{d}}\,i$, $\xi' = \sqrt{x - y\sqrt{d}}\,i$. Then $K = \mathbf{Q}(\xi)$ is a CM-field of degree 4. We define $\sigma \in J_K$ by $\xi^\sigma = \xi'$. Then $\Phi = \{\mathrm{id}, \sigma\}$ is a CM-type of K. Let $\mathfrak{q}$ be an integral ideal of K and we consider Grössencharacters of K such that

$$\lambda_{a,b}^{(1)}((\alpha)) = \left(\frac{\alpha^\rho}{|\alpha|}\right)^a \left(\frac{\alpha^{\sigma\rho}}{|\alpha^\sigma|}\right)^b, \qquad \alpha \equiv 1 \mod {}^\times\mathfrak{q},$$

$$\lambda_{a,b}^{(2)}((\alpha)) = \left(\frac{\alpha^\rho}{|\alpha|}\right)^a \left(\frac{\alpha^{\sigma}}{|\alpha^{\sigma\rho}|}\right)^b, \qquad \alpha \equiv 1 \mod {}^\times\mathfrak{q}.$$

Here a and b are positive integers such that $a \equiv b \mod 2$. We put

$$Q = \frac{L(1, \lambda_{2,2}^{(1)})L(1, \lambda_{2,2}^{(2)})}{\pi^8 g_K(\mathrm{id}, \mathrm{id})^4 g_{K^\sigma}(\mathrm{id}, \mathrm{id})^4},$$

$$Q_1 = \frac{L(1, \lambda_{4,2}^{(1)})L(1, \lambda_{4,2}^{(2)})}{\pi^{10} g_K(\mathrm{id}, \mathrm{id})^8 g_{K^\sigma}(\mathrm{id}, \mathrm{id})^4},$$

$$Q_2 = \frac{L(1, \lambda_{2,4}^{(1)})L(1, \lambda_{2,4}^{(2)})}{\pi^{10} g_K(\mathrm{id}, \mathrm{id})^4 g_{K^\sigma}(\mathrm{id}, \mathrm{id})^8}.$$

Assume that $\mathfrak{q} = (1)$, $h_K = 1$. Then, for $\alpha \in \mathrm{Gal}(\overline{\mathbf{Q}}/\mathbf{Q})$, Conjecture C states that

$$Q^\alpha = \zeta Q,$$

$$Q_1^\alpha = \zeta_1 Q_1, \quad Q_2^\alpha = \zeta_2 Q_2, \qquad \text{if } \alpha|F = \mathrm{id},$$

$$Q_1^\alpha = \zeta_1 Q_2, \quad Q_2^\alpha = \zeta_2 Q_1, \qquad \text{if } \alpha|F \neq \mathrm{id},$$

where $\zeta = \zeta(\alpha)$, $\zeta_1 = \zeta_1(\alpha)$ and $\zeta_2 = \zeta_2(\alpha)$ are roots of unity which depend on α.

Now take $F = \mathbf{Q}(\sqrt{29})$. Then $\epsilon_0 = \frac{5+\sqrt{29}}{2}$ is the fundamental unit and $\epsilon = \epsilon_0^2 = \frac{27+5\sqrt{29}}{2}$ is the totally positive fundamental unit. The narrow class number of F is 1. We take (1) as the representative for the narrow ideal class. We take $C_1 = C(1, \epsilon)$, $C_2 = C(1)$ for the cone decomposition. Let $K = \mathbf{Q}\left(\sqrt{\frac{9+\sqrt{29}}{2}}\, i\right)$. The conductor of K over F is $\mathfrak{f}\infty_1\infty_2$, $\mathfrak{f} = \left(\frac{9+\sqrt{29}}{2}\right)$. We have $h_K = 1$, $N(\mathfrak{f}) = 13$. By numerical computation, we find

$$Q = \frac{2^4 \cdot 3}{29^2},$$

$$Q_1 = \epsilon^{9/13} \cdot \frac{2935 + 503\sqrt{29}}{13 \cdot 29^2}, \qquad Q_2 = \epsilon^{-9/13} \cdot \frac{2935 - 503\sqrt{29}}{13 \cdot 29^2}.$$

Both sides coincide to the 50th decimal place. We have $C_\mathfrak{f} = \{c_1, c_2\}$, $c_1 = 1$. For the V-term, we have

$$V(c_1) - V(c_2) = -\frac{1}{13}\sqrt{29}\log\epsilon$$

and the congruence

$$9 - \sqrt{29} \equiv 0 \mod \mathfrak{f}', \qquad \mathfrak{f}' = \left(\frac{9 - \sqrt{29}}{2}\right).$$

In this example, we have $\zeta(\alpha) = 1$,

$$\zeta_1(\alpha) = (\epsilon^{9/13})^\alpha / \epsilon^{9/13} \quad \text{if } \alpha|F = \mathrm{id},$$

$$\zeta_1(\alpha) = (\epsilon^{9/13})^\alpha \epsilon^{9/13} \quad \text{if } \alpha|F \neq \mathrm{id},$$

$\zeta_2(\alpha) = \zeta_1(\alpha)^{-1}$. We see that they are 13th roots of unity.

§6. *Reflection on periods and units*

In this section, we will show that the invariant $X(c)$ contains information of CM-periods and units of number fields.

Let us first recall a conjecture of Stark–Shintani. Let F be a totally real algebraic number field of degree $n \geq 2$. We assume that ∞_1 corresponds to the identical embedding of F into $\mathbf{R}$. Let M be an abelian extension of F whose conductor is of the form $\mathfrak{f}\infty_2 \cdots \infty_n$. In other words, if σ_i is the embedding $F \subset \mathbf{R}$ corresponding to ∞_i and if $\tilde{\sigma}_i$ is any extension of σ_i to an embedding $M \subset \mathbf{C}$, we have $\tilde{\sigma}_1(M) \subset \mathbf{R}$ and $\tilde{\sigma}_i(M) \not\subset \mathbf{R}$ for $2 \leq i \leq n$. For $\sigma \in \mathrm{Gal}(M/F)$, let

$$\zeta(s,\sigma) = \sum_{\mathfrak{a},\left(\frac{M/F}{\mathfrak{a}}\right)=\sigma} N(\mathfrak{a})^{-s}$$

be the partial zeta function attached to σ. Here $\left(\frac{M/F}{\mathfrak{a}}\right)$ denotes the Artin symbol.

Conjecture of Stark-Shintani. *In the above situation, there exists a unit $u \in E_M$ such that*

$$\exp(-2\zeta_F'(0,\sigma)) = u^\sigma \qquad \text{for every } \sigma \in \mathrm{Gal}(M/F).$$

Now let F be a real quadratic field and $\mathfrak{f}$ be an integral ideal. For $i = 1, 2$, we can choose $\nu_i \in \mathcal{O}_F$ so that $\nu_i \equiv 1 \mod \mathfrak{f}$, $\nu_i^{(i)} < 0 < \nu_i^{(j)}$ if $j \neq i$. Let $s_i \in C_\mathfrak{f}$ be the class of the ideal (ν_i). A character χ of $C_\mathfrak{f}$ ramifies at ∞_i if and only if $\chi(s_i) = -1$.

Theorem 6.1. *Let ϵ_0 be the fundamental unit of F. If $N(\epsilon_0) = -1$, we have*

$$X(c;\{C_j\},\mathfrak{a}_\mu) + X(cs_2;\{C_j\},\mathfrak{a}_\mu) = a\log\epsilon_0$$

with $a \in \mathbf{Q}$. If $N(\epsilon_0) = 1$, write $F = \mathbf{Q}(\sqrt{d})$ with a square free integer d. Then we have

$$X(c;\{C_j\},\mathfrak{a}_\mu) + X(cs_2;\{C_j\},(\sqrt{d})\mathfrak{a}_\mu) = a\log\epsilon_0$$

with $a \in \mathbf{Q}$.

Note that the narrow ideal class of s_2 is trivial if and only if $N(\epsilon_0) = -1$. We can give an explicit formula for the value of a. For the proof, see [Y4], Chapter III, Theorems 5.8 and 5.12. For $c \in C_{\mathfrak{f}}$, we put

$$X(c) = X(c; \{C_j\}, \mathfrak{a}_\mu),$$

$$X(cs_2) = \begin{cases} X(cs_2; \{C_j\}, \mathfrak{a}_\mu) & \text{if } N(\epsilon_0) = -1, \\ X(cs_2; \{C_j\}, (\sqrt{d})\mathfrak{a}_\mu) & \text{if } N(\epsilon_0) = 1. \end{cases}$$

Now let L be the maximal ray class field modulo $\mathfrak{f}\infty_1\infty_2$. Then we have $\mathrm{Gal}(L/F) \cong C_{\mathfrak{f}}$. (We do not assume that the conductor of L over F is equal to $\mathfrak{f}\infty_1\infty_2$.) We can show that the maximal totally real subfield F_0 of L corresponds to the subgroup generated by s_1 and s_2. There exists the maximal CM-subfield K of L if and only if there exists a character of $C_{\mathfrak{f}}$ which ramifies at ∞_1 and ∞_2. In this case, K corresponds to the subgroup generated by $s_1 s_2$.

Let M_i be the fixed field of s_i for $i = 1, 2$. Then ∞_i is unramified in M_i. Assume that the conductor of M_1 over F is $\mathfrak{f}\infty_2$. For $c \in C_{\mathfrak{f}}$, the Stark–Shintani conjecture says

$$\exp(-2(\zeta_F'(0, c) + \zeta_F'(0, cs_1))) = u^{\left(\frac{M_1/F}{c}\right)} \tag{6.1}$$

with a unit u of M_1. Let σ be the nontrivial automorphism of F. Let E stand for the term of the form $a \log \epsilon_0$ with $a \in \mathbf{Q}$. Put

$$X(c^\sigma) = X(c^\sigma; \{C_j\}, \mathfrak{a}_\mu^\sigma), \qquad c \in C_{\mathfrak{f}}.$$

Using Theorem 4.2, we can show that

$$X(c) + X(c^\sigma) = \zeta_F'(0, c) + E.$$

Now (6.1) can be written as

$$X(c) + X(cs_1) + X(c^\sigma) + X(c^\sigma s_1^\sigma) = -\frac{1}{2} \log u^{\left(\frac{M_1/F}{c}\right)} + E.$$

Here the sum of the last two terms of the left hand side is of the form E by Theorem 6.1. Therefore, assuming (6.1), we obtain for $c \in C_{\mathfrak{f}}$ that

$$X(c) + X(cs_1) = -\frac{1}{2} \log u^{\left(\frac{M_1/F}{c}\right)} + E, \tag{6.2}$$

$$X(c) + X(cs_2) = E. \tag{6.3}$$

On the other hand, it is not difficult to see that L contains the maximal CM-subfield if and only if $s_1 \neq 1$. In this case, our Conjecture B says that the quantity $X(c) + X(cs_1 s_2)$ is related to CM-periods. Thus our invariants $X(c)$ contain information of CM-periods and the Stark–Shintani units in a clear-cut manner. On reflecting this situation, we conclude that the periods and units (algebraic numbers) are tied together very closely. This will become even more clear if we develop the p-adic theory.

Remark. A careful reader may be curious about the asymmetry of ∞_1 and ∞_2. This is because we consider F as a subfield of $\mathbf{R}$, hence ∞_1 is the privileged infinite place.

Part II. p-Adic Theory

§7. *A p-adic analogue of Shintani's formula (Kashio's formula)*

Let p be a prime. For simplicity, we assume that p is odd. Let $\mathbf{C}_p$ be the completion of an algebraic closure $\overline{\mathbf{Q}}_p$ of $\mathbf{Q}_p$. We normalize the absolute value $|\ \ |_p$ of $\mathbf{C}_p$ so that $|p|_p = p^{-1}$. Put

$$U = \{u \in \mathbf{C}_p \mid |u|_p = 1\}, \qquad D = \{u \in \mathbf{C}_p \mid |u-1|_p < 1\}.$$

Let V be the subgroup of U consisting of all roots of unity whose orders are prime to p. Then we have the direct product decomposition

$$U = V \times D.$$

For $u \in U$, let $\theta_p(u)$ be the projection of u to V. Let φ be a function on $\mathbf{Z}_p^r$ taking values in $\mathbf{C}_p$. We define the p-adic integration of φ by

$$I(\varphi) = \lim_{l_1 \to \infty} \cdots \lim_{l_r \to \infty} \frac{1}{p^{l_1 + \cdots + l_r}} \sum_{n_1=0}^{p^{l_1}-1} \cdots \sum_{n_r=0}^{p^{l_r}-1} \varphi(n_1, \ldots, n_r)$$

whenever the limit exists. Let $\mathcal{O}_p = \{x \in \mathbf{C}_p \mid |x|_p \leq 1\}$ be the ring of integers of $\mathbf{C}_p$. Take $\omega = (\omega_1, \ldots, \omega_r)$, $\omega_i \in \mathcal{O}_p$, $\omega_i \neq 0$, $x \in \mathcal{O}_p$. We consider a linear form

$$f(X) = \sum_{i=1}^{r} \omega_i X_i + x, \qquad X = (X_1, \ldots, X_r)$$

and for $s \in \mathbf{Z}_p$, define a $\mathbf{C}_p$-valued function on $\mathbf{Z}_p^r$ by

$$f^{(0)}(X,s) = \begin{cases} F(X)^r(\theta_p(f(X))^{-1}f(X))^{-s} & \text{if } f(X) \in \mathcal{O}_p, \\ 0 & \text{if } f(X) \notin \mathcal{O}_p. \end{cases}$$

We define the p-adic r-ple zeta function by

$$\zeta_{p,r}(s,\omega,x) = \frac{(-1)^r I(f^{(0)}(X,s))}{(r-s)(r-1-s)\cdots(1-s)\omega_1\cdots\omega_r}.$$

In analogy to the complex case, we define the p-adic logarithmic r-ple gamma function by

$$L\Gamma_{p,r}(x,\omega) = \frac{\partial}{\partial s}\zeta_{p,r}(s,\omega,x)\bigg|_{s=0}.$$

We have

$$L\Gamma_{p,1}(x,1) = \log_p \Gamma_p(x).$$

Where $\log_p$ is the Iwasawa p-adic logarithmic function and Γ_p is the Morita p-adic gamma function. We can extend the domain of definition of $\zeta_{p,r}(s,\omega,x)$ and $L\Gamma_{p,r}(x,\omega)$ to $x \in \mathbf{C}_p$, $\omega \in (\mathbf{C}_p^\times)^r$ in the following way. Suppose that $x \notin \mathcal{O}_p$ or there exists ω_i such that $\omega_i \notin \mathcal{O}_p$. We take $c \in \mathbf{C}_p^\times$ so that

$$cx \in \mathcal{O}_p, \qquad c\omega_i \in \mathcal{O}_p, \quad 1 \le i \le r$$

and that $|c|_p$ takes the maximum value among the elements satisfying this condition. We write

$$c = p^a u, \qquad a \in \mathbf{Q}, \quad u \in U$$

and put

$$\zeta_{p,r}(s,\omega,x) = (\theta_p(u)^{-1}u)^s \zeta_{p,r}(s,c\omega,cx).$$

This is well defined and we obtain the extension of the domain of definition. A similar procedure can be applied to $L\Gamma_{p,r}(x,\omega)$.

Now let F be a totally real algebraic number field. Let $\mathfrak{f}$ be an integral ideal of F which is divisible by every prime factor of (p). For $c \in C_{\mathfrak{f}}$, taking an ideal $\mathfrak{a} \in c$, we put

$$\omega(c) = \theta_p(N(\mathfrak{a})).$$

As shown by Barsky, Cassou-Noguès [Ca] and Deligne–Ribet [DR], there exists the p-adic partial zeta function $\zeta_{p,F}(s,c)$. It is analytic except for a possible simple pole at $s=1$ and characterized by the interpolation property

$$\zeta_{p,F}(-m,c)=\omega(c)^{-m}\zeta_F(-m,c), \qquad 0\le m\in\mathbf{Z}.$$

We fix an embedding $F\subset\overline{\mathbf{Q}}\subset\mathbf{C}_p$.

Theorem 7.1 (Kashio [K]). *Let $\mathfrak{f}$ be an integral ideal of F which is divisible by every prime factor of (p). Let $\widetilde{F}$ be the normal closure of F over $\mathbf{Q}$ and let $\epsilon_1,\ldots,\epsilon_m$ be generators of $E^+_{\widetilde{F}}$. Write*

$$\zeta'_F(0,c)=\sum_{\sigma\in J_F}\sum_{j\in J}\sum_{z\in R(C_j,c)}\log\frac{\Gamma_{r(j)}(z^\sigma,v_j^\sigma)}{\rho_{r(j)}(v_j^\sigma)}+\sum_{i=1}^{m}a_i\log\epsilon_i-\log N(\mathfrak{a}_\mu\mathfrak{f})\zeta_F(0,c)$$

with $a_i\in\widetilde{F}$. Here $v_j=(v_{j1},\ldots,v_{jr(j)})=A_j^{(1)}$ and $v_j^\sigma=(A_j^{(1)})^\sigma$. Then we have

$$\zeta'_{p,F}(0,c)=\sum_{\sigma\in J_F}\sum_{j\in J}\sum_{z\in R(C_j,c)}L\Gamma_{p,r(j)}(z^\sigma,v_j^\sigma)+\sum_{i=1}^{m}a_i\log_p\epsilon_i-\log_p N(\mathfrak{a}_\mu\mathfrak{f})\zeta_{p,F}(0,c).$$

§8. *p-adic absolute CM-period symbol*

We fix an embedding $F\subset\mathbf{C}_p$ and let $\mathfrak{p}$ be the prime ideal of F induced by this embedding. Let $\mathfrak{f}$ be an integral ideal of F. For $c\in C_{\mathfrak{f}}$, we define

$$G_p(c)=\sum_{j\in J}\sum_{z\in R(C_j,c)}L\Gamma_{p,r(j)}(z,v_j). \tag{8.1}$$

Taking generators $\epsilon_1,\ldots,\epsilon_{n-1}$ of E^+_F, write $V(c)$ in the form $V(c)=\sum_{i=1}^{n-1}a_i\log\epsilon_i$ with $a_i\in F$. By a theorem of Baker, this expression is unique. We define

$$(8.2) \qquad V_p(c) = \sum_{i=1}^{n-1} a_i \log_p \epsilon_i.$$

For a fractional ideal $\mathfrak{a}$ of F, we define $\log_p(\mathfrak{a})$ and $\log(\mathfrak{a})$ in the following way. Let h_0 be the class number of F in narrow sense. Let $\mathfrak{q}$ be a prime ideal. We write $\mathfrak{q}^{h_0} = (\beta)$ with a totally positive element β. Then we define

$$\log_p(\mathfrak{q}) = \frac{1}{h_0} \log_p \beta, \qquad \log(\mathfrak{q}) = \frac{1}{h_0} \log \beta.$$

For $\sigma \in J_F$, we formally define

$$\log_p(\mathfrak{q}^\sigma) = \frac{1}{h_0} \log_p \beta^\sigma, \qquad \log(\mathfrak{q}^\sigma) = \frac{1}{h_0} \log \beta^\sigma.$$

(It may happen that $\mathfrak{q} = \mathfrak{q}^\sigma$, but $\log \mathfrak{q} \neq \log \mathfrak{q}^\sigma$. We regard $\log \mathfrak{q}^\sigma$ as the symbol attached to $\mathfrak{q}$ and σ.) We extend this definition to fractional ideals so that $\log_p(\mathfrak{a}^\sigma \mathfrak{b}^\sigma) = \log_p(\mathfrak{a}^\sigma) \log_p(\mathfrak{b}^\sigma)$ holds, and similarly to log.

Now taking $\mathfrak{a}_\mu$ so that c and $\mathfrak{a}_\mu \mathfrak{f}$ belong to the same narrow ideal class, we define the W_p-term by

$$(8.3) \qquad W_p(c) = -\log_p(\mathfrak{a}_\mu) \cdot (\zeta_F(0, c) - \zeta_F(0, \underline{\mathfrak{p}}^{-1} c)).$$

Here $\underline{\mathfrak{p}}$ is the class of $\mathfrak{p}$ in $C_\mathfrak{f}$ if $\mathfrak{p} \nmid \mathfrak{f}$; if $\mathfrak{p} \mid \mathfrak{f}$, the term $\zeta_F(0, \underline{\mathfrak{p}}^{-1} c)$ should be dropped. For $W_p(c)$, this definition leaves the ambiguity of the addition of elements of the form $a \log_p \epsilon$, $a \in \mathbf{Q}$, $\epsilon \in E_F^+$. We put

$$(8.4) \qquad X_p(c) = G_p(c) + V_p(c) + W_p(c).$$

If $\mathfrak{f}$ is divisible by every prime factor of (p), Theorem 7.1 can be written as

$$(8.5) \qquad \zeta'_{p,F}(0, c) = \sum_{\sigma \in J_F} X_p(c^\sigma) - \log_p N(\mathfrak{f}) \zeta_{p.F}(0, c).$$

Now let K be a CM-field which is abelian over F. We fix an embedding $K \subset \mathbf{C}_p$ and let $\mathfrak{P}$ be the prime ideal induced by this embedding. We put

$$\mathfrak{P}^{h_K} = (\alpha).$$

We define the (logarithmic) p-adic absolute CM-period symbol by

$$(8.6) \quad \begin{aligned} lg_{p,K}(\mathrm{id},\tau) = & -\frac{\mu(\tau)}{2}\log_p(\mathfrak{p}) \\ & + \frac{1}{|G|}\sum_{\chi\in\hat{G}_-}\frac{\chi(\tau)}{L(0,\chi)}\sum_{c\in C_{\mathfrak{f}(\chi)\mathfrak{p}}}\chi(c)X_p(c). \end{aligned}$$

To state a precise conjecture on $lg_{p,K}(\mathrm{id},\tau)$, we define the logarithmic version of the absolute CM-period symbol:

$$(8.7) \quad \begin{aligned} lg_K(\mathrm{id},\tau;\{C_j\},\{\mathfrak{a}_\mu\}) = & -\frac{\mu(\tau)}{2}\log\pi \\ & + \frac{1}{|G|}\sum_{\chi\in\hat{G}_-}\frac{\chi(\tau)}{L(0,\chi)}\sum_{c\in\mathfrak{f}(\chi)}\chi(c)X(c;\{C_j\},\{\mathfrak{a}_\mu\}). \end{aligned}$$

Here we modify the W-term in the definition of $X(c)$ by

$$W(c) = -\log(\mathfrak{a}_\mu\mathfrak{f})\zeta_F(0,c).$$

From [Y4], Chapter III, §3.6, §3.7, it follows that

$$(8.8) \quad lg_K(\mathrm{id},\tau;\{C_j\},\{\mathfrak{a}_\mu\mathfrak{p}\}) - lg_K(\mathrm{id},\tau;\{C_j\},\{\mathfrak{a}_\mu\}) = \sum_{i=1}^{n-1} a_i\log\epsilon_i$$

with generators ϵ_i of E_F^+ and $a_i \in \mathbf{Q}$. Now we can state

Conjecture P. *Let $\epsilon_1, \epsilon_2, \ldots, \epsilon_{n-1}$ be generators of E_F^+. If $\mathfrak{p}$ is completely decomposed in K, then we have*

$$lg_{p,K}(\mathrm{id},\tau) = \frac{1}{2h_K}\log_p(\alpha^{\tau^{-1}\rho}/\alpha^{\tau^{-1}}) + \sum_{i=1}^{n-1} a_i\log_p\epsilon_i$$

with $a_i \in \mathbf{Q}$ defined by (8.8).

Remark. We can define another version of the p-adic absolute CM-period symbol, replacing $\sum_{c \in C_{\mathfrak{f}(\chi)\mathfrak{p}}}$ by $\sum_{c \in C_{\mathfrak{f}(\chi)}}$. Both symbols have almost same properties, but the version defined by (8.6) is somewhat simpler.

Now let us recall a conjecture of Gross [G], which can be regarded as a p-adic analogue of the Stark-Shintani conjecture.

Conjecture of Gross. *Let F, K and the other notations be as above. Let $\mathfrak{p}_1, \mathfrak{p}_2, \ldots, \mathfrak{p}_r$ be all the prime ideals of F lying above p. We assume that $\mathfrak{p} = \mathfrak{p}_1$ splits completely in K. Then, for $\chi \in \widehat{G}_-$, we have*

$$\frac{L_p'(0, \omega\chi)}{L(0, \chi)} = \frac{1}{2h_K} \prod_{i=2}^{t} (1 - \chi(\mathfrak{p}_i)) \sum_{\tau \in G} \chi(\tau) \log_p (N_{K_{\mathfrak{P}}/\mathbf{Q}_p}(\alpha^{\tau\rho}/\alpha^{\tau})).$$

We can show that Conjecture P implies the Gross conjecture. Moreover it produces interesting algebraic numbers (Weil numbers) even when $L_p'(0, \omega\chi) = 0$.

§9. *Numerical examples (p-adic case)*

As in §5, we take $F = \mathbf{Q}(\sqrt{29})$, $K = \mathbf{Q}\left(\sqrt{\frac{9+\sqrt{29}}{2}}\, i\right)$ and use the same notation. Let χ be the character of $C_{\mathfrak{f}}$ which corresponds to the extension K/F. We write $\mathfrak{p} = (\alpha_0)$ with a totally positive α_0. Since $L(0, \chi) = 2$, (8.6) can be written as

$$lg_{p,K}(\mathrm{id}, \mathrm{id}) = -\frac{1}{2} \log_p \alpha_0 + \frac{1}{4} \sum_{c \in C_{\mathfrak{f}\mathfrak{p}}} \chi(c) X_p(c). \tag{9.1}$$

First take $p = 79$. Then p remains prime in F and decomposes completely in K as $p\mathcal{O}_K = (\alpha)(\alpha^\rho)$ with

$$\alpha = \frac{1+\sqrt{29}}{2} \left(\sqrt{\frac{9+\sqrt{29}}{2}}\, i + \frac{3+\sqrt{29}}{2} \right) - (5 + 2\sqrt{29}).$$

By (9.1) and $\alpha_0 = p$, $\log_p p = 0$, we have

$$lg_{p,K}(\mathrm{id}, \mathrm{id}) = \frac{1}{4} \sum_{c \in \mathfrak{f}(p)} \chi(c) X_p(c).$$

We can check easily that the correction term (8.8) is 0. Hence Conjecture P states that

$$lg_{p,K}(\mathrm{id},\mathrm{id}) = \frac{1}{2}\log_p(\alpha^\rho/\alpha). \tag{9.2}$$

Both sides are elements of $\mathcal{O}_{F_p}$ and we verified, by numerical computation, that they are equal modulo p^{50}. Let σ is the nontrivial automorphism of F. Since σ gives a continuous automorphism of F_p, (9.2) yields

$$\sum_{c\in\mathfrak{f}(p)} \chi(c)X_p(c) = 2\log_p(\alpha^\rho/\alpha), \quad \sum_{c\in\mathfrak{f}(p)} \chi(c)X_p(c^\sigma) = 2\log_p(\alpha^{\rho\sigma}/\alpha^\sigma).$$

By (8.5), we have

$$L_p'(0,\chi\omega) = \sum_{c\in\mathfrak{f}(p)} \chi(c)(X_p(c) + X_p(c^\sigma)).$$

Therefore Conjecture P gives

$$L_p'(0,\chi\omega) = 2\log_p(N_{K_\mathfrak{P}/\mathbf{Q}_p}(\alpha^\rho/\alpha)).$$

This is what the Gross conjecture states in this case. Conjecture P is finer since it gives a factorization of $\log_p(N_{K_\mathfrak{P}/\mathbf{Q}_p}(\alpha^\rho/\alpha))$.

Next let $p = 23$; p decomposes in F as $p\mathcal{O}_F = \mathfrak{p}\mathfrak{p}'$ with

$$\mathfrak{p} = \left(\frac{11+\sqrt{29}}{2}\right), \qquad \mathfrak{p}' = \left(\frac{11-\sqrt{29}}{2}\right).$$

Then both of $\mathfrak{p}$ and $\mathfrak{p}'$ decompose completely in K. The Gross conjecture states

$$L_p(0,\omega\chi) = L_p'(0,\omega\chi) = 0,$$

which is a special case of a theorem proved in [K]. We may take

$$\alpha = \sqrt{\frac{9+\sqrt{29}}{2}}\,i + 1, \qquad \alpha_0 = \frac{11+\sqrt{29}}{2}.$$

We can show that the correction term (8.8) is equal to

$$-\frac{1}{8}\log\epsilon\sum_{c\in C_{\mathfrak{f}}}\chi(c)\sum_{z\in R(C(1,\alpha_0),c)}\mathrm{Tr}_{F/\mathbf{Q}}(\zeta_2(0,(1,\alpha_0),z))$$
$$=-\frac{109}{2\cdot 13\cdot p}\log\epsilon.$$

Hence Conjecture P states that

$$(9.3)\qquad lg_{p,K}(\mathrm{id},\mathrm{id})=\frac{1}{2}\log_p(\alpha^\rho/\alpha)-\frac{109}{2\cdot 13\cdot p}\log_p\epsilon.$$

Both sides are elements of $\mathbf{Z}_p$; by numerical computation, we verified that the equality holds modulo p^{50}.

§10. *Interpretation of $lg_{p,K}$ in terms of p-adic periods*

We wish to know the nature of the p-adic absolute period symbol $lg_{p,K}(\mathrm{id},\tau)$. Conjecture P clarifies it when $\mathfrak{p}$ is completely decomposed in K. Now our problem is to drop this assumption.

For the formulation of the conjecture in general case, it seems most advantageous to consider motives attached to Grössencharacters of type A_0. In fact, in the complex case, Conjecture C is formulated with respect to the critical values of the L-functions attached to such Grössencharacters, which eventually turns out to be amenable to numerical tests.

Let K be a CM-field. Let χ be a Grössencharacter of K of conductor $\mathfrak{f}$ such that

$$\chi((\alpha))=\prod_{\sigma\in J_K}\sigma(\alpha)^{l_\sigma},\qquad \alpha\equiv 1 \mod{}^\times\mathfrak{f}.$$

Here l_σ are integers such that $l_\sigma+l_{\sigma\rho}$ are independent of σ. Put

$$E=\mathbf{Q}(\chi(\mathfrak{a})\mid \mathfrak{a}\text{ is a fractional ideal prime to }\mathfrak{f}).$$

Then E is an algebraic number field of finite degree.

There exists a motive $M=M(\chi)$ over K with coefficients in E; M is characterized by the property

$$L(M,s)=(L(s,\sigma(\chi)))_{\sigma\in J_E},\qquad J_E=\mathrm{Hom}(E,\mathbf{C}).$$

We consider four realizations of M; de Rham, p-adic, Betti and crystalline.

First M has the de Rham realization $H_{\mathrm{DR}}(M)$ which is a free $E \otimes_{\mathbf{Q}} K$-module of rank 1. For a finite place λ of E, M has the λ-adic realization $H_\lambda(M)$ which is an E_λ-module of rank 1. We have an E_λ-linear action of $\mathrm{Gal}(\overline{K}/K)$ on $H_\lambda(M)$. It is related to the p-adic realization by

$$H_p(M) = \oplus_{\lambda|p} H_\lambda(M).$$

Let $\overline{\mathbf{Q}}$ be the algebraic closure of $\mathbf{Q}$ in $\mathbf{C}$. We fix an embedding of $\overline{\mathbf{Q}}$ into $\mathbf{C}_p$ and let $\mathfrak{P}$ be the prime ideal of K induced by this embedding. Below we take $K_{\mathfrak{P}}$ as the basic local field.

Using $K \subset \overline{\mathbf{Q}} \subset \mathbf{C}$, we have the Betti realization $H_B(M)$ which is an E-module of rank 1. We have the canonical isomorphism

$$\text{(10.1)} \qquad i_p : H_B(M) \otimes_{\mathbf{Q}} \mathbf{Q}_p \cong H_p(M)$$

as $E \otimes_{\mathbf{Q}} \mathbf{Q}_p$-modules of rank 1. The de Rham and p-adic realizations are related in the following way. Let B_{DR} be the discrete valuation field introduced by Fontaine [Fo]. We have

$$\text{(10.2)} \qquad I_{\mathrm{DR}} : H_p(M) \otimes_{\mathbf{Q}_p} B_{\mathrm{DR}} \cong H_{\mathrm{DR}}(M) \otimes_K B_{\mathrm{DR}}.$$

Both sides are isomorphic as free $E \otimes_{\mathbf{Q}} B_{\mathrm{DR}}$-modules of rank 1. The isomorphism is compatible with the action of $\mathrm{Gal}(\overline{\mathbf{Q}}_p/K_{\mathfrak{P}})$ and filtrations.

Suppose that χ is unramified at $\mathfrak{P}$. Then M has the crystalline realization $H_{\mathrm{cris}}(M)$. Let k be the residue field of $K_{\mathfrak{P}}$ and $K_{\mathfrak{P},0}$ be the quotient field of $W(k)$, the ring of Witt vectors over k. We identify $K_{\mathfrak{P},0}$ with the maximal unramified extension of $\mathbf{Q}_p$ contained in $K_{\mathfrak{P}}$. $H_{\mathrm{cris}}(M)$ is a free $E \otimes_{\mathbf{Z}} W(k) \cong E \otimes_{\mathbf{Q}} K_{\mathfrak{P},0}$ -module of rank 1. We have the isomorphism

$$\text{(10.3)} \qquad I_{\mathrm{cris}} : H_p(M) \otimes_{\mathbf{Q}_p} B_{\mathrm{cris}} \cong H_{\mathrm{cris}}(M) \otimes_{W(k)} B_{\mathrm{cris}}.$$

Both sides are isomorphic as free $E \otimes_{\mathbf{Q}} B_{\mathrm{cris}}$-modules of rank 1. The isomorphism is compatible with the action of $\mathrm{Gal}(\overline{\mathbf{Q}}_p/K_{\mathfrak{P}})$ and

Frobenius. Here $B_{\rm cris}$ denotes the subring of $B_{\rm DR}$ introduced by Fontaine.

Let $\varphi \in \mathrm{Gal}(\overline{\mathbf{Q}}_p/\mathbf{Q}_p)$ be a Frobenius and let $\varphi_{\mathfrak{P}}$ be a Frobenius of $\mathfrak{P}$ which lies in $\mathrm{Gal}(\overline{\mathbf{Q}}_p/K_{\mathfrak{P}})$. We may take $\varphi_{\mathfrak{P}} = \varphi^f$ where f denotes the degree of $\mathfrak{P}$ over $\mathbf{Q}$. The action of $\varphi_{\mathfrak{P}}$ on $H_{\rm cris}(M)$ is given by

$$\varphi_{\mathfrak{P}} \mid H_{\rm cris}(M) = \chi(\mathfrak{P}) \otimes 1. \tag{10.4}$$

There is a $K_{\mathfrak{P},0}$-structure on $H_{\rm DR}(M) \otimes_K K_{\mathfrak{P}}$ which we denote by $H_{\rm DR}(M/K_{\mathfrak{P},0})$. We have

$$I_0 : H_{\rm cris}(M) \otimes_{W(k)} K_{\mathfrak{P},0} \cong H_{\rm DR}(M/K_{\mathfrak{P},0}), \tag{10.5}$$

$$I_{\rm DR} = (I_0)_{\rm DR} \circ (I_{\rm cris})_{\rm DR}. \tag{10.6}$$

Here $(I_{\rm cris})_{\rm DR}$ denotes the isomorphism obtained by taking $\otimes_{B_{\rm cris}} B_{\rm DR}$ in (10.3) and $(I_0)_{\rm DR}$ denotes the isomorphism by taking $\otimes_{K_{\mathfrak{P},0}} B_{\rm DR}$ in (10.5).

Hereafter we assume that $K_{\mathfrak{P},0} = K_{\mathfrak{P}}$ for simplicity. This condition is satisfied when $\mathfrak{P}$ is unramified over $\mathbf{Q}$.

We take $0 \neq c_B \in H_B(M)$. Then $i_p(c_B)$ is a generator of $H_p(M)$ as an $E \otimes_{\mathbf{Q}} \mathbf{Q}_p$-module. Take a generator $c_{\rm DR} \in H_{\rm DR}(M)$ as an $E \otimes_{\mathbf{Q}} K$-module. Then we define the period $P(\chi)$ by

$$I_{\rm DR}(i_p(c_B) \otimes 1) = P(\chi)(c_{DR} \otimes 1). \tag{10.7}$$

We have $P(\chi) \in (E \otimes_{\mathbf{Q}} B_{\rm DR})^\times$; it is determined up to the multiplication of elements of $(E \otimes_{\mathbf{Q}} K)^\times$. Here we regard $E \otimes_{\mathbf{Q}} K$ as a subring of $E \otimes_{\mathbf{Q}} B_{\rm DR}$. $P(\chi)$ is the p-adic period in the standard sense.

We can choose a generator $c_{\rm cris}$ of $H_{\rm cris}(M)$ as an $E \otimes_{\mathbf{Z}} W(k)$-module so that

$$I_0(c_{\rm cris} \otimes 1) = c_{\rm DR} \otimes 1. \tag{10.8}$$

We define $\widetilde{P}(\chi) \in (E \otimes_{\mathbf{Q}} B_{\rm cris})^\times$ by

$$I_{\rm cris}(i_p(c_B) \otimes 1) = \widetilde{P}(\chi)(c_{\rm cris} \otimes 1). \tag{10.9}$$

Since $B_{\mathrm{cris}} \subset B_{\mathrm{DR}}$, we can regard $\widetilde{P}(\chi) \in (E \otimes_{\mathbf{Q}} B_{\mathrm{DR}})^{\times}$. Then, by (10.6) and (10.9), we have

$$P(\chi) = \widetilde{P}(\chi). \tag{10.10}$$

Since $H_{\mathrm{cris}}(M, W(k)) \otimes_{W(k)} K_{\mathfrak{P},0}$ is a free $E \otimes_{\mathbf{Q}} K_{\mathfrak{P},0}$-module of rank 1, we can write

$$\varphi^i(c_{\mathrm{cris}} \otimes 1) = Q^{(i)}(c_{\mathrm{cris}} \otimes 1), \qquad 1 \leq i \in \mathbf{Z} \tag{10.11}$$

with $Q^{(i)} \in (E \otimes_{\mathbf{Q}} K_{\mathfrak{P},0})^{\times}$. Applying the Frobenius i-times on (10.9), we obtain

$$\Phi^i_{\mathrm{cris}}(\widetilde{P}(\chi))\varphi^i(c_{\mathrm{cris}} \otimes 1) = \widetilde{P}(\chi)(c_{\mathrm{cris}} \otimes 1).$$

Here Φ_{cris} denotes the action of Frobenius on B_{cris}. Therefore we obtain

$$\Phi^i_{\mathrm{cris}}(\widetilde{P}(\chi))Q^{(i)} = \widetilde{P}(\chi), \qquad 1 \leq i \in \mathbf{Z}. \tag{10.12}$$

We put $Q = Q^{(1)}$. In particular, we have

$$\Phi_{\mathrm{cris}}(\widetilde{P}(\chi))Q = \widetilde{P}(\chi). \tag{10.13}$$

Applying Φ_{cris} on both sides noting that Φ_{cris} acts on $K_{\mathfrak{P},0}$ by the absolute Frobenius φ, we get

$$\Phi^2_{\mathrm{cris}}(\widetilde{P}(\chi))\varphi(Q) = \Phi_{\mathrm{cris}}(\widetilde{P}(\chi)) = \widetilde{P}(\chi)Q^{-1}.$$

Hence we have

$$Q^{(2)} = \varphi(Q)Q.$$

Repeating this process, we get

$$Q^{(i)} = \varphi^{i-1}(Q) \cdots \varphi(Q)Q, \qquad 1 \leq i \in \mathbf{Z}. \tag{10.14}$$

We note that

$$Q^{(f)} = \chi(\mathfrak{P}) \otimes 1 \in (E \otimes_{\mathbf{Q}} K_{\mathfrak{P},0})^{\times} \tag{10.15}$$

which follows from (10.14).

Using $Q^{(i)}$, we can predict the nature of $lg_{p,K}$ in the general case. Suppose that K is abelian over a totally real number field F as in our construction of $lg_{p,K}$. Take $\tau \in \mathrm{Gal}(K/F)$. We write $lg_{p,K/F}(\mathrm{id},\tau)$ for $lg_{p,K}(\mathrm{id},\tau)$. We have

$$Q^{(i)} \in (E \otimes_{\mathbf{Q}} K_{\mathfrak{P},0})^\times \subset (E \otimes_{\mathbf{Q}} \overline{\mathbf{Q}_p})^\times \cong \prod_{\sigma: E \subset \overline{\mathbf{Q}}} \overline{\mathbf{Q}_p}^\times .$$

Let $Q^{(i)}(\sigma) \in \overline{\mathbf{Q}_p}^\times$ denote the σ-component of $Q^{(i)}$. Let $\mathfrak{p}$ be the prime ideal of F below $\mathfrak{P}$ and let f_0 be the degree of $\mathfrak{p}$ over $\mathbf{Q}$. We take a Hecke character χ of the form

$$\chi((\alpha)) = (\tau(\alpha)/\rho\tau(\alpha))^l, \qquad \alpha \equiv 1 \mod {}^\times \mathfrak{f}$$

with $1 \le l \in \mathbf{Z}$. Let $\widetilde{K}$ be the normal closure of K over $\mathbf{Q}$. We can take l and χ so that $E \subset \widetilde{K}$. Then $E \otimes_{\mathbf{Q}} \widetilde{K} \cong \prod_{\sigma \in J_E} \widetilde{K}$.

Conjecture Q. *We have*

$$lg_{p,K/F}(id,\tau^{-1}) = -\frac{1}{2l}\log_p Q^{(f_0)}(\mathrm{id}) + \sum_i a_i \log_p \epsilon_i + \log_p b$$

with $a_i \in F$, $\epsilon_i \in E_F^+$, $b \in \widetilde{K}$.

Since $\widetilde{P}(\chi)$ is determined up to the multiplication of elements in $(E \otimes_{\mathbf{Q}} K)^\times$, $Q^{(f_0)}$ is determined up to the multiplication of elements of the form $\varphi^{f_0}(c)/c$, $c \in (E \otimes_{\mathbf{Q}} K)^\times$ by (10.12). Therefore the validity of Conjecture Q does not depend on choices of c_B, c_{cris} and c_{DR}.

We can show, using (10.15), that Conjecture P implies that Conjecture Q is true with $b = 1$ if $\mathfrak{p}$ is completely decomposed in K. The p-adic periods can be understood more explicitly in this case, but we will not go into details here.

Remark. What is remarkable in the p-adic case is that the nature of the p-adic absolute period symbol $lg_{p,K/F}(\mathrm{id},\tau)$ depends strongly on F. (This is not so when $\mathfrak{p}$ is completely decomposed in K.) In the complex case, $p_K(\mathrm{id},\tau)^{2l}$ appears in place of $Q^{(f_0)}(\mathrm{id})^{-1}$.

The author thanks Professor Don Blasius for discussions which are useful in writing §10.

References

[A] W. Anderson, Logarithmic derivatives of Dirichlet L-functions and the periods of abelian varieties, Comp. Math. 45(1982), 315–332.

[Ba] E. W. Barnes, On the theory of the multiple gamma function, Trans. Cambridge Philos. Soc. 19 (1904), 374–425.

[Bl1] D. Blasius, On the critical values of Hecke L-series, Ann. of Math. 124(1986), 23–63.

[Bl2] D. Blasius, A p-adic property of Hodge classes on abelian variety, Proc. Symposia Pure Math. 55(1994), Part 2, 293–308.

[BO] P. Berthlot and A. Ogus, F-isocrystals and De Rham cohomology I, Inv. Math., 72(1983), 159–199.

[Ca] P. Cassou-Noguès, Valeurs aux entiers négative des fonction zêta et fonction zêta p-adiques, Inv. Math. 51(1979), 29–59.

[Co] P. Colmez, Périods des variétés abéliennes à multiplication complexe, Ann. of Math. 138(1993), 625–683.

[DR] P. Deligne and K. A. Ribet, Values of abelian L-function at negative integers over totally real fields, Inv. Math. 59(1980), 227–286.

[DS] E. de Shalit, On monomial relations between p-adic periods, Journal für reine und angew. Math., 374(1987), 193–207.

[Fa] G. Faltings, Crystalline cohomology and p-adic Galois representations, Algebraic ananlysis, and Number theory (J. I. Igusa, ed.), Johns Hopkins Univ. Press, 1990, pp. 25–79.

[Fo] J. -M. Fontaine, Sur certains types de représentations p-adiques du grou-pe de Galois d'un corps local; construction d'un anneaux de Barsotti-Tate, Ann. of Math., 115(1982), 529–577.

[Gi] R. Gillard, Relations monomiales entre périodes p-adiques, Invent. Ma-th. 93(1988), 355–381.

[Gr] B. H. Gross, p-adic L-series at $s = 0$, J. Fac. Sci. Univ. Tokyo 28(1981), 979–994.

[GK] B. H. Gross and N. Koblitz, Gauss sums and the p-adic Γ-function, Ann. Math., 109(1979), 569–581.

[I] K. Iwasawa, Lectures on p-adic L-functions, Annals of Math. Studies, 74, Princeton University Press, 1972.

[JKS] U. Jannsen, S. Kleiman and J-P. Serre (eds.), Motives, Proc. Symposia Pure Math. 55(1994), Part 1 and 2.

[K] T. Kashio, On a p-adic analogue of Shintani's formula, J. Math. Kyoto Univ. 45(2005), 99-128.

[Kat] N. Katz, p-adic L-functions for CM-fields, Invent. Math. 49(1978), 199-297.

[KY1] T. Kashio and H. Yoshida, On the p-adic absolute CM-period symbol, Algebra and Number theory (ed. R. Tandon), Proceedings of the Silver Jubilee Conference, University of Hyderabad, 2005, 359–399, Hindustan Book Agency.

[KY2] T. Kashio and H. Yoshida, On p-adic absolute CM-periods, in preparation.

[O] A. Ogus, A p-adic analogue of the Chowla-Selberg formula, Lecture notes in Math., 1454, 319–341, Springer Verlag, 1990.

[S1] G. Shimura, Automorphic forms and periods of abelian varieties, J. Math. Soc. Japan 31(1979), 561–592 (=Collected Papers III, [79a]).

[S2] G. Shimura, The arithmetic of certain zeta functions and automorphic forms on orthogonal groups, Ann. of Math. 111(1980), 313–375 (=Collected Papers III, [80]).

[S3] G. Shimura, Abelian varieties with complex multiplication and modular functions, Princeton Mathematical Series 46, Princeton University Press, 1998.

[Sh1] T. Shintani, On evaluation of zeta functions of totally real algebraic number fields at non-positive integers, J. Fac. Sci. Univ. Tokyo 23(1976), 393–417.

[Sh2] T. Shintani, On values at $s = 1$ of certain L functions of totally real algebraic number fields, in Algebraic Number Theory, Proc. International Symp., Kyoto, 1976, 201–212, Kinokuniya, 1977.

[Sh3] T. Shintani, On certain ray class invariants of real quadratic fields, J. Math. Soc. Japan 30(1978), 139–167.

[St] H. M. Stark, L-functions at $s = 1$, I, II, III, IV, Advances in Math. 7(1971), 301–343; 17(1975), 60–92; 22(1976), 64–84; 35(1980), 197–235.

[W] A. Weil, Sur les périodes des intégrales abéliennes, Comm. on Pure and Appl. Math. XXIX (1976), 813–819 (=Collected Papers III, [1976b]).

[Y1] H. Yoshida, On the zeta functions of Shimura varieties and periods of Hilbert modular forms, Duke Math. J. 75(1994), 121–191.

[Y2] H. Yoshida, On absolute CM–periods, Proc. Symposia Pure Math. 66, Part 1, 1999, 221–278.

[Y3] H. Yoshida, On absolute CM-periods II, Amer. J. Math. 120(1998), 1199–1236.

[Y4] H. Yoshida, Absolute CM-periods, Mathematical Surveys and Monographs 106, American Mathematical Society, 2003.

Hiroyuki YOSHIDA
Department of Mathematics,
Faculty of Science,
Kyoto University,
Kyoto 606-8502, Japan
Email: yoshida@math.kyoto-u.ac.jp

Special Zeta Values in Positive Characteristic

Jing YU

1 Introduction

This is the story of certain zeta values in positive characteristic. The arithmetic here starts with the ring $\mathbf{A} := \mathbb{F}_q[\theta]$, where $\mathbb{F}_q$ is the finite field of q elements with q a power of prime number p, and θ is a variable. This ring $\mathbf{A}$ plays the role of $\mathbb{Z}$. It sits discretely inside $\mathbb{F}_q((\frac{1}{\theta}))$ with respect to the $\frac{1}{\theta}$-adic topology just as $\mathbb{Z}$ is discrete subset on the real line $\mathbb{R}$. Moreover one has that $\mathbb{F}_q((\frac{1}{\theta}))/\mathbf{A}$ is compact as against the fact that $\mathbb{Z}$ is cocompact inside $\mathbb{R}$. The number $q-1$ is particularly relevant since there are $q-1$ signs (i.e. units in $\mathbf{A}$) in this positive characteristic world. In the simplest case $q=2$, it is a world where sign does not matter. On the other hand for $q>2$, one also needs the set of all monic polynomials in $\mathbf{A}$ to be denoted by $\mathbf{A}_+$.

In 1935 L. Carlitz [4] initiated the study of the following zeta values:

$$(1.1) \qquad \zeta_{\mathbf{A}}(n) := \sum_{a \in \mathbf{A}_+} \frac{1}{a^n} \in \mathbb{F}_q\left(\left(\frac{1}{\theta}\right)\right), \quad n = 1, 2, \ldots.$$

These special values are natural analogues of the classical Riemann zeta values $\zeta(n)$ at integer $n > 1$. They contain very interesting information about deeper structures of the arithmetic of function fields, cf. [7] and [9]. Carlitz made the discovery that there is a constant $\pi_{\mathbf{A}}$ algebraic over $\mathbb{F}_q((\frac{1}{\theta}))$ such that $\zeta_{\mathbf{A}}(n)/\pi_{\mathbf{A}}^n := b_n$ falls in $k := \mathbb{F}_q(\theta)$ if n is divisible by $q-1$. This constant $\pi_{\mathbf{A}}$ is later shown to be transcendental over k by Wade [10]. Moreover Carlitz has explicit formulas for these rationals b_n in case n is divisible by $q-1$. If $q=2$ this explains perfectly the nature of $\zeta_{\mathbf{A}}(n)$ for all

$n > 0$. For $q > 2$ it also leads to the more interesting question of explaining the nature of $\zeta_{\mathbf{A}}(n)$ for all n which are **odd**, in the sense that n is not divisible by $q-1$.

A central feature of Carlitz theory is the introduction of the so-called Carlitz module. It plays the role of $\mathbb{G}_m$ for arithmetic of the rational field k. The crucial transcendental element $\pi_{\mathbf{A}}$ above occurs as a fundamental period of this Carlitz module. We shall survey Carlitz theory in Section 2, together with the Euler-Carlitz relations : $\zeta_{\mathbf{A}}(n) = b_n \pi_{\mathbf{A}}^n$ for all n divisible by $q-1$. As in the classical story of Riemann zeta values $\zeta(n)$ for integer $n > 1$, one is tempting to conjecture that all $\zeta_{\mathbf{A}}(n)$ are transcendental over k and these Euler-Carlitz relations generates all the non-trivial relations among these zeta values. Because we are now in the characteristic p world, p-th power (Frobenius) relations are always there for all positive integers n and m :

$$\zeta_{\mathbf{A}}(p^m n) = \zeta_{\mathbf{A}}(n)^{p^m}. \tag{1.2}$$

These relations certainly are not considered as non-trivial.

Half century after Carlitz's discovery, a breakthrough towards understanding these zeta values takes place. First Anderson-Thakur [2] is able to relate the zeta values $\zeta_{\mathbf{A}}(n)$ to a special point on the n-th higher Carlitz module for all positive integer n. As a result we have proved [12] the transcendence of the zeta values $\zeta_{\mathbf{A}}(n)$ for all positive integer n, in particular for those **odd** n not divisible by $q-1$. Later in [13], we are also able to prove that the Euler-Carlitz relations are the only linear relations (with algebraic coefficients) among these Carlitz zeta values and powers of $\pi_{\mathbf{A}}$. In Section 3 we will sketch this theory of Anderson-Thakur and Yu, with n-th higher Carlitz module playing the central figure. These so-called t-modules can be interpreted as the tensor powers of the old Carlitz module.

The grand final question is to determine all the algebraic relations among the values $\zeta_{\mathbf{A}}(n), n = 1, 2, 3, \ldots$. That is to prove that the Euler-Carlitz relations and the p-th power relations, account for all the existing algebraic relations. With the recent advances on transcendence theory of positive characteristic made by Anderson-Brownawell-Papanikolas [3] and Papanikolas [8], this is completely settled by Chang-Yu 2005 [6]:

Theorem. *The transcendence degree of* $\mathbb{F}_q(\theta)(\pi_{\mathbf{A}}, \zeta_{\mathbf{A}}(1), \ldots, \zeta_{\mathbf{A}}(n))$ *over* $\mathbb{F}_q(\theta)$ *is* :

$$n - \lfloor n/p \rfloor - \lfloor n/(q-1) \rfloor + \lfloor n/(p(q-1)) \rfloor + 1.$$

We will give a very brief account of this latest development. This round, the central figure on stage changes to t-motives (introduced by Anderson [1]) which is a structure dual to the previously predominant t-module structure. In Section 4, we outline the construction of t-motives whose periods are related to the values $\zeta_{\mathbf{A}}(n)$. The breakthrough in passing from linear independence to algebraic independence, is achieved by developing an analogue of classical differential Galois theory and relating it to the motivic Galois groups of the t-motives (cf. Papanikolas [8]). This last episode can be viewed philosophically as realizing a function field version of Grothendieck's conjecture on periods of abelian varieties. Short sketch of proof our main theorem will be given in the final Section 5.

2 Carlitz Theory

Recall the Riemann zeta function:

$$\zeta(s) = \sum_{n=1}^{\infty} \frac{1}{n^s}, \qquad \mathrm{Re}(s) > 1.$$

For even $m > 1$, Euler showed

$$\zeta(m) = \sum_{n=1}^{\infty} \frac{1}{n^m} = \frac{-(2\pi\sqrt{-1})^m B_m}{2\ m!},$$

where Bernoulli numbers B_m are given by

$$\frac{z}{e^z - 1} = \sum_{m=0}^{\infty} B_m \frac{z^m}{m!}.$$

Here e^z is the exponential map for the algebraic group $\mathbb{G}_m$:

$$\begin{array}{ccc} \mathbb{C} & \stackrel{e^z}{\rightarrow} & \mathbb{G}_m(\mathbb{C}) = \mathbb{C}^\times \\ n(\cdot) \downarrow & & \downarrow (\cdot)^n \\ \mathbb{C} & \stackrel{e^z}{\rightarrow} & \mathbb{G}_m(\mathbb{C}) = \mathbb{C}^\times. \end{array}$$

Fundamental period of $\mathbb{G}_m$ is $2\pi\sqrt{-1}$ which satisfies

$$0 \to \mathbb{Z}\,2\pi\sqrt{-1} \to \mathbb{C} \overset{e^z}{\to} \mathbb{C}^\times \to 1.$$

For arithmetic of the function field k, Carlitz observed that instead of $\mathbb{G}_m$ one should let the key role be played by the additive group $\mathbb{G}_a$ together with the following t-action

$$\phi_C(t) : x \longmapsto \theta x + x^q.$$

This structure is now called the Carlitz module. One has the Carlitz exponential map to linearize the non-linear t-action:

$$\begin{array}{ccc} \mathbb{C}_\infty & \overset{\exp_C}{\to} & \mathbb{G}_a(\mathbb{C}_\infty) = \mathbb{C}_\infty \\ \theta(\cdot)\downarrow & & \downarrow \phi_C(t) \\ \mathbb{C}_\infty & \overset{\exp_C}{\to} & \mathbb{G}_a(\mathbb{C}_\infty) = \mathbb{C}_\infty. \end{array}$$

Here $\mathbb{C}_\infty$ is the completion of a fixed algebraic closure of $k_\infty := \mathbb{F}_q((1/\theta))$, on which the chosen absolute value will be denoted by $|\cdot|_\infty$.

The commutative diagram above amounts to the following functional equation for the exponential:

$$\exp_C(\theta z) = \theta \exp_C(z) + \exp_C(z)^q.$$

Solving the recursive relation for the Taylor coefficients inductively gives the expansion

$$\exp_C(z) = \sum_{h=0}^{\infty} \frac{z^{q^h}}{D_h}, \quad \text{for all } z \in \mathbb{C}_\infty,$$

where

$$\begin{aligned} D_0 &:= 1 \\ D_i &:= \prod_{j=0}^{i-1} (\theta^{q^i} - \theta^{q^j}). \end{aligned}$$

Note that D_i is in fact the product of all monic polynomials in $\mathbf{A}$ with degree i.

One also has the exact sequence

$$0 \to \mathbf{A}\pi_{\mathbf{A}} \to \mathbb{C}_\infty \overset{\exp_C}{\to} \mathbb{C}_\infty \to 0,$$

as a result of the following basic theorem of Carlitz:

Theorem 2.1. *One has the product expansion*

$$\exp_C(z) = z \prod_{a \neq 0 \in \mathbf{A}} \left(1 - \frac{z}{a\pi_{\mathbf{A}}}\right), \quad \text{for all } z \in \mathbb{C}_\infty, \tag{2.1}$$

where

$$\pi_{\mathbf{A}} = \theta(-\theta)^{\frac{1}{q-1}} \prod_{i=1}^{\infty} \left(1 - \theta^{1-q^i}\right)^{-1}. \tag{2.2}$$

We will fix from now on a choice of $(-\theta)^{\frac{1}{q-1}}$ so that $\pi_{\mathbf{A}}$ is a well-defined element in $\overline{k_\infty}$.

Proof. From the Carlitz t-action one sees that the zero set of $\exp_C(z)$ is a discrete $\mathbf{A}$-submodule of rank one inside $\mathbb{C}_\infty$ with respect to scalar multiplication by $\mathbf{A}$. Thus (2.1) is established by checking that all zeros are simple. We sketch here a proof of (2.2) from [2]. Consider the following generating function in variable t:

$$E(t) := \sum_{m=0}^{\infty} \exp_C \left(\frac{\pi_{\mathbf{A}}}{\theta^{m+1}}\right) t^m.$$

For each m, substitute the product expansion (2.1) of $\exp_C(\frac{\pi_{\mathbf{A}}}{\theta^{m+1}})$ into the above Taylor series defining $E(t)$. We see that this generating function can be written as a function analytic at $t = \theta$ plus the dominating term :

$$\sum_{m=0}^{\infty} \frac{\pi_{\mathbf{A}}}{\theta} \left(\frac{t}{\theta}\right)^m = \frac{-\pi_{\mathbf{A}}}{t - \theta}.$$

Thus $E(t)$ has residue $-\pi_{\mathbf{A}}$ at $t = \theta$. On the other hand we also contend that $E(t)$ has product expansion

$$E(t) = (-\theta)^{\frac{1}{q-1}} \prod_{i=0}^{\infty} \left(1 - \frac{t}{\theta^{q^i}}\right)^{-1}. \tag{2.3}$$

Formula (2.2) results from computing residue at θ using this product expansion.

To prove the infinite product (2.3), let $F(t)$ be the function on the right-hand side. "Twisting" the series for $E(t)$ to

$$E^{1)}(t) := \sum_{m=0}^{\infty} \exp_C\left(\frac{\pi_{\mathbf{A}}}{\theta^{m+1}}\right)^q t^m$$

$$= \sum_{m=0}^{\infty}\left(\exp_C\left(\frac{\pi_{\mathbf{A}}}{\theta^m}\right) - \theta\exp_C\left(\frac{\pi_{\mathbf{A}}}{\theta^{m+1}}\right)\right)t^m = (t-\theta)E(t),$$

we find it satisfies the same equation as twisting $F(t)$ to $F^{(1)}(t)$. It follows that we must have $E(t) = F(t)$. □

Carlitz also introduced "factorials" for the ring $\mathbf{A}$. To define, say Γ_n for positive integer n, just write down the q-adic expansion $\sum_{i=0}^{\infty} n_i q^i$ of $n-1$, and let

$$\Gamma_n := \prod_{i=0}^{\infty} D_i^{n_i}\ .$$

With these Carlitz factorials, the Bernoulli-Carlitz "numbers" c_n in $\mathbf{A}$ are then given by the following series expansion

$$\frac{z}{\exp_C(z)} = \sum_{n=0}^{\infty} \frac{c_n}{\Gamma_{n+1}} z^n.$$

The important point is that both the fundamental period $\pi_{\mathbf{A}}$ and the Bernoulli-Carlitz numbers come from the Carlitz module structure.

By taking logarithmic derivative of (2.1), Carlitz ([5]) arrives immediately at the Euler-Carlitz relations:

Theorem 2.2. *For all positive integer n divisible by $q-1$, one has*

$$\zeta_{\mathbf{A}}(n) = \frac{c_n}{\Gamma_{n+1}} \pi_{\mathbf{A}}^n. \tag{2.4}$$

It should be noted that as in the classical arithmetic, both the Carlitz factorials and the Bernoulli-Carlitz numbers enjoy good divisibility or congruence properties, cf. [7] and [9]. There are lots of arithmetic information behind the Euler-Carlitz relations.

The formal inverse to the Carlitz exponential is the Carlitz logarithm. It is the series

$$\log_C(z) := \sum_{k=0}^{\infty} \frac{z^{q^k}}{L_k},$$

where

$$L_0 := 1,$$
$$L_i := \prod_{j=1}^{i}(\theta - \theta^{q^j}) \ (i = 1, 2, \ldots).$$

This is no longer an entire series. It only converges for all $z \in \mathbb{C}_\infty$ satisfying $|z|_\infty < |\theta|_\infty^{\frac{q}{q-1}}$. It also satisfies the functional equation

$$\theta \log_C(z) = \log_C(\theta z) + \log_C(z^q)$$

whenever these values in question are defined.

As pointed out by Thakur, cf. [2] Sec. 3, Carlitz already noticed the following identity connecting logarithm with the zeta value at 1:

Theorem 2.3.

$$\zeta_{\mathbf{A}}(1) = \log_C(1). \tag{2.5}$$

Proof Let y be a variable and consider the generating function:

$$\exp_C(y \log_C(z)) = \sum_{d=0}^{\infty} B_d(y) z^{q^d},$$

where

$$B_d(y) = \sum_{i=0}^{d} \frac{y^{q^i}}{D_i L_{d-i}^{q^i}} \in k[y]$$

satisfying, for all $a \in \mathbb{F}_q[t]$, that

$$\phi_C(a)(x) = \sum_d B_d(a) x^{q^d}.$$

Here $a \mapsto \phi_C(a)$ is the $\mathbb{F}_q[t]$-action on $\mathbb{G}_a$ determined by the Carlitz t-action.

Let $\mathbf{A}_{d+}$ denote the set of monic polynomials in $\mathbf{A}$ having degree d. One checks that $1 - B_d(y)$ has elements in $\mathbf{A}_{d+}$ as simple zeros. Hence, by taking logarithmic derivatives, we have

$$\sum_{a \in \mathbf{A}_{d+}} \frac{1}{y-a} = -\frac{B_d'(0)}{1 - B_d(y)}.$$

The identity (2.5) follows by setting $y = 0$, in fact one gets even:

$$\sum_{a \in \mathbf{A}_{d+}} \frac{1}{a} = \frac{1}{L_d}.$$

This identity brings us to the starting point toward understanding the Carlitz zeta value at integers not divisible by $q-1$.

Let $\bar{k}$ be the algebraic closure of k in $\mathbb{C}_\infty$. We shall say that an element in $\mathbb{C}_\infty$ is transcendental if it is not in $\bar{k}$. After giving ad hoc proof of the transcendence of $\pi_{\mathbf{A}}$, Wade went on in [11] to prove the following analogue of Hermite-Lindemann theorem for the Carlitz exponential:

Theorem 2.4. *For every $\alpha \neq 0 \in \bar{k}$, $\exp_C(\alpha)$ is transcendental.*

This implies not only the transcendence of the fundamental period $\pi_{\mathbf{A}}$ but also the transcendence of $\zeta_{\mathbf{A}}(1)$ because of (2.5).

3 Anderson-Thakur Theory

To understand those zeta values $\zeta_{\mathbf{A}}(n)$ for $n > 1$ not divisible by $q-1$, Anderson introduced in [2] the concept of n-th higher Carlitz module. This is the n-dimensional vector group $\mathbb{G}_a^n$ together with t-action given by the polynomial map

$$\Phi_n(t) : \begin{pmatrix} x_1 \\ \vdots \\ x_n \end{pmatrix} \longmapsto \begin{pmatrix} \theta & 1 & 0 \\ & \ddots & 1 \\ 0 & & \theta \end{pmatrix} \begin{pmatrix} x_1 \\ \vdots \\ x_n \end{pmatrix} + \begin{pmatrix} 0 & \cdots & 0 \\ \vdots & & \vdots \\ 1 & \cdots & 0 \end{pmatrix} \begin{pmatrix} x_1^q \\ \vdots \\ x_n^q \end{pmatrix}.$$

Still one has exponential map $\exp_n$ to linearize the t-action

$$\begin{array}{ccc} \mathbb{C}_\infty^n & \stackrel{\exp_n}{\longrightarrow} & \mathbb{G}_a^n(\mathbb{C}_\infty) = \mathbb{C}_\infty^n \\ d\Theta_n(t) \downarrow & & \downarrow \Phi_n(t) \\ \mathbb{C}_\infty^n & \stackrel{\exp_n}{\longrightarrow} & \mathbb{G}_a^n(\mathbb{C}_\infty) = \mathbb{C}_\infty^n. \end{array}$$

When $n = 1$ it is just the Carlitz module with $\exp_1(z) = \exp_C(z)$. Moreover for any n, the entire analytic map $\exp_n(z_1, \ldots, z_n)$ satisfying the above commutative diagram can be easily solved as everywhere convergent power series with matrix coefficients:

$$\exp_n \begin{bmatrix} z_1 \\ \vdots \\ z_n \end{bmatrix} = \sum_{i=0}^{\infty} P_i \begin{bmatrix} z_1^{q^i} \\ \vdots \\ z_n^{q^i} \end{bmatrix}, \quad P_i \in \mathrm{M}_n(k).$$

The polynomial map $\Phi_n(t)$ determines, via composition and addition, an action of $\mathbb{F}_q[t]$ on $\mathbb{G}_a^n$ with constants in $\mathbb{F}_q$ acting as scalars. Such a structure is called a t-module. The kernel of the exponential map is called the period lattice of the t-module. Regarding $\mathbb{C}_\infty^n$ as tangent space at the origin of $\mathbb{G}_a^n$, Anderson-Thakur was able to prove for n-th higher Carlitz module the following fundamental

Theorem 3.1. *For all n one has exact sequence*

$$d\Phi_n(\mathbb{F}_q[t])\Omega_n \rightarrowtail \mathbb{C}_\infty^n \overset{\exp_n}{\to} \mathbb{C}_\infty^n \to 0$$

$$\textit{where } \Omega_n = (\ldots, \pi_{\mathbf{A}}^n) \in \mathbb{C}_\infty^n.$$

Writing down the compositional inverse to $\exp_n$ as formal series

$$\log_n \begin{bmatrix} z_1 \\ \vdots \\ z_n \end{bmatrix} = \sum_{i=0}^{\infty} Q_i \begin{bmatrix} z_1^{q^i} \\ \vdots \\ z_n^{q^i} \end{bmatrix}$$

with matrix coefficients from $Q_i \in \mathrm{M}_n(k)$, one finds that this series converges for all $(z_1, \ldots, z_n) \in \mathbb{C}_\infty^n$ satisfying

$$|z_i|_\infty < |\theta|_\infty^{i-n+\frac{nq}{q-1}}.$$

Moreover one has the following particular simple formula:

$$\log_n \begin{bmatrix} 0 \\ \vdots \\ 0 \\ z \end{bmatrix} = \begin{bmatrix} * \\ \vdots \\ * \\ \sum_{i=0}^{\infty} \frac{z^{q^i}}{L_i^n} \end{bmatrix}. \tag{3.1}$$

We call the n-th Carlitz polylogarithm the series

$$\mathrm{polylog}_n(z) := \sum_{k=0}^{\infty} \frac{z^{q^k}}{L_k^n}$$

which converges for all $z \in \mathbb{C}_\infty$ with $|z|_\infty < |\theta|_\infty^{\frac{nq}{q-1}}$. Its value at particular $z = \alpha \neq 0$ is called the n-th polylogarithm of α. As can be easily checked these polylogarithms always has no zero value. Also when $n = 1$, one has $\mathrm{polylog}_1(z) = \log_C(z)$. In transcendence theory we are interested in those polylogarithms of $\alpha \in \bar{k}$, as analogues of classical logarithms of algebraic numbers.

In [12], we have an analogue of Hermite-Lindemann Theorem for n-th higher Carlitz module:

Theorem 3.2. *If* $\mathbf{v} = (l_1, \ldots, l_n) \neq \mathbf{0}$ *in* $\mathbb{C}_\infty^n$, *and* $\exp_n(\mathbf{v}) \in \overline{k}^n$, *then* l_n *must be transcendental.*

Given $u \in \mathbb{C}_\infty$, if there exists elements $l_1, \ldots, l_{n-1}$ such that $\exp_{C^{\otimes n}}(l_1, \ldots, l_{n-1}, u) = \mathbf{p}$ is in $\overline{k}^n$, then we also shows that the algebraic point $\mathbf{p} = \mathbf{p}(u)$ on $\mathbb{G}_a^n$ is determined completely by u. We call such u a last coordinate logarithm of algebraic point on the n-th higher Carlitz module. The key result of Anderson-Thakur in [2] is

Theorem 3.3. *For all* $n \geq 1$, $\zeta_{\mathbf{A}}(n)$ *is last coordinate logarithm of certain algebraic point on the* n*-th higher Carlitz module.*

Applying Theorem 3.2 to the algebraic points of Theorem 3.3, we obtain immediately the transcendence of all $\zeta_{\mathbf{A}}(n)$. The **special** algebraic point on the n-th higher Carlitz module corresponds to $\zeta_{\mathbf{A}}(n)$ can be written down explicitly. They are torsion points for t-action precisely when q is even, i.e. $n \equiv 0 \pmod{q-1}$. Writing down these special points amounts to deriving nice "integral" formulas for $\zeta_{\mathbf{A}}(n)$. To do this, let $\mathbf{y}_{ni}$ be the column vector of length n with last coordinate θ^i and all other coordinates zero. Define the algebraic point

$$\mathbf{p}_n := \sum_i \Phi_n(h_{n,i})(\mathbf{y}_{ni}),$$

where the $h_{n,i} \in \mathbf{A}$ come from a generating function identity with variables x, y:

$$\left(1 - \sum_{j=0}^{\infty} \frac{\prod_{i=1}^{j}(\theta^{q^j} - y^{q^i})}{D_j} x^{q^j}\right)^{-1} := \sum_{n=0}^{\infty} \frac{H_n(y)}{\Gamma_{n+1}} x^n,$$

$$H_{n-1}(y) := \sum_i h_{n,i} y^i \in \mathbf{A}[y].$$

What Anderson-Thakur established is that $\Gamma_n \zeta_{\mathbf{A}}(n)$ is the last coordinate logarithm of this point $\mathbf{p}_n$.

By checking that the coefficients h_{ni} vanish for $i \geq \frac{nq}{q-1}$, Anderson-Thakur also obtained the following formula connecting $\zeta_{\mathbf{A}}(n)$ with polylogarithms of appropriate powers of θ.

Theorem 3.4. *Given positive integer n, one can find non-negative integer $l_n < \frac{nq}{q-1}$ such that $h_{n,l_n} \neq 0$, and the following identity holds*

$$\Gamma_n \zeta_{\mathbf{A}}(n) = \sum_{i=0}^{l_n} h_{n,i} \operatorname{polylog}_n(\theta^i). \tag{3.2}$$

In other words, $\zeta_{\mathbf{A}}(n)$ can be written as linear combination of n-th polylogarithms of powers of θ with rational coefficients from k. This fact eventually enables us to determine all the algebraic relations among the zeta values in question.

Example. For $1 \leq n \leq q-1$, one has simply the formula

$$\zeta_{\mathbf{A}}(n) = \operatorname{polylog}_n(1).$$

4 t-Motives

Following Anderson [1] and Papanikolas [8] we now consider t-motives.

Let $\bar{k}[t, \sigma]$ be the twisted polynomial ring such that for all $c \in \bar{k}$:

$$ct = tc, \quad \sigma t = t\sigma, \quad \sigma c = c^{1/q}\sigma.$$

An Anderson t-motive is a left $\bar{k}[t, \sigma]$-module M which is free and finitely generated both as left $\bar{k}[t]$-module and left $\bar{k}[\sigma]$-module, and satisfying for N sufficiently large

$$(t - \theta)^N M \subset \sigma M.$$

Let $\mathbf{m} \in \mathrm{M}_{r\times 1}(M)$ be a $\bar{k}[t]$-basis of M. Then the multiplication by σ on M is given as

$$\sigma(\mathbf{m}) = \Phi\mathbf{m},$$

for some matrix $\Phi \in \mathrm{M}_r(\bar{k}[t])$.

We also need the Tate algebra $\mathbb{T}$ in variable t. This is the subring in $\mathbb{C}_\infty[[t]]$ consisting of power series which converge on $\mid t \mid_\infty \leq 1$. The field of fractions of $\mathbb{T}$ is denoted by $\mathbb{L}$. There is a twisting automorphism on $\mathbb{T}$ defined by:

$$f^{(-1)} = \sum_i a_i^{q^{-1}} t^i, \text{if } f = \sum_i a_i t^i \in \mathbb{T}.$$

This twisting leaves invariant various natural subrings of $\mathbb{L}$, e.g. $\bar{k}(t)$.

It is convenient to introduce left $\bar{k}(t)[\sigma, \sigma^{-1}]$-modules which are finite dimensional over $\bar{k}(t)$. These are called pre-t-motives. They form an abelian category. Each Anderson t-motive M corresponds to a pre-t-motive $\bar{k}(t) \otimes_{\bar{k}[t]} M$, with

$$\sigma(f \otimes m) := f^{(-1)} \otimes \sigma m.$$

A given pre-t-motive of dimension r over $\bar{k}(t)$ is said to be rigid analytically trivial if there exists $\Psi \in \mathrm{GL}_r(\mathbb{L})$ such that

$$\sigma(\Psi) = \Psi^{(-1)} = \Phi\Psi, \tag{4.1}$$

where Φ represents multiplication by σ with respect to a basis of this pre-t-motive. This Ψ is said to be a rigid analytic trivialization of the matrix Φ.

Rigid analytically trivial pre-t-motives that can be constructed from Anderson t-motives using direct sums, subquotients, tensor products, duals and internal Hom's, are called t-motives. These t-motives form a neutral Tannakian category $\mathcal{T}$. By Tannakian duality, for each t-motive M of dimension r, the Tannakian subcategory generated by M is equivalent to the category of finite-dimensional representations over $\mathbb{F}_q(t)$ of some algebraic subgroup $\Gamma_M \subset \mathrm{GL}_r$ defined over $\mathbb{F}_q(t)$. This algebraic group Γ_M is called the Galois group of the t-motive M.

Let M be a t-motive, with $\Phi \in \mathrm{GL}_r(\bar{k}(t)) \cap \mathrm{M}_r(\bar{k}[t])$ represents multiplication by σ on M. It can be shown that entries of Ψ are in fact everywhere convergent power series in t with coefficients lie in finite extensions of k_∞. One is allowed to substitute θ for the variable t in Ψ. Call $\Psi(\theta)^{-1} \in \mathrm{GL}_r(\bar{k}_\infty)$ a period matrix of the motive M. The entries of $\Psi(\theta)$ generate a field over $\bar{k}$ whose transcendence degree over $\bar{k}$ is an analytic invariant of the motive M.

Let $\mathbb{E} \subset \mathbb{T}$ be the set of entire power series with coefficients lie in finite extensions of k_∞. A fundamental Theorem of Papanikolas [8] relates the analytic invariant of the t-motive coming from Ψ with the algebraic Galois group Γ_M:

Theorem 4.1. *Let M and Φ be given as above such that* $\det \Phi = c(t-\theta)^s$, *with $c \in \bar{k}^\times$. Let Ψ be a rigid analytic trivialization of Φ in $\mathrm{GL}_r(\mathbb{T}) \cap \mathrm{M}_r(\mathbb{E})$ and let L be the field generated by the entries of $\Psi(\theta)$ over $\bar{k}$. Then*

$$\dim \Gamma_M = \mathrm{tr.deg}_{\bar{k}} L.$$

One can view equation (4.1) as analogue of system of linear differential equations with column vectors of Ψ as solutions. The solution space is of dimension r over $\mathbb{F}_q(t)$. The motivic Galois group Γ_M, as algebraic subgroup of GL_r over $\mathbb{F}_q(t)$ can be interpreted as "differential" Galois group associated to this linear system (4.1).

Example 1. The Carlitz motive C. Let $C = \bar{k}[t]$ with σ-action:

$$\sigma(f) = (t-\theta)f^{(-1)}, \quad f \in C.$$

Here $\Phi = (t-\theta)$ and $r = 1$. An analytic solution Ψ of (4.1) is

$$\Psi_C(t) = (-\theta)^{-q/(q-1)} \prod_{i=1}^{\infty}(1 - t/\theta^{q^i}).$$

Period of Carlitz motive is $\Psi_C(\theta)^{-1} = -\pi_{\mathbf{A}}$, where $\pi_{\mathbf{A}}$ is fundamental period of the Carlitz module. One computes that $\Gamma_C = \mathbb{G}_m$ thereby gets transcendence of $\pi_{\mathbf{A}}$.

Example 2. The s-th tensor power of the Carlitz motive $C^{\otimes s}$. Let s be a positive integer and

$$C^{\otimes s} = C \otimes_{\bar{k}[t]} \cdots \otimes_{\bar{k}[t]} C$$

with σ-action:

$$\sigma(f) = (t-\theta)^s f^{(-1)}, \quad f \in C^{\otimes s}.$$

Here $\Phi = (t-\theta)^s$ and $r = 1$. An analytic solution Ψ of (4.1) is Ψ_C^s. Period of this motive is $\Psi_C^n(\theta)^{-1} = (-\pi_{\mathbf{A}})^s$. One also computes that $\Gamma_{C^{\otimes s}} = \mathbb{G}_m$.

Example 3. The t-motives for s-th Carlitz polylogarithms (case $s = 1$ is given by Papanikolas [8]). Given $\alpha_1, \ldots, \alpha_m \in \bar{k}^\times$ satisfying $|\alpha_i|_\infty < |\theta|_\infty^{sq/q-1}$. Set

$$\Phi_s(\alpha_1, \ldots, \alpha_m) = \begin{pmatrix} (t-\theta)^s & 0 & \cdots & 0 \\ \alpha_1^{(-1)}(t-\theta)^s & 1 & \cdots & 0 \\ \vdots & \vdots & \ddots & \vdots \\ \alpha_m^{(-1)}(t-\theta)^s & 0 & \cdots & 1 \end{pmatrix}.$$

This defines a t-motive M_{Φ_s} which is an extension of $\mathbf{1}^m$ by the s-th tensor power Carlitz motive $C^{\otimes s}$. Here $\mathbf{1}$ is the trivial rank one t-motive.

Analytic solution to (4.1):

$$\Psi_s(\alpha_1, \ldots, \alpha_m) = \begin{pmatrix} (\Psi_C)^s & 0 & \cdots & 0 \\ L_{\alpha_1,s}(\Psi_C)^s & 1 & \cdots & 0 \\ \vdots & \vdots & \ddots & \vdots \\ L_{\alpha_m,s}(\Psi_C)^s & 0 & \cdots & 1 \end{pmatrix},$$

where

$$L_{\alpha_i,s}(t) = \alpha_i + \sum_{j=1}^{\infty} \frac{\alpha_i^{q^j}}{(t-\theta^q)^s \cdots (t-\theta^{q^j})^s}.$$

One has

$$L_{\alpha,s}^{(-1)} = \alpha^{(-1)} + \frac{L_{\alpha,s}}{(t-\theta)^s}.$$

Moreover for all i, one has

$$L_{\alpha_i,s}(\theta) = \mathrm{polylog}_s(\alpha_i).$$

To compute the Galois group Γ_{Φ_s} for the t-motive M_{Φ_s}, consider

$$G = \left\{ \begin{pmatrix} \star & 0 \\ \star & I_m \end{pmatrix} \in \mathrm{GL}_{m+1} \right\}.$$

One checks that $\Gamma_\Phi \subset G$, and $\Gamma_\Phi \twoheadrightarrow \mathbb{G}_m$ canonically. Let V be kernel of the later map. Then V can be identified as a linear subspace of $\mathbb{G}_a^m$ over $\mathbb{F}_q(t)$. This observation leads to the following

Theorem 4.2. *Suppose* $\mathrm{polylog}_s(\alpha_1), \ldots, \mathrm{polylog}_s(\alpha_m)$ *and* $\pi_{\mathbf{A}}{}^s$ *are linearly independent over* k*. Then the Galois group* Γ_{ϕ_s} *is an extension of* $\mathbb{G}_a^m$ *by* $\mathbb{G}_m$ *and* $\dim \Gamma_{\Phi_s} = m+1$.

It follows that $\pi_{\mathbf{A}}{}^s$ and $\mathrm{polylog}_s(\alpha_1), \ldots, \mathrm{polylog}_s(\alpha_m)$ must be algebraically independent over $\bar{k}$ if they are linearly independent over k.

5 Algebraic Independence of the Special Zeta Values

Given n not divisible by $q-1$, we let

$$N_n := k\text{-span}\{\mathrm{polylog}_n(1), \mathrm{polylog}_n(\theta), \ldots, \mathrm{polylog}_n(\theta^{l_n})\}.$$

By Theorem 3.4 we have $\zeta_{\mathbf{A}}(n) \in N_n$. Since $\zeta_{\mathbf{A}}(n)$ and $\pi_{\mathbf{A}}{}^n$ are linearly independent over k_∞, they are also linearly independent over k. For each such n we fix once for all a finite set of distinct exponents

$$0 \le i_0(n), \ldots, i_{m_n}(n) \le l_n$$

so that both

$$\{\mathrm{polylog}_n(\theta^{i_0(n)}), \ldots, \mathrm{polylog}_n(\theta^{i_{m_n}(n)})\}$$

and

$$\{\zeta_{\mathbf{A}}(n), \mathrm{polylog}_n(\theta^{i_1(n)}), \ldots, \mathrm{polylog}_n(\theta^{i_{m_n}(n)})\}$$

are bases of N_n over k.

Let M_{Φ_n} be the t-motive defined by the matrix

$$\Phi_n = \Phi_n(\theta^{i_0(n)}, \ldots, \theta^{i_{m_n}(n)}). \tag{5.1}$$

The Galois group of this t-motive M_{Φ_n} has dimension $m_n + 2$ by Theorem 4.2 On the other hand, a rigid analytic trivialization of Φ_n is given by

$$\Psi_n := \begin{bmatrix} \Psi_C^n & 0 & \cdots & 0 \\ \Psi_C^n L_{\theta^{i_0(n)},n} & 1 & \cdots & 0 \\ \vdots & \vdots & \ddots & \vdots \\ \Psi_C^n L_{\theta^{i_{m_n}(n)},n} & 0 & \cdots & 1 \end{bmatrix}.$$

Hence by Theorem 4.1, m_n+2 also equals to the transcendence degree over $\bar{k}$ of the field

$$\bar{k}({\pi_{\mathbf{A}}}^n, \mathrm{polylog}_n(\theta^{i_0(n)}), \ldots, \mathrm{polylog}_n(\theta^{i_{m_n}(n)}))$$

$$= \bar{k}({\pi_{\mathbf{A}}}^n, \zeta_{\mathbf{A}}(n), \mathrm{polylog}_n(\theta^{i_1(n)}), \ldots, \mathrm{polylog}_n(\theta^{i_{m_n}(n)})).$$

In particular, the elements

$${\pi_{\mathbf{A}}}^n, \zeta_{\mathbf{A}}(n), \mathrm{polylog}_n(\theta^{i_1(n)}), \ldots, \mathrm{polylog}_n(\theta^{i_{m_n}(n)})$$

are algebraically independent over $\overline{k}$. Therefore, ${\pi_{\mathbf{A}}}^n$ and $\zeta_{\mathbf{A}}(n)$ are algebraically independent over $\overline{k}$. This completes the proof of the following

Theorem 5.1. *For any positive integer n satisfying $q-1 \nmid n$, $\pi_{\mathbf{A}}$ and $\zeta_{\mathbf{A}}(n)$ are algebraically independent over $\overline{k}$.*

Given positive integer s, we set $U(s) := \{1 \leq n \leq s \mid p \nmid n,\ q-1 \nmid n, \}$.

Define diagonal block matrices

$$\Phi(s) := \oplus_{n \in U(s)} \Phi_n.$$

The matrix $\Phi(s)$ defines a t-motive $M(s) := M_{\Phi(s)}$ which is the direct sum of the t-motives M_{Φ_n} with $n \in U(s)$. The algebraic Galois group $\Gamma_{\Phi(s)}$ can be computed, in particular it gives:

Theorem 5.2. *Fix any $s \in \mathbb{N}$, we have*

$$\dim \Gamma_{\Phi(s)} = 1 + \sum_{n \in U(s)} (m_n + 1).$$

Writing down rigid analytic trivilization $\Psi(s)$ of $\Phi(s)$ and applying Theorem 4.1 to $\Gamma_{\Phi(s)}$, we find that $1+\sum_{n\in U(s)}(m_n+1)$ is exactly the transcendence degree over $\bar{k}$ of the following field:

$$\bar{k}\left(\tilde{\pi}, \bigcup_{n\in U(s)} \{\mathrm{polylog}_n(\theta^{i_0(n)}), \ldots, \mathrm{polylog}_n(\theta^{i_{m_n}(n)})\}\right)$$

$$= \bar{k}\left(\tilde{\pi}, \bigcup_{n\in U(s)} \{\zeta_{\mathbf{A}}(n), \mathrm{polylog}_n(\theta^{i_1(n)}), \ldots, \mathrm{polylog}_n(\theta^{i_{m_n}(n)})\}\right).$$

It follows that the set

$$\bigcup_{n\in U(s)} \{\zeta_{\mathbf{A}}(n), \mathrm{polylog}_n(\theta^{i_1(n)}), \ldots, \mathrm{polylog}_n(\theta^{i_{m_n}(n)})\}$$

is algebraically independent over $\bar{k}$, hence also $\{\zeta_{\mathbf{A}}(n) \mid n \in U(s)\}$ is algebraically independent over $\bar{k}$. Counting cardinality of $U(s)$ we obtain

Theorem 5.3. *The transcendence degree of $\overline{k}(\pi_{\mathbf{A}}, \zeta_{\mathbf{A}}(1), \ldots, \zeta_{\mathbf{A}}(n))$ over $\overline{k}$ is:*

$$n - \lfloor n/p \rfloor - \lfloor n/(q-1) \rfloor + \lfloor n/(p(q-1)) \rfloor + 1.$$

Finally it remains to compute the Galois group $\Gamma_{\Psi(s)}$.

Let $l(s) := \sum_{n\in U(s)}(m_n + 2)$. Define $G(s)$ to be the algebraic subgroup of $\mathrm{GL}_{l(s)}$ over $\mathbb{F}_q(t)$ which consists of all block diagonal matrices of the form

$$\begin{bmatrix} 1 & 0 \\ * & 1 \end{bmatrix} \oplus \cdots \oplus \begin{bmatrix} 1 & 0 & \cdots & 0 \\ * & 1 & \cdots & 0 \\ \vdots & \vdots & \ddots & \vdots \\ * & 0 & \cdots & 1 \end{bmatrix}$$

where the block matrix at the position corresponding to $n \in U(s)$ has the size $m_n + 2$. Thus $G(s)$ is isomorphic to direct product of $\sum_{n\in U(s)}(m_n + 1)$ copies of $\mathbb{G}_a$ canonically.

Since the Carlitz motive C is contained in M_{Φ_1}, it is also contained in $M(s)$ for any positive integers s. There is a surjection of algebraic groups over $\mathbb{F}_q(t)$

$$\mathrm{pr} : \Gamma_{\Psi(s)} \twoheadrightarrow \mathbb{G}_m$$

which coincides with the natural projection on the upper left corner position of the first 2×2 block matrix. Let $V(s)$ be the kernel of pr so that one has an exact sequence of algebraic groups over $\mathbb{F}_q(t)$

$$1 \to V(s) \to \Gamma_{\Psi(s)} \twoheadrightarrow \mathbb{G}_m \to 1. \tag{5.2}$$

From the equation $\Psi(s)^{(-1)} = \Phi(s)\Psi(s)$ we find that $V(s) \subseteq G(s)$. We claim that in fact $V(s) = G(s)$.

To prove this claim we intoduce a $\mathbb{G}_m$-action on $G(s)$. On the block matrix at the position corresponding to $n \in U(s)$ it is defined by

(5.3)

$$a \cdot \begin{bmatrix} 1 & 0 & \cdots & 0 \\ * & 1 & \cdots & 0 \\ \vdots & \vdots & \ddots & \vdots \\ * & 0 & \cdots & 1 \end{bmatrix} \mapsto \begin{bmatrix} 1 & 0 & \cdots & 0 \\ a^n * & 1 & \cdots & 0 \\ \vdots & \vdots & \ddots & \vdots \\ a^n * & 0 & \cdots & 1 \end{bmatrix}, \text{ for } a \in \mathbb{G}_m(\overline{\mathbb{F}_q(t)}).$$

Since $\oplus_{n\in U(s)}C^{\otimes n}$ is canonically contained in $M(s)$, writing down coordinates for the Galois group implies that the restriction of the above $\mathbb{G}_m$-action to $V(s)$ agrees with the conjugation action of $\mathbb{G}_m$ on $V(s)$ coming from (5.2). Taking into consideration of the $\mathbb{G}_m$-action, we are able to establish:

Theorem 5.4. *Fix any $s \in \mathbb{N}$. Then we have exact sequence of algebraic groups over $\mathbb{F}_q(t)$:*

$$1 \to V(s) \to \Gamma(s) \twoheadrightarrow \mathbb{G}_m \to 1$$

where $V(s)$ is isomorphic to the vector group

$$\prod_{n\in U(s)} \mathbb{G}_a^{m_n+1}.$$

Furthermore, the conjugation action by $\mathbb{G}_m$ *on* $V(s)$ *has multi-weight* $(n)_{n \in U(s)}$.

References

[1] Greg W. Anderson, *t-mptives*, Duke. Math. J **53** (1986), 457-502.

[2] Greg W. Anderson and Dinesh S.Thakur, *Tensor powers of the Carlitz module and zeta values.* Ann. of Math. **132** (1990), 159-191.

[3] Greg W. Anderson, W. Dale Brownnawell and Matthew A. Papanikolas, *Determination of the algebraic relations among special Gamma-values in positive characteristic*, Ann. of Math. **160** (2004), 237-313.

[4] L. Carlitz, *On certain functions connected with polynomials in a Galois field*, Duke. Math. J **1** (1935), 137-168.

[5] L. Carlitz, *An analogue of the von Staudt-Clausen theorem*, Duke. Math. J **3** (1937), 503-517.

[6] Chieh-Yu Chang and Jing Yu, *Determination of algebraic relations among special zeta values in positive characteristic*, preprint, 2006.

[7] David Goss, *Basic structures of function field arithmetic* , Springer-Verlag, Berlin, 1996.

[8] Matthew A. Papanikolas, *Tannakian duality for Anderson-Drinfeld motives and algebraic independence of Carlitz logarithms*, http://arxiv.org/abs/math/0506078, 2005.

[9] Dinesh S. Thakur, *Function field arithmetic*, World Scientific Publishing, River Edge NJ, 2004.

[10] L. I. Wade, *Certain quantities transcendental over GF(pn,x)*, Duke. Math. J **8** (1941), 701-720.

[11] L. I. Wade, *Transcendence properties of the Carlitz ψ-functions*, Duke. Math. J **13** (1946), 71-78.

[12] Jing Yu, *Transcendence and special zeta values in characteristic p*, Ann. of Math. (2) **134** (1991), 1-23.

[13] Jing Yu, *Analytic homomorphisms into Drinfeld modules*, Ann. of Math. (2) **145** (1997), 215-233.

Jing YU
Department of Mathematics,
National Tsing-Hua University
Hsinchu, Taiwan 30043
Email: yu@math.cts.nthu.edu.tw

Automorphic Forms, Eisenstein Series and Spectral Decompositions

Lin WENG

In Memory of Serge

And Serge did this sort of thing, through the decades, with many of the young: he would proffer to them gracious, yet demanding, invitations to engage as a genuine colleague – not teacher to student – but mathematician to mathematician; he did all this naturally, and with extraordinary generosity and success.

From *For Serge*, by B. Mazur

This note is prepared for the reader who wants to learn Langlands' fundamental results about Eisenstein series and spectral decompositions, say using Moeglin and Waldspurger's Cambridge tract. It results from the six lectures given at a special seminar on Automorphic Forms and Eisenstein Series at Department of Mathematics, University of Toronto in 2005. (Once a week, one and a half hours starting from October 25, after another set of six beautiful lectures on automorphic forms given by Henry Kim.) The author wishes to thank Arthur and Kim very much for their kind invitation and support which, together with a JSPS fund, make his long visit possible.

Contents

Day One: Basics of Automorphic Forms

1 Basic Decompositions

Of central importance here are the following two decompositions:
(i) (**Langlands Decomposition**) $G(\mathbb{A}) = U(\mathbb{A}) \cdot A_{M(\mathbb{A})} \cdot M(\mathbb{A})^1 \cdot K$; and
(ii) (**Reduction Theory**) $G(\mathbb{A}) = G(F) \cdot S(\omega; t_0) = P(F) \cdot S^P(\omega; t_0)$.

1.1 Langlands Decomposition

1.1.1 Reductive Groups

Let G be a connected reductive group defined over a number field F. Denote by $G(\mathbb{A})$ the associated adelic group.
Recall that by definition,
(i) an algebraic group is called *reductive*, if its unipotent radical is trivial;
(ii) if F_v is the v-adic completion of F at its place v (with $\mathcal{O}_v$ the v-adic ring of integers when v is finite), then the associated *adelic ring* $\mathbb{A}$ of F is defined to be the restricted product $\mathbb{A} = \prod' F_v$ of F_v with respect to (the maximal compact) subgroup $\mathcal{O}_v$. In other words, $m = (m_v) \in \prod_v F_v$ belongs to $\mathbb{A}$ if and only if for almost all v, $m_v \in \mathcal{O}_v$. Similarly, in order to introduce the *adelic group* $G(\mathbb{A})$, we need to first introduce maximal compact subgroups $K_v := G(\mathcal{O}_v)$ for almost all places v, then define $G(\mathbb{A})$ as the restricted product $\prod'_v G(F_v)$ with respect to K_v. So the difficulty here is to introduce ring valued groups $G(\mathcal{O}_v)$ and can be, say, overcome as follows: First of all, there is no problem for this to be done for the general linear group GL. Then being an algebraic group, there exists a natural embedding i'_G of G into the general linear group GL_n for certain n. So by taking care of the inverse operation as well, the morphism $i_G : G \to GL_{2n}$ defined by $g \mapsto \mathrm{diag}(i_G(g), i_G(g)^{-t})$ will do the job.

1.1.2 Parabolic Subgroups

Let P be a parabolic subgroup of G defined over F. (Recall that this means that as algebraic varieties, P is defined over F and G/P is complete, Grothendieck's notion in algebraic geometry generalizing the common concept of separated and compact.) We have the so-called *Levi decomposition* $P = MU = UM$ for P, where U is the *unipotent radical* of P and M is what we call a *Levi subgroup* of P. For later use, fix a minimal parabolic subgroup P_0 of G and a minimal Levi subgroup M_0 of P, both defined over F, and call a parabolic subgroup P and a Levi subgroup M *standard* if they are defined over F and contain P_0 and M_0 respectively. Denote by U_0 the unipotent radical of P_0. It is known that

(i) *every parabolic subgroup P of G defined over F conjugates by an element of $G(F)$ to a standard one;* and

(ii) *every standard parabolic subgroup P contains a unique standard Levi subgroup of G.*

Ex. (a) With $G = GL_n$, parabolic subgroups P correspond to partitions $n = r_1 + r_2 + \cdots + r_{|P|}$ of n. Normally, we choose the minimal parabolic subgroup P_0 as the Borel subgroup, i.e., the subgroup consisting of upper triangle matrices and the minimal Levi subgroup M_0 as the subgroup consisting of diagonal matrices. As such, the associated unipotent radical U_0 is the subgroup consisting of upper triangle matrices whose diagonal entries are all 1. In terms of partitions, P_0 corresponding to the partition $I_0 = (1, 1, \ldots, 1)$ resulting from $n = 1 + 1 + \cdots + 1$;

(b) More generally, let $I = (n_1, n_2, \ldots, n_N)$ be the partition resulting from $n = n_1 + n_2 + \cdots + n_N$. Then the corresponding standard parabolic subgroup P_I is the subgroup of G consisting of the blocked upper triangle matrices, for which the diagonal blocks are square matrices of sizes $n_1, n_2, \ldots,$ and n_N respectively. Consequently, the corresponding standard parabolic subgroup M_I consists of blocked diagonal matrices in P_I, while the unipotent radical U_I consists of blocked upper triangle matrices whose diagonal matrices are identity matrices of sizes $n_1, n_2, \ldots, n_N$ respectively.

1.1.3 Maximal Compact Subgroups

Recall that even in the defintion of $G(\mathbb{A})$, we need a maximal open compact subgroup $K = \prod_v K_v$ of $G(\mathbb{A})$, for which $K_v = G(\mathcal{O}_v)$ for almost all v. For our purpose, we also require that K satisfies the following conditions:

(i) (**Langlands Decomposition**) $G(\mathbb{A}) = P(\mathbb{A}) \cdot K = U(\mathbb{A}) \cdot M(\mathbb{A}) \cdot K$; and
(ii) (**Admissible Condition**) $P(\mathbb{A}) \cap K = \Big(U(\mathbb{A}) \cap K\Big) \cdot \Big(M(\mathbb{A}) \cap K\Big)$.

Ex. With $G = SL_n$, $K_{\mathbb{R}} = SO(n)$ and $K_{\mathbb{C}} = SU(n)$.

1.1.4 Decomposition $M(\mathbb{A}) = A_{M(\mathbb{A})} \cdot M(\mathbb{A})^1$

This is supposed to be the first difficulty the reader meets. The basic idea is to separate infinite places from others so as to get a commutative real Lie group with which the unbounded quotient space we encounter later will admit a relative simpler description.

Starting point is the decomposition of idele group $\mathbb{I}$ of F. (As a group, $\mathbb{I}$ can be understood as $\mathbb{A}^* = \mathbb{G}_m(\mathbb{A}) = \mathrm{GL}_1(\mathbb{A})$, where $\mathbb{A}^*$ stands for the collection of invertible elements of the ring $\mathbb{A}$, $\mathbb{G}_m$ stands for the algebraic group representing the so-called multiplicative group.)

We have the subgroup $\mathbb{I}^0$ of the so-called norm 1 ideles, in which multiplicative group F^* consisting of non-zero elements of F is naturally embedded as discrete subgroup via the diagonal map $f \mapsto i(f) := (\ldots, f, \ldots, f, \ldots)$. (For $f \in F^*$, there are only finitely many v such that $|f|_v \neq 1$, so $i(f) \in \mathbb{I}$, while by the product formula, the norm of $i(f)$ is simply one, hence, $i(f) \in \mathbb{I}^0$.) As such essential points are that
(i) $F^* \backslash \mathbb{I}^0 = \mathrm{GL}_1(F) \backslash \mathrm{GL}_1(\mathbb{A})$ is compact; and
(ii) $\mathbb{I}^0 \backslash \mathbb{I} \simeq \mathbb{R}_+^*$ given by mapping g to the idele defined as follows: each component at a finite place is trivially 1, while for each infinite place the entry is simply $\|g\|^{\frac{1}{[F:\mathbb{Q}]}}$.

Now a generalization to general reductive group is quite obvious: We need to

(i) introduce spaces corresponding to $\mathbb{I}^0$ and $\mathbb{R}_{>0}$, which are simply $M(\mathbb{A})^1$ and $A_{M(\mathbb{A})}$; and
(ii) introcuce a generalization of the logarithmic map.

Let

$$\mathrm{Rat}\,(M) := \Big\{\chi : M \to \mathbb{G}_m\Big\}$$

be the collection of morphisms between M and $\mathbb{G}_m$ (as algebraic groups). This is a finitely generated $\mathbb{Z}$-module. Set

$$\mathfrak{a}_M^* := \mathrm{Rat}(M) \otimes_{\mathbb{Z}} \mathbb{C}, \qquad \text{and} \qquad \mathfrak{a}_M^* := \mathrm{Hom}_{\mathbb{Z}}\Big(\mathrm{Rat}\,(M), \mathbb{C}\Big)$$

which contain the real subspaces

$$\mathrm{Re}\,\mathfrak{a}_M^* := \mathrm{Rat}(M) \otimes_{\mathbb{Z}} \mathbb{R}, \qquad \text{and} \qquad \mathrm{Re}\,\mathfrak{a}_M^* := \mathrm{Hom}_{\mathbb{Z}}\Big(\mathrm{Rat}\,(M), \mathbb{R}\Big)$$

respectively.

For $\chi \in \mathrm{Rat}(M)$, by definition, χ is a character of M, i.e., a morphism from the algebraic group M to $\mathbb{G}_m$. Thus in particular, over the local field F_v, χ induces a morphism $\chi_v = \chi|_{G(F_v)}$: $G(F_v) \to \mathbb{G}_m(F_v) = F_v^*$. Accordingly, set $|\chi| : M(\mathbb{A}) \to \mathbb{R}_{>0}$ to be the continuous map defined by $m^{|\chi|} := (m_v)^{|\chi|} := \prod_v |m_v|_v^{\chi_v}$ where $m = (m_v) \in M(\mathbb{A})$. With this, let

$$M(\mathbb{A})^1 := \cap_{\chi \in \mathrm{Rat}\,(M)} \mathrm{Ker}\,|\chi|.$$

Next, we define $A_{M(\mathbb{A})}$.

(**a**) $A_{M_0(\mathbb{A})}$ for the **minimal** Levi M_0. Here $P_0 = U_0 M_0$ is a minimal parabolic subgroup. Denote by Z_{M_0} the center of M_0. Then Z_{M_0} contains a maximal F-split torus T_0 (of G), and there exists a natural number R, the *semi-simple rank* of G, such that

$$T_0 \simeq \mathbb{G}_m^R.$$

As such, with the injection $\mathbb{R}_{>0} \hookrightarrow \mathbb{I} = \mathbb{G}_m(\mathbb{A})$ defined by $t \mapsto (1, \ldots, 1; t, \ldots, t)$, i.e., the idele whose finite components are all 1 while whose Archemidean components are all t, we obtain a natural inclusion

$$\mathbb{R}_{>0}^R \hookrightarrow \mathbb{G}_m(\mathbb{A}) \simeq T_0(\mathbb{A})$$

which is known to be split (as an entension of groups, since $\mathbb{R}_{>0}$ is simply connected.) Therefore there exists a unique subgroup $A_{M_0(\mathbb{A})}$ of $T_0(\mathbb{A}) \subset Z_{M_0(\mathbb{A})}$, the image of the above morphism, such that it projects back onto $\mathbb{R}^R_{>0}$.
(**b**) $A_{M(\mathbb{A})}$ for **general** (standard) Levi M. Being standard Levi, $M_0(\mathbb{A}) \subset M(\mathbb{A})$. Consequently, $Z_{M(\mathbb{A})} \subset Z_{M_0(\mathbb{A})}$ (bigger group admits smaller center). As such, set $A_{M(\mathbb{A})} := A_{M_0(\mathbb{A})} \cap Z_{M(\mathbb{A})}$.
Theorem. *For any standard parabolic subgroup $P = MU$ of G,*
(a) $M(\mathbb{A}) = M(\mathbb{A})^1 \cdot A_{M(\mathbb{A})} = A_{M(\mathbb{A})} \cdot M(\mathbb{A})^1$; and
(b) (**Langlands Decomposition**) $G(\mathbb{A}) = U(\mathbb{A}) \cdot M(\mathbb{A})^1 \cdot A_{M(\mathbb{A})} \cdot K$.

The reader should understand that while (b) is a structural result, (a) is merely a simple decomposition for our convenience.

1.1.5 Logarithmic Map

Naturally associated to the decomposition $M(\mathbb{A}) = M(\mathbb{A})^1 \cdot A_{M(\mathbb{A})}$ is the so-called *logarithmic map*

$$\begin{array}{cccc} \log_M : & M(\mathbb{A}) & \to & \mathrm{Re}\,\mathfrak{a}_M \\ & m & \mapsto & \log_M(m) \end{array}$$

characterized by $\langle \chi, \log_M m \rangle := \log m^{|\chi|}$ or the same,

$$m^{|\chi|} =: e^{\langle \chi, \log_M m \rangle}, \qquad \forall m \in M(\mathbb{A}) \ \&\ \chi \in \mathrm{Rat}(M).$$

Clearly
(i) $\log_M$ factors through $M(\mathbb{A})^1 \backslash M(\mathbb{A})$; and
(ii) $\log_M \Big(M(\mathbb{A})^1 \backslash M(\mathbb{A}) \Big) = \mathrm{Re}\,\mathfrak{a}_M$.

For convenience, define also a map

$$m_P : G(\mathbb{A}) \to M(\mathbb{A})^1 \backslash M(\mathbb{A}) \qquad \text{by} \qquad g \mapsto M(\mathbb{A})^1 \cdot m(g)$$

where $g = u \cdot m(g) \cdot k$ is a Langlands decomposition with $u \in U(\mathbb{A}), m(g) \in M(\mathbb{A})$ and $k \in K$. (m_P is well-defined since $M(\mathbb{A}) \cap K$ is contained in $M(\mathbb{A})^1$ and $M(\mathbb{A})^1$ is a normal subgroup of $M(\mathbb{A})$.) Accordingly, we are able to talk about $H_P(g) := \log_M m_P(g)$ an element in $\mathrm{Re}\, a_M$ associated to $g \in G(\mathbb{A})$, a notation used commonly in literature.

1.1.6 Decomposition $\mathfrak{a}_{M_0}^* = \mathfrak{a}_M^* \oplus \left(\mathfrak{a}_{M_0}^M\right)^*$

At the level of associated Lie algebras, there is a natural decomposition

$$\mathfrak{a}_{M_0}^* = \mathfrak{a}_M^* \oplus \left(\mathfrak{a}_{M_0}^M\right)^*.$$

(**a**) *The space* $\mathfrak{a}_{M_0}^*$. While defined as $\mathrm{Rat}\,(M) \otimes_{\mathbb{Z}} \mathbb{C}$, it may also be identified with the space generated by the root system $R(T_0, G)$. (Recall that the adjoint representation gives a natural action of the group T_0 on the Lie algebra $\mathfrak{g}$ associated to the reductive group G. Being commutative, the non-trivial common eigenspaces of T_0 in $\mathfrak{g}$ give the elements, the so-called roots, in $R(T_0, G)$.) As such, with respect to the parabolic subgroup P_0, we then further get the set of positive roots $R^+(T_0, G)$ in $R(T_0, G)$ and also the set of simple roots $\Delta_0 := \Delta_{M_0} \subset R^+(T_0, G)$. As a complex vector space, $\mathfrak{a}_{M_0}^*$ admits a basis Δ_0;
(**b**) *The space* $\mathfrak{a}_M^*$. Similarly, being commutative, maximal F-split torus T_M in the centre Z_M of M acts naturally on the Lie algebra $\mathfrak{g}$ of G. Denote by $R(T_M, G)$ the set of 'roots' of G with respect to this action of T_M, i.e., these with non-trivial eigenspaces. In general, $R(T_M, G)$ is not a root system. Set $\Delta_M :=$ the set consisting of elements of $R(T_M, G)$ obtained from the non-trivial restrictions to $T_M(\subset T_0)$ of elements in Δ_0. (And similarly for $R^+(T_M, G)$.) $\mathfrak{a}_M^*$ admits a natural basis Δ_M;
(**c**) *The space* $\left(\mathfrak{a}_{M_0}^M\right)^*$. The adjoint action for T_{M_0} on the Lie algebra associated to M gives the set $R(T_0, M)$, the root of M with respect to T_0. Set $\Delta_{M_0}^M := \Delta_0 \cap R(T_0, M)$ be a subset of Δ_0, consisting of only elements which come from the M-level, i.e., those operating only within M. The space $\left(\mathfrak{a}_{M_0}^M\right)^*$ admits a natural basis $\Delta_{M_0}^M$.

As all Δ_0, Δ_M and $\Delta_{M_0}^M$ are defined over $\mathbb{R}$, we have the decomposition

$$\mathrm{Re}\,\mathfrak{a}_{M_0}^* = \mathrm{Re}\,\mathfrak{a}_M^* \oplus \mathrm{Re}\left(\mathfrak{a}_{M_0}^M\right)^*.$$

Similarly, we get decompositions of their duals

$$\mathfrak{a}_{M_0} = \mathfrak{a}_M \oplus \mathfrak{a}_{M_0}^M \qquad \text{and} \qquad \mathrm{Re}\,\mathfrak{a}_{M_0} = \mathrm{Re}\,\mathfrak{a}_M \oplus \mathrm{Re}\,\mathfrak{a}_{M_0}^M.$$

Set accordingly

$$\rho_0 = \frac{1}{2} \sum_{\alpha \in R^+(T_0, G)} \alpha \qquad \text{and} \qquad \rho_P = \frac{1}{2} \sum_{\alpha \in R^+(T_M, G)} \alpha.$$

Ex. In the case $G = GL_n$ and $P = P_I$. The subset $\Delta_{M_0}^M$ of Δ_0 consists of simple roots which operate within the small blocks associated to M, while Δ_M consists of restrictions of simple roots in Δ_0 which operate exactly among different blocks associated to M.

1.1.7 Haar Measures

Key is that all have to be compactible with the Langlands decomposition

$$G(\mathbb{A}) = U_0(\mathbb{A}) \cdot A_{M_0(\mathbb{A})} \cdot M_0(\mathbb{A})^1 \cdot K = U(\mathbb{A}) \cdot A_{M(\mathbb{A})} \cdot M(\mathbb{A})^1 \cdot K.$$

For compact groups K, there is no problem. Even for unipotent radicals U, since $U(F)$ is discrete and $U(F)\backslash U(\mathbb{A})$ is compact, it is obvious. With this, for all compactly supported continuous functions f on $G(\mathbb{A})$,

$$f \mapsto \int_{U_0(\mathbb{A}) \cdot A_{M_0(\mathbb{A})} \cdot M_0(\mathbb{A})^1 \cdot K} f(uamk)\, dk \left(a^{-2\rho_0} da\right) dm\, du$$

defines a Haar measure dg on $G(\mathbb{A})$ where $u \in U_0(\mathbb{A}), a \in A_{M_0(\mathbb{A})}, m \in M_0(\mathbb{A})$ and $k \in K$. Replacing G with M, we also obtain a Haar measure dm on $M(\mathbb{A})$. Among all these Haar measures, we have the following

Compatibility Condition. *For any compactly supported continuous function f on $G(\mathbb{A})$,*

$$\int_{G(\mathbb{A})} f(g)\, dg = \int_{U(\mathbb{A}) \cdot A_{M(\mathbb{A})} \cdot M(\mathbb{A})^1 \cdot K} f(uamk)\, dk\, a^{-2\rho_P} da\, dm\, du$$

where $u \in U(\mathbb{A}), a \in A_{M(\mathbb{A})}, m \in M(\mathbb{A})^1$ and $k \in K$.

1.2 Reduction Theory: Siegel Sets

Topologically, $P(F)$ is a discrete subgroup of $G(\mathbb{A})$. We want to understand the quotient space $P(F)\backslash G(\mathbb{A})$, which classically is realized via the so-called fundamental domain. A typical example is for the upper half plane $\mathcal{H}$ under the action of $SL(2,\mathbb{Z})$, or equivalently, for the quotient space $SL(2,\mathbb{Z})\backslash SL(2,\mathbb{C})/SO(2)$, a model for $SL(2,\mathbb{Q})\backslash SL(2,\mathbb{A})$ modulo certain maximal compact subgroup K.

According to Langlands decomposition, $G(\mathbb{A}) = U(\mathbb{A}) \cdot A_{M(\mathbb{A})} \cdot M(\mathbb{A})^1 \cdot K$ and the Levi decomposition $P = UM$, we have

$$P(F)\backslash G(\mathbb{A}) = U(F)\backslash U(\mathbb{A}) \cdot A_{M(\mathbb{A})} \cdot M(F)\backslash M(\mathbb{A})^1 \cdot K.$$

Note that being unipotent radical, $U(F)\backslash U(\mathbb{A})$, being essentially copies of $F\backslash\mathbb{A}$, is compact. (In fact, here, a filtration and its associated graded quotients have to be used.) Consequently, the problem of reduction mainly reduces to the one for $M(F)\backslash M(\mathbb{A})$, or the same for $A_{M(\mathbb{A})} \cdot M(F)\backslash M(\mathbb{A})^1$, with K being compact.

As such, then we need to know that $M(F)\backslash M(\mathbb{A})^1$ is of finite volume, while $A_{M(\mathbb{A})}$ is a commutative real Lie group. With all this, then we can go as follows:
(**a**) The case where $P = P_0 = U_0M_0$ is minimal parabolic. Then M_0 is a torus. So $M_0(F)\backslash M_0(\mathbb{A})^1$ is simple, being essentially copies of compact space $F^*\backslash\mathbb{I}^0$ (at least when G is split). In any case, here, the reduction theory does not offer us any new thing as to a large extent both $A_{M_0(\mathbb{A})}$ and $M_0(F)\backslash M_0(\mathbb{A})$ are left untouched;
(**b**) The case where $P = G$. There is nothing we can say about $A_{G(\mathbb{A})}$. But for $G(F)\backslash G(\mathbb{A})^1$, we use the maximal infinitesimal space $\mathfrak{a}_{M_0}^G$, or the same $A_{M_0(\mathbb{A})}$. So let ω be a compact subset of $P_0(\mathbb{A})$ and $t_0 \in M_0(\mathbb{A})$. Then set

$$A_{M_0(\mathbb{A})}(t_0) := \Big\{a \in A_{M_0(\mathbb{A})} \,\big|\, a^\alpha > t_0^\alpha \ \forall\alpha \in \Delta_0\Big\},$$

which is a unbounded subset of $A_{M_0(\mathbb{A})}$. Accordingly, set the associated *Siegel set* to be

$$S(\omega; t_0) := S := \Big\{pak \,\big|\, p \in \omega, a \in A_{M_0(\mathbb{A})}(t_0), k \in K\Big\}.$$

Reduction Theory$_0$. *We may choose big enough ω and small enough t_0 such that $G(\mathbb{A}) = G(F) \cdot S(\omega; t_0)$.*
(Here t_0 small enough means that for all $\alpha \in \Delta_0$, t_0^α are very close to zero.) As such, Siegel sets $S(\omega; t_0)$, whose structure is rather simple, may be viewed as approximations of usually complicated fundamental domains. (For many purposes in the theory of automorphic forms, exact fundamental domains, which are very hard to obtain, only play a secondary role, while much simpler structured Siegel sets are already sufficient.)
Ex. $G = SL(2)$. We are lead to consider $SL(2,\mathbb{Z})\backslash\mathcal{H}$. Then the canonical fundamental domain is essentially given by

$$D := \left\{ z = x + iy \in \mathcal{H} \,\middle|\, |z| \geq 1 \, |x| \leq \frac{1}{2} \right\},$$

an intersection of the strip defined by a linear inequality $|x| \leq \frac{1}{2}$ and a quadratic inequality $|z| \geq 1$. On the other hand, a Siegel domain may be chosen to be a much simpler

$$S := \left\{ z = x + iy \in \mathcal{H} \,\middle|\, |x| \leq \delta, y \geq c \right\}$$

for all $\delta \geq \frac{1}{2}$ and $0 < c < \frac{\sqrt{3}}{2}$ defined by two linear inequalities $|x| \leq \delta$ and $y \geq c$.
(**c**) The case where $P = UM$ is standard parabolic. As said, we only need to study the space $A_{M(\mathbb{A})} \times M(F)\backslash M(\mathbb{A})^1$. Now the difficulty is that unlike the case $F^*\backslash\mathbb{I}^0$, $M(F)\backslash M(\mathbb{A})^1$ is no longer compact (otherwise the life would be very easy) – it is only of finite volume.

Recall now the infinitesimal decomposition $\mathfrak{a}_{M_0}^* = \left(\mathfrak{a}_{M_0}^M\right)^* \oplus \mathfrak{a}_M^*$. Clearly, reduction theory is supposed to offer us a clear way to understand $A_{M(\mathbb{A})} \times M(F)\backslash M(\mathbb{A})^1$. Thus being very simple, we will leave the A-part $A_{M(\mathbb{A})}$ untouched. Or in terms of infinitesimal structure, leave $\mathfrak{a}_M^*$ wild. As for the part $M(F)\backslash M(\mathbb{A})^1$, then we use infinitesimal space $\left(\mathfrak{a}_{M_0}^M\right)^*$ which admits a basis $\Delta_{M_0}^M$. That is to say, set

$$A_{M_0(\mathbb{A})}^P(t_0) := \left\{ a \in A_{M_0(\mathbb{A})} \,\middle|\, a^\alpha > t_0^\alpha \; \forall \alpha \in \Delta_{M_0}^M \right\}.$$

This is a unbounded subset of $A_{M_0(\mathbb{A})}$ which contains the entire subspace corresponding to $\mathfrak{a}_M$ or the same $\mathfrak{a}_M^*$. Accordingly, let the P-level *Siegel set* to be

$$S^P(\omega;t_0) := S^P := \Big\{pak \,\big|\, p \in \omega, a \in A^P_{M_0(\mathbb{A})}(t_0), k \in K\Big\}.$$

Reduction Theory$_P$. *We may choose big enough ω and small enough t_0 such that $G(\mathbb{A}) = P(F) \cdot S^P(\omega;t_0)$.*

We will not say anything about proof. Instead we point out that theory of automorphic forms may be understood from $\mathfrak{a}_{M_0}^* = \left(\mathfrak{a}_{M_0}^M\right)^* \oplus \mathfrak{a}_M^*$ or the same from $A_{M(\mathbb{A})} \times M(F)\backslash M(\mathbb{A})^1$. Indeed, reduction theory uses only $\left(\mathfrak{a}_{M_0}^M\right)^*$ for $M(F)\backslash M(\mathbb{A})^1$, while $\mathfrak{a}_M^*$ gives new dimension of freedom. Say, automorphic representations are parametrized by $\mathfrak{a}_M^*$; Eisenstein series is with variable $\lambda \in \mathfrak{a}_M^*$ in which residue can be taken; pseudo-Eisenstein series, being integrations of ESes over $\mathrm{Im}\,\mathfrak{a}_M^*$, are square integrable, etc. Moreover, such freedom is dominated by a huge Hecke algebra and a nice algebra of Langlands operators (both of which are associated to $\mathfrak{a}_M^*$).

1.3 Moderate Growth and Rapidly Decreasing

1.3.1 Heights

Being unbounded, we need to measure the size of elements of $G(\mathbb{A})$ or $P(F)\backslash G(\mathbb{A})$. Borrowing a concept from Diophantine geometry, we have the so-called (Weil) heights. There are many ways to define them, but it is a well-known theorem due to Weil that up to $O(1)$, all these heights are the same. Thus it is of less importance for many purpose to figure out the exact definition. However to start with, for any $g \in G(\mathbb{A})$, define its *height* by

$$\|g\| := \prod_v \sup\Big\{|g_{st}|_v : s,t = 1,2,\ldots,2n\Big\}$$

where $(g_{st}) = i_G(g)$ is the element in GL_{2n} corresponding to g under the map $i_G : G \hookrightarrow GL_{2n}$ in 1.1.1.

Lemma. (a) (**Naturalness**) *There exist $c, r > 0$ such that, for any g, g_1, $g_2 \in G(\mathbb{A})$,*

(i) $\|g\| \geq c$;
(ii) $\|g_1 g_2\| \leq c\|g_1\| \cdot \|g_2\|$; and
(iii) $\|g^{-1}\| \leq c \cdot \|g\|^r$.
(b) (**Log**) (i) *There exist* $c_1, c_2 > 0$ *and* $c \in \mathbb{R}$ *such that, for any* $a \in A_{M_0(\mathbb{A})}$,

$$c_1 \|\log_{M_0} a\| \leq \log \|a\| + c \leq c_2 \|\log_{M_0} a\|;$$

(ii) *For any* $\lambda \in \operatorname{Re} \mathfrak{a}^*_{M_0}$, *there exists* $c, r > 0$ *such that*

$$m_{P_0}(g)^\lambda \leq c \cdot \|g\|^r \qquad \forall g \in G(\mathbb{A});$$

(iii) *there exist* $\lambda_0 \in \operatorname{Re} \mathfrak{a}^*_{M_0}$ *and* $c > 0$ *such that*

$$\|g\| \leq c \cdot m_{P_0}(g)^{\lambda_0} \qquad \forall g \in G(\mathbb{A})^1 \cap S;$$

(c) (**Arith**) *There exists* $c > 0$ *such that*

$$\|g\| \leq c \cdot \|\gamma g\| \qquad \forall \gamma \in G(F),\ g \in S;$$

(d) (**Geo**) *There exist* $c, r_1, r_2 > 0$ *such that*

$$c \cdot \|a\|^{r_1} \cdot \|g\|^{r_2} \leq \|ag\| \qquad \forall a \in A_{G(\mathbb{A})},\ g \in G(\mathbb{A})^1.$$

1.3.2 Moderate Growth

Denote by $\mathcal{U}$ the universal enveloping algebra of the complexified Lie algebra $\mathrm{Lie}G(\mathbb{A}_\infty) \otimes_{\mathbb{R}} \mathbb{C}$, where $\mathbb{A}_\infty$ denotes the Archimedean components of the adeles.

By definition, a function $f : G(\mathbb{A}) \to \mathbb{C}$ is called *smooth*, if for any $g \in G(\mathbb{A})$ with the decomposition $g = g_f \cdot g_\infty$ according to $G(\mathbb{A}) = G(\mathbb{A}_f) \times G(\mathbb{A}_\infty)$, there exist a neighborhood V_∞ of g_∞ in $G(\mathbb{A}_\infty)$, a neighborhood V_f of g_f in $G(\mathbb{A}_f)$ and a C^∞-function $f_\infty : V_\infty \to \mathbb{C}$ such that

$$f(g'_\infty g'_f) = f_\infty(g'_\infty) \qquad \forall g'_\infty \in V_\infty,\ g'_f \in V_f.$$

For many purposes, a good control on growth of functions is crucial. We will need two types of them, that is, moderate growth and rapidly decreasing.

(**a**) *G-Level.* A function $f : G(\mathbb{A}) \to \mathbb{C}$ is called of *moderate growth* if there exist $c, r \in \mathbb{R}$ such that

$$|f(g)| \leq c \cdot \|g\|^r, \qquad \forall g \in G(\mathbb{A}).$$

(Simply put, f is of moderate growth if $f(g)$ is about the size of g.) Moreover a smooth function $f : G(\mathbb{A}) \to \mathbb{C}$ is called of *uniformly moderate growth* if there exist $r \in \mathbb{R}$ and for all $X \in \mathcal{U}$, there exists a constant $c_X > 0$ such that

$$\big|(\delta(X)f)(g)\big| \leq c_X \cdot \|g\|^r, \qquad \forall g \in G(\mathbb{A}).$$

(**b**) *P-Level* (for $P = UM$ a standard parabolic). A function $f : U(\mathbb{A})M(F)\backslash G(\mathbb{A}) \to \mathbb{C}$ is called of *moderate growth* if there exist $c, r \in \mathbb{R}$ and $\lambda \in \mathrm{Re}(\mathfrak{a}_{M_0}^M)^*$ such that

$$|f(amk)| \leq c \cdot \|a\|^r \cdot m_{P_0}(m)^\lambda, \ \forall a \in A_{M(\mathbb{A})},\ m \in M(\mathbb{A})^1 \cap S^P,\ k \in K.$$

Moreover a smooth function $f : U(\mathbb{A})M(F)\backslash G(\mathbb{A}) \to \mathbb{C}$ is called of *uniformly moderate growth* if there exist $r \in \mathbb{R}$ and for all $X \in \mathcal{U}$, there exists a constant $c_X > 0$ and $\lambda_X \in \mathrm{Re}(\mathfrak{a}_{M_0}^M)^*$ such that

$$\big|(\delta(X)f)(amk)\big| \leq c_X \cdot \|a\|^r \cdot m_{P_0}(m)^{\lambda_X},$$

for all $a \in A_{M(\mathbb{A})}, m \in M(\mathbb{A})^1 \cap S^P, k \in K$.

As it stands, moderate growth condition is the one on $M(\mathbb{A}) = A_{M(\mathbb{A})} \cdot M(\mathbb{A})^1$-component, i.e., always in terms of $\|a\|^r$ for all $a \in A_{M(\mathbb{A})}$ with a fixed r, and $m_{P_0}(m)^{\lambda_X}$ for $m \in M(\mathbb{A})^1$ with a fixed λ or λ_X. (Simply put, moderate growth looks like polynomial growth.)

1.3.3 Rapidly Decreasing

(**a**) *G-Level.* A function $f : G(F)\backslash G(\mathbb{A}) \to \mathbb{C}$ is called *rapidly decreasing* if there exists $r > 0$ and for any $\lambda \in \mathrm{Re}\mathfrak{a}_{M_0}^*$ there exists $c > 0$ such that

$$|f(ag)| \leq c \cdot \|a\|^r \cdot m_{P_0}(g)^\lambda, \qquad \forall a \in A_{G(\mathbb{A})},\ g \in G(\mathbb{A})^1 \cap S.$$

(**b**) *P-Level.* A function $f : U(\mathbb{A})M(F)\backslash G(\mathbb{A}) \to \mathbb{C}$ is called rapidly decreasing if there exists $r > 0$ and for all $\lambda \in \mathrm{Re}(\mathfrak{a}_{M_0}^M)^*$ there exists $c > 0$ such that

$$|f(ag)| \le c \cdot \|a\|^r \cdot m_P(g)^\lambda, \qquad \forall a \in A_{M(\mathbb{A})},\ g \in M(\mathbb{A})^1 \cap S^P.$$

Here $A_{M(\mathbb{A})}$-part plays only a secondary role. Simply put, the rapidly decreasing condition is the one on $M(\mathbb{A})^1$-part, where functions decay like exponential functions, or better, decay faster than any rational functions.

1.3.4 Natural Pairing

Let $\phi, \varphi : U(\mathbb{A})M(F)\backslash G(\mathbb{A}) \to \mathbb{C}$ be two functions, one is of moderate growth and one is rapidly decreasing. Then we have the following
Lemma. *The integral*

$$\langle \phi, \varphi \rangle := \int_{Z_{M(\mathbb{A})}U(\mathbb{A})M(F)\backslash G(\mathbb{A})} \overline{\phi(g)} \cdot \varphi(g)\, dg$$

converges absolutely.
Sketch of a proof. After writing out using reduction theory, the integration looks like $\int_c^\infty Q(t)e^{-t}dt$ where Q is a polynomial of t and $c \ge \delta > 0$, (since essentially moderate growth means polynomial growth while rapidly decreasing means exponential decay). Clearly, this latest integration converges.

1.4 Automorphic Forms

1.4.1 Automorphic Forms

Let $P = UM$ be a standard parabolic subgroup of G. For any $k \in K$ and a function $f : U(\mathbb{A})M(F)\backslash G(\mathbb{A}) \to \mathbb{C}$, define a new function $f_k : M(F)\backslash M(\mathbb{A}) \to \mathbb{C}$ by setting

$$f_k(m) := m^{-\rho_P} \cdot f(mk) \qquad m \in M(F)\backslash M(\mathbb{A}).$$

With this, f is said to be *K-finite*, if the space of $\mathbb{C}$-linear span of $\Big\{ f_k \,\Big|\, k \in K \Big\}$ is finite dimensional. Similarly, f is called $\mathfrak{z}$*-finite* if the space of $\mathbb{C}$-linear span of $\Big\{ \delta(X) f \,\Big|\, X \in \mathfrak{z} \Big\}$ is finite dimensional.

Definition. A function $\phi : U(\mathbb{A})M(F)\backslash G(\mathbb{A}) \to \mathbb{C}$ is called an *automorphic form* if
(1) it is of moderate growth;
(2) it is smooth;
(3) it is K-finite; and
(4) it is $\mathfrak{z}$-finite.

Denote by $A\Big(U(\mathbb{A})M(F)\backslash G(\mathbb{A})\Big)$ the collection of automorphic forms on the space $U(\mathbb{A})M(F)\backslash G(\mathbb{A})$, the so-called *$P$-level automorphic forms.* In particular, we have the G-level automorphic forms $A\Big(G(\mathbb{A})\backslash G(\mathbb{A})\Big)$ hence also the *M-level automorphic forms* $A\Big(M(\mathbb{A})\backslash M(\mathbb{A})\Big)$.

Bridge Lemma. $\phi \in A\Big(U(\mathbb{A})M(F)\backslash G(\mathbb{A})\Big) \Leftrightarrow \phi$ *is smooth, K-finite, and for all $k \in K$, $\phi_k \in A\Big(M(\mathbb{A})\backslash M(\mathbb{A})\Big)$.*

Key to the proof. Let $\mathfrak{z}_M$ be the center of the universal enveloping algebra of the complixified Lie algebra of $M(\mathbb{A}_\infty)$. Then there a natural morphism $\mathfrak{z} \to \mathfrak{z}_M$ which makes $\mathfrak{z}_M$ a finitely generated $\mathfrak{z}$-module. So the $\mathfrak{z}_M$ finiteness becomes natural. As such the only thing left is about the moderate growth condition. This is essentially a direct consequence of the definition since $A_{M_0(\mathbb{A})} = A_{M(\mathbb{A})} \cdot A^M_{M_0(\mathbb{A})}$, and the $A^M_{M_0(\mathbb{A})}$ part is already a part of the definiton of M-level, while the $A_{M(\mathbb{A})}$-part is a direct consequence of $\mathfrak{z}_M$-finiteness.

1.4.2 Constant Terms and Cusp Forms

For a function $f : G(F)\backslash G(\mathbb{A}) \to \mathbb{C}$, its *constant term along P* is defined by the integration

$$f_P(g) := \int_{U(F)\backslash U(\mathbb{A})} f(ug)\, du.$$

Let $\phi \in A\Big(U(\mathbb{A})M(F)\backslash G(\mathbb{A})\Big)$ be a P-level automorphic form. Then by definition ϕ is called *cuspidal* (or a *cusp form*) if for any standard parabolic subgroup P' of G which is properly contained in P we have $\phi_{P'} = 0$. Denote by $A_0\Big(U(\mathbb{A})M(F)\backslash G(\mathbb{A})\Big)$ the space of P-level cusp forms on $U(\mathbb{A})M(F)\backslash G(\mathbb{A})$. Similarly, we have the space $A_0\Big(M(\mathbb{A})\backslash M(\mathbb{A})\Big)$ of M-level cusp forms.

2 Structural Results

2.1 Moderate Growth and Rapid Decreasing

Here we want to construct rapidly decreasing functions out of (uniformly) moderate growth functions. As a by-product, we show that there is a natural pairing between automorphic forms and cusp forms.

Theorem A. *Let $f : G(F)\backslash G(\mathbb{A}) \to \mathbb{C}$ be a smooth K-finite function. If f and its derivatives have uniformly moderate growth, then*

$$\begin{array}{rccc} [f]: & S & \to & \mathbb{C} \\ & g & \mapsto & [f](g) := \sum_P (-1)^{\mathrm{rank}\big(Z_M/Z_G\big)} \cdot f_P(g) \end{array}$$

is rapidly decreasing.

Key to the proof is that unipotent groups have simple structures. Essentially $U(F)\backslash U(\mathbb{A})$ is a certain copies of $F\backslash\mathbb{A}$. As such, we only need to understand what happens for $\mathbb{Z}\backslash\mathbb{R}$: it is then well known that over the compact space $\mathbb{Z}\backslash\mathbb{R}$, non-constant terms of the Fourier expansion of a smooth function is rapidly decreasing.

Ex. Let $\phi \in A_0\Big(U(\mathbb{A})M(F)\backslash G(\mathbb{A})\Big)$ be a cusp form, then $[\phi] : S^P \to \mathbb{C}$ is simply ϕ itself by definition, since if $P = G$, $\phi_P = \phi$ while if $P \neq G$, by the cuspidality, $\phi_P = 0$. As such, cusp forms are rapidly decreasing if we can somehow show that not only they are of moderate growth, their derivatives are also of moderate growth. (This is almost the case. See e.g., Lem I.4.1 [MW].) Consequently, we have the following

Natural Pairing. *There is a natural pairing between automorphic and cusp forms:*

$$\begin{array}{cccc} \langle\cdot,\cdot\rangle : & A\Big(U(\mathbb{A})M(F)\backslash G(\mathbb{A})\Big) \times A_0\Big(U(\mathbb{A})M(F)\backslash G(\mathbb{A})\Big) & \to & \mathbb{C} \\ & (\varphi,\phi) & \mapsto & \langle\varphi,\phi\rangle \end{array}$$

defined by

$$\langle \varphi, \phi \rangle := \int_{\left(Z_{M(\mathbb{A})}U(\mathbb{A})M(F)\right)\backslash G(\mathbb{A})} \overline{\varphi(g)} \cdot \phi(g)\, dg.$$

2.2 Semi-Simpleness

This is a fundamental result due to Harish-Chandra. Note that $A\Big(M(F)\backslash M(\mathbb{A})\Big)$ is not invariant under the action of $M(\mathbb{A}_\infty)$ even we have the $\mathfrak{z}_M$-finiteness. So it is not an $M(\mathbb{A})$-module. However, a weak version holds correctly: the space can be viewed as an $M(\mathbb{A}_f) \times \Big(\mathrm{Lie}M(\mathbb{A}_\infty) \otimes_{\mathbb{R}} \mathbb{C}, M(\mathbb{A}) \cap K\Big)$-module. This in literature is called the theory of $(\mathfrak{g} - K)$-modules. The center of this theory is in fact not about automorphic forms, rather it is about cusp forms. More precisely, we have the following

Theorem B. (Harish-Chandra) *As an* $M(\mathbb{A}_f) \times \Big(\mathrm{Lie}M(\mathbb{A}_\infty) \otimes_{\mathbb{R}} \mathbb{C}, M(\mathbb{A}) \cap K\Big)$*-module,* $A_0\Big(M(F)\backslash M(\mathbb{A})\Big)$ *is semi-simple.*

To understand this result, let us fix an irreducible representation σ of $\mathfrak{z}$ and an irreducible representation μ of K, and set $A(\sigma,\mu)$ to be the isotypic submodule of $A_0\Big(M(F)\backslash M(\mathbb{A})\Big)$ of type (σ,μ). Then we have the following

Theorem B′. (a) *Each space* $A(\mu,\sigma)$ *is finite dimensional*; and (b) $A_0\Big(M(F)\backslash M(\mathbb{A})\Big) = \oplus_{\sigma,\mu} A(\sigma,\mu)$.

With this, over P-level, we obtain a natural morphism

$$\begin{array}{rccc} \bullet^{\mathrm{cusp}}: & A\Big(U(\mathbb{A})M(F)\backslash G(\mathbb{A})\Big) & \to & A_0\Big(U(\mathbb{A})M(F)\backslash G(\mathbb{A})\Big) \\ & \varphi & \mapsto & \varphi^{\mathrm{cusp}}, \end{array}$$

such that for any cusp form $\phi_0 \in A_0\Big(U(\mathbb{A})M(F)\backslash G(\mathbb{A})\Big)$,

$$\langle \varphi, \phi_0 \rangle = \langle \varphi^{\mathrm{cusp}}, \phi_0 \rangle.$$

2.3 $\mathfrak{z}$-Finiteness

Key to the $\mathfrak{z}$-finiteness in the definition of automorphic forms is that $\mathfrak{z}$ is Noetherian for reductive groups. Say in the case of GL_n, they are in fact polynomial rings. Clearly, it is then natural to understand this finiteness in more details, say to make the finite dimensional space even one dimensional. To give a good approximation to this, the first thing we want to do is to introduce a center character ξ_M of $Z_{M(\mathbb{A})}$ and set $A\Big(M(F)\backslash M(\mathbb{A})\Big)_{\xi_M} :=$

$$\Big\{\phi \in A\Big(M(F)\backslash M(\mathbb{A})\Big) \,|\, \phi(zm) = z^{\xi_M}\phi(m)\ \forall z \in Z_{M(\mathbb{A})}, m \in M(\mathbb{A})\Big\}$$

and accordingly

$$A_0\Big(U(\mathbb{A})M(F)\backslash G(\mathbb{A})\Big)_{\xi_M} := \Big\{\phi \in A_0\Big(U(\mathbb{A})M(F)\backslash G(\mathbb{A})\Big) \\ |\phi(zg) = z^{\rho_P} \cdot z^{\xi_M} \cdot \phi(g)\ \forall z \in Z_{M(\mathbb{A})}, g \in G(\mathbb{A})\Big\}.$$

(In fact in Theorem B′, we should work on $A_0\Big(U(\mathbb{A})M(F)\backslash G(\mathbb{A})\Big)_{\xi_M}$ for a fixed ξ_M.)

The idea is that now we may consider a weaker condition: being $Z_{M(\mathbb{A})}$-finite, or the same being $A_{M(\mathbb{A})}$-finite. As such the problem then becomes simpler since the group involved is $A_{M(\mathbb{A})}$ which is commutative. Consequently its infinitesimal version will be a polynomial ring over a vector space. So the corresponding finiteness condition means that for every automorphic form ϕ there exists a finitely codimensional ideal (of this polynomial ring) which annihilates ϕ. Consequently, using all the non-trivial polynomials in the resulting quotient space, which is of finite dimensional, we can write our automorphic form as a certain combination, (which is far from being canonical), in terms of these polynomials whose coefficients are again automorphic. These later automorphic forms are then expected to be much simpler in the sense that this time they behave according to certain fixed central characters. Put this in a more theoretic form, we have the follows:

First, for any $Q \in \mathbb{C}[\mathrm{Re}\,\mathfrak{a}_M]$ and $\phi \in A_{(0)}\Big(U(\mathbb{A})M(F)\backslash G(\mathbb{A})\Big)_{\xi_M}$,

$$g \mapsto Q\big(\log_M m_P(g)\big) \cdot \phi(g)$$

is again a P-level automorphic form. Then, let

$$A_{(0)}\Big(U(\mathbb{A})M(F)\backslash G(\mathbb{A})\Big)_Z := \sum_{\xi_M\in \mathrm{Hom}(Z_{M(\mathbb{A})},\mathbb{C}^*)} A_{(0)}\Big(U(\mathbb{A})M(F)\backslash G(\mathbb{A})\Big)_{\xi_M}.$$

Theorem C. (**Global Version**) *The above natural map induces an isomorphism*

$$\mathbb{C}[\mathrm{Re}\,\mathfrak{a}_M]\times A_{(0)}\Big(U(\mathbb{A})M(F)\backslash G(\mathbb{A})\Big)_Z \simeq A_{(0)}\Big(U(\mathbb{A})M(F)\backslash G(\mathbb{A})\Big).$$

Set $\Pi_0(M(\mathbb{A}))_{\xi_M} :=$ the set of isomorphism classes of irreducible representations of $M(\mathbb{A}_f)\times\Big(\mathrm{Lie}M(\mathbb{A}_\infty)\otimes_{\mathbb{R}}\mathbb{C}, M(\mathbb{A})\cap K\Big)$, or simply, of $M(\mathbb{A})$, occuring as submodule of $A_0\Big(M(F)\backslash M(\mathbb{A})\Big)_{\xi_M}$, and

$$\Pi_0(M(\mathbb{A}))_{\xi_M} := \cup_{\xi_M\in\mathrm{Hom}(Z_{M(\mathbb{A})},\mathbb{C}^*)}\Pi_0(M(\mathbb{A}))_{\xi_M}.$$

Then for $\pi\in\Pi_0(M(\mathbb{A}))_{\xi_M}$, (or better using (V,π) with V the realization space of π), set $A_0\Big(M(F)\backslash M(\mathbb{A})\Big)_\pi :=$ the isotypic component of type π in $A_0\Big(M(F)\backslash M(\mathbb{A})\Big)_{\xi_M}$, that is, the image of $V\times \mathrm{Hom}_{M(\mathbb{A})}\Big(V, A_0(M(F)\backslash M(\mathbb{A}))_{\xi_M}\Big)$. Set also the P-level space by

$$A_0\Big(U(\mathbb{A})M(F)\backslash G(\mathbb{A})\Big)_\pi := \Big\{\phi\in A_{(0)}\Big(U(\mathbb{A})M(F)\backslash G(\mathbb{A})\Big)_{\xi_M} \mid \phi_k\in A_0\Big(M(F)\backslash M(\mathbb{A})\Big)_\pi, \forall k\in K\Big\}.$$

Theorem B″. *We have the natural decompositions*

$$A_0\Big(M(F)\backslash M(\mathbb{A})\Big) = \oplus_{\pi\in\Pi_0(M(\mathbb{A}))} A_0\Big(M(F)\backslash M(\mathbb{A})\Big)_\pi,$$
$$A_0\Big(U(\mathbb{A})M(F)\backslash G(\mathbb{A})\Big) = \oplus_{\pi\in\Pi_0(M(\mathbb{A}))} A_0\Big(U(\mathbb{A})M(F)\backslash G(\mathbb{A})\Big)_\pi.$$

Theorem C′. (**Micro Version**) *For $\phi\in A_0\Big(U(\mathbb{A})M(F)\backslash G(\mathbb{A})\Big)$, there exists a finite set*

$$D(M,\phi)\subset \mathbb{C}[\mathrm{Re}\,\mathfrak{a}_M]\times\Pi_0(M(\mathbb{A}))\times A_0\Big(U(\mathbb{A})M(F)\backslash G(\mathbb{A})\Big)$$

such that
(i) *if* $(Q,\pi,\varphi) \in D(M,\phi)$, $\varphi \in A_0\Big(U(\mathbb{A})M(F)\backslash G(\mathbb{A})\Big)_\pi$; and
(ii) $\phi(g) = \sum_{(Q,\pi,\varphi)\in D(M,\phi)} Q(\log_M m_P(g)) \cdot \varphi(g)$.
Usually we call $D(M,\phi)$ the *cuspidal data* for ϕ and set $\Pi_0(M(\mathbb{A}),\phi)$ be the projection of $D(M,\phi)$ to $\Pi_0(M(\mathbb{A}))$, called the *cuspidal support* of ϕ.

In particular, if now $\phi \in A\Big(G(F)\backslash G(\mathbb{A})\Big)$ is a G-level automorphic form, then its constant term ϕ_P along P is a P-level automorphic form. So it makes sense to talk about ϕ_P^{cusp} the corresponding P-level cusp form introduced in 4.2 and hence also $\Pi_0\Big(M(\mathbb{A}),\phi_P^{\mathrm{cusp}}\Big)$. Set $\Pi_0(M,\phi) := \Pi_0\Big(M(\mathbb{A}),\phi_P^{\mathrm{cusp}}\Big)$ and call it the *cuspidal support* of ϕ.

2.4 Philosophy of Cusp Forms

This result will play a key role in the study follows: It exposes the reason why cusp forms are really the 'heart' of that for automorphic forms.

Let $C_0\Big(U(\mathbb{A})M(F)\backslash G(\mathbb{A})\Big) :=$ the space of functions on the quotient $U(\mathbb{A})M(F)\backslash G(\mathbb{A})$ which are linear combinations of functions of the form $g \mapsto b(m_P(g)) \cdot \phi(g)$ where b are compactly supported C^∞-functions on $M(\mathbb{A})^1\backslash M(\mathbb{A})$ and $\phi \in A_0\Big(U(\mathbb{A})M(F)\backslash G(\mathbb{A})\Big)$ are P-level cusp forms.

Now let $f : G(F)\backslash G(\mathbb{A}) \to \mathbb{C}$ be a function of moderate growth. By definiton, the *cuspidal P-form of f* is the linear form on the space $C_0\Big(U(\mathbb{A})M(F)\backslash G(\mathbb{A})\Big)$ given by

$$\varphi \mapsto \int_{U(\mathbb{A})M(F)\backslash G(\mathbb{A})} \overline{f_P(g)} \cdot \varphi(g)\, dg.$$

Note that here P is allowed to be taken as the entire group G itself.

Theorem D. *Let $f : G(F)\backslash G(\mathbb{A}) \to \mathbb{C}$ be of moderate growth. Then the following two conditions are equivalent:*
(i) *For all standard parabolic subgroup P of G, the cuspidal P-form of ϕ along P are zero;*

(ii) $f = 0$ *almost everywhere.*

While important, the proof is a simple manipulation of integrations using a unfolding argument and a standard approximation argument. We leave it to the reader.

Consequently, $\phi : G(F)\backslash G(\mathbb{A}) \to \mathbb{C}$ *is an automorphic form (in the sense of almost everywhere) if and only if all its cuspidal* P*-forms can be induced by automorphic forms.*

2.5 L^2-Automorphic Forms

Recall that by Theorems B and C, if ϕ is an automorphic form, then

$$\phi(uamk) = \sum_{j=1}^{l} Q_j(H_P(a)) \cdot e^{\langle \lambda_j + \rho_P, H_P(a) \rangle} \cdot \varphi_j(mk)$$
$$\forall u \in U(\mathbb{A}),\ a \in A_{M(\mathbb{A})},\ m \in M(\mathbb{A})^1,\ k \in K,$$

where $H_P(a) := \log_M m_P(a)$, and φ_j is an automorphic form of level P on which the $A_{M(\mathbb{A})}$-action is simply given by $\varphi_j(zg) = \varphi_j(g)$ for all $z \in Z_{M(\mathbb{A})}$, $g \in G(\mathbb{A})$. (Here, ρ_P is added for normalization, and the most important point is that z is an element of $Z_{M(\mathbb{A})}$, not just in $Z_{G(\mathbb{A})}$ which is much smaller in gereral.) We call $\lambda + \rho_P \in \Pi_0(M(\mathbb{A}))$ the *exponents* of ϕ. Since ϕ is of moderate growth, so is φ_j on $M(F)\backslash M(\mathbb{A})^1$. Thus by the finiteness of the volume $M(F)\backslash M(\mathbb{A})^1$, φ_j's are relatively easy. Consequently, the condition that ϕ is square integrable is really one for Q and the exponents λ_j. Note that among two factors $e^{\langle \lambda + \rho_P, H_P(a) \rangle}$ and $Q(H_P(a))$, the formal one behaves like an exponential function while the second like a polynomial. Hence we are lead to $\int_{\{a \in A_{G(\mathbb{A})} \backslash A_{M_0(\mathbb{A})} : a^\alpha > t_0^\alpha \forall \alpha \in \Delta_0\}} Q\big(H_P(a)\big) \cdot a^{2(\mathrm{Re}\,\lambda + \rho_P) - 2\rho_P} da$, where $-2\rho_P$ comes form Haar measure $a^{-2\rho_P} da$ on the $A_{M(\mathbb{A})}$ part. So it is a question of exponential decay competing with polynomial growth. As such we are ready to understand the following

Theorem E. **(Exponents of L^2-Automorphic Forms)** *Let* ξ *be a unitary central character of* $Z_{G(\mathbb{A})}$*. Then* $\phi \in A\Big(G(F)\backslash G(\mathbb{A})\Big)_\xi$ *is* $L^2 \Leftrightarrow \forall \pi \in \Pi_0(M, \phi)$ *with* $\mathrm{Re}\,\pi = \sum_{\alpha \in \Delta_M} x_\alpha \cdot \alpha$*, we have* $x_\alpha < 0$*.*

Day Two: Eisenstein Series

3 Definition

3.1 Equivalence Classes of Automorphic Representations

Let $P = UM$ be a standard parabolic subgroup of G. Then the space $A\Big(M(F)\backslash M(\mathbb{A})\Big)$ is a $M(\mathbb{A}_f) \times \Big(\mathrm{Lie}M(\mathbb{A}_\infty) \otimes_{\mathbb{R}} \mathbb{C}, M(\mathbb{A}) \cap K\Big)$-module and $A\Big(U(\mathbb{A})M(F)\backslash G(\mathbb{A})\Big)$ is a $G(\mathbb{A}_f) \times \Big(\mathrm{Lie}G(\mathbb{A}_\infty) \otimes_{\mathbb{R}} \mathbb{C}, K\Big)$-module. For simplicity, we say that they are $M(\mathbb{A})$ and $G(\mathbb{A})$-modules respectively.

Suppose that π_0 is an irreducible cuspidal representation, i.e.,

$$A\Big(M(F)\backslash M(\mathbb{A})\Big)_{\pi_0} \cap A_0\Big(M(F)\backslash M(\mathbb{A})\Big) \neq \{0\}.$$

Basic Hypothesis. *If π_0 is cuspidal, then*

$$A\Big(M(F)\backslash M(\mathbb{A})\Big)_{\pi_0} \subset A_0\Big(M(F)\backslash M(\mathbb{A})\Big)$$

and

$$A\Big(U(\mathbb{A})M(F)\backslash G(\mathbb{A})\Big)_{\pi_0} \subset A_0\Big(U(\mathbb{A})M(F)\backslash G(\mathbb{A})\Big).$$

Note that if $\phi \in A\Big(M(F)\backslash M(\mathbb{A})\Big)_{\pi_0}$ and $\lambda \in \mathfrak{a}_M^*$, then the function

$$\lambda\phi : m \mapsto m^\lambda \cdot \phi(g)$$

is again an M-level automorphic form. Denote by

$$A\Big(M(F)\backslash M(\mathbb{A})\Big)_{\pi_0 \otimes \lambda} := \Big\{\lambda\phi | \phi \in A\Big(M(F)\backslash M(\mathbb{A})\Big)_{\pi_0}\Big\},$$

which defines an automorphic representation of $M(\mathbb{A})$ as well. Similarly, at the P-level, if $\phi \in A\Big(U(\mathbb{A})M(F)\backslash G(\mathbb{A})\Big)_{\pi_0}$, then the function $\lambda\phi : g \mapsto m_P(g)^\lambda \cdot \phi(g)$ is again a P-level automorphic form.

Denote by

$$A\Big(U(\mathbb{A})M(F)\backslash G(\mathbb{A})\Big)_{\pi_0\otimes\lambda} := \Big\{\lambda\phi \,|\, \phi \in A\Big(U(\mathbb{A})M(F)\backslash G(\mathbb{A})\Big)_{\pi_0}\Big\}.$$

One checks that this space is the same as the P-level $G(\mathbb{A})$-module induced from the M-level automorphic representation $\pi_0 \otimes \lambda$. For this reason, we call two automorphic representations π_1, π_2 are *equivalent* if there exists $\lambda \in \mathfrak{a}_M^*$ such that $\pi_2 = \pi_1 \otimes \lambda$. Quite often, we use $\mathfrak{P}$ as a running symbol for equivalence classes of automorphic representations. Clearly, if $\pi \in \mathfrak{P}$, then $\mathfrak{P} = \pi \otimes \mathfrak{a}_M^*$.

As such, for a fixed automorphic form ϕ, we get a family of automorphic forms by setting $\phi_\lambda(g) = m_P(g)^\lambda \cdot \phi(g)$ parametrized by $\lambda \in \mathfrak{a}_M^*$. Roughly speaking, an Eisenstein series associated to ϕ with variable λ is an average of such a function over the coset $P(F)\backslash G(F)$. Before going to that, let us give the following easy calculation on the change of central characters with respect to the shift from ϕ to $\lambda\phi$:

$$\begin{aligned}(\lambda\phi)(zg) =& m_P(zg)^\lambda \phi(zg) = (z^\lambda \cdot m_P(g)^\lambda) \cdot (z^{\rho_P+\xi_M}\phi(g))\\ =& z^{\rho_P} \cdot z^{\xi_M+\lambda} \cdot \phi(g), \qquad \forall z \in Z_{M(A)},\ g \in G(\mathbb{A}).\end{aligned}$$

i.e., $\lambda\phi \in A(U(\mathbb{A})M(F)\backslash G(\mathbb{A}))_{\xi_M+\lambda}$ if $\phi \in A(U(\mathbb{A})M(F)\backslash G(\mathbb{A}))_{\xi_M}$. On the other hand, for $\xi : Z_{G(\mathbb{A})} \to \mathbb{C}^*$, and we set

$$\begin{aligned}A(U(\mathbb{A})M(F)\backslash G(\mathbb{A}))_\xi :=&\{\phi \in A(U(\mathbb{A})M(F)\backslash G(\mathbb{A}))\,|\\ &\phi(zg) = z^\xi \cdot \phi(g),\ \forall z \in Z_{G(\mathbb{A})},\ g \in G(\mathbb{A})\}.\end{aligned}$$

Then by the fact that $z^\lambda = 1$ for all $z \in Z_{G(\mathbb{A})}$ and $\lambda \in \mathfrak{a}_M^*$, we conclude that if $\phi \in A(U(\mathbb{A})M(F)\backslash G(\mathbb{A}))_\xi$ then so is $\lambda\phi$. So the reader should distinguish ξ and ξ_M as well as their associated spaces of automorphic spaces: In general, since $Z_{G(\mathbb{A})} \subset Z_{M(\mathbb{A})}$ for a fixed ξ there are many extensions ξ_M to $Z_{M(\mathbb{A})}$ and the space for ξ is simply the summations of the corresponding spaces for ξ_M for all possible ξ_M. As such clearly, when we talk about $G(\mathbb{A})$-modules, we should work over a fixed ξ instead of a fixed ξ_M. For our own convenience, an automorphic representation is called ξ-*admissible*, if $A(U(\mathbb{A})M(F)\backslash G(\mathbb{A}))_\pi \subset A(U(\mathbb{A})M(F)\backslash G(\mathbb{A}))_\xi$.

3.2 Eisenstein Series and Intertwining Operators

Fix a central character ξ and let π_0 be a ξ-admissible automorphic representation. Fix $\phi_{\pi_0} \in A\Big(U(\mathbb{A})M(F)\backslash G(\mathbb{A})\Big)_{\pi_0}$ and write $\phi_\pi := \phi_{\pi_0 \otimes \lambda} := \lambda\phi_{\pi_0}$ a family of automorphic forms parametrized by $\pi = \pi_0 \otimes \lambda \in \mathfrak{P}$ (or equivalently by $\lambda \in \mathfrak{a}_M^*$). View π (or λ) as variable.

Definition (1) The *Eisenstein series associated to* ϕ_π is defined by the following series

$$\begin{aligned} & E(\phi_\pi, \pi)(g) := E(\phi_{\pi_0}; \lambda)(g) \\ := & \sum_{\delta \in P(F)\backslash G(F)} m_P(\delta g)^\lambda \cdot \phi_{\pi_0}(\delta g) \\ = & \sum_{\delta \in P(F)\backslash G(F)} \Big(\lambda \cdot \phi_{\pi_0}\Big)(\delta g), \qquad \forall g \in G(\mathbb{A}); \end{aligned}$$

(2) For any $w \in G(F)$ and $\pi \in \mathfrak{P}$, the associated *intertwining operator* is defined by

$$\phi_\pi \mapsto \Big(M(w,\pi)\phi_\pi\Big)(g) := \int_{\big(U'(F)\cap wU(F)w^{-1}\big)\backslash U'(F)} \phi_\pi(w^{-1}ug)\, du$$

where $\phi_\pi \in A\Big(U(\mathbb{A})M(F)\backslash G(\mathbb{A})\Big)_\pi$, U' is the unipotent radical of the parabolic subgroup associated to the Levi subgroup $M' := wMw^{-1}$.

Quite often we also write $M(w;\lambda) = M(w, \pi_0 \otimes \lambda)$ when π_0 is fixed to indicate that λ is a variable. Naturally, for any automorphic representation, we have a corresponding automorphic representation $w\pi$ of wMw^{-1}.

So far, we have not said anything about the convergence of the series and the integration above. This will be dealt with later. Assume this, easily we have:

Theorem B. (1) $E(\phi_\pi, \pi) \in A\Big(G(F)\backslash G(\mathbb{A})\Big)_\xi$; and

(2) *The intertwining operator gives a morphism*

$$M(w,\pi) : A\Big(U(\mathbb{A})M(F)\backslash G(\mathbb{A})\Big)_\pi \to A\Big(U'(\mathbb{A})M'(F)\backslash G(\mathbb{A})\Big)_{w\pi}.$$

3.3 Convergence

Here we want to establish the convergence of $E(\phi_{\pi_0};\lambda)$.
Theorem A. *There is a cone $\mathcal{C} \subset \mathrm{Re}\,\mathfrak{a}_M^*$ such that if $\mathrm{Re}\,\lambda \in \mathcal{C}$, then the series*

$$E(\phi_{\pi_0};\lambda)(g) := \sum_{\delta\in P(F)\backslash G(F)} m_P(\delta g)^\lambda \cdot \phi_{\pi_0}(\delta g)$$

converges absolutely for any fixed $g \in G(\mathbb{A})$. Moreover if g is contained in a compact subset, the series converges uniformly.
Sketch of a proof. There are two steps. In the first we show that when $M = M_0$ is minimal, and ϕ is an M_0-level cusp form, the associated Eisenstein series converges as claimed. In the second, we show that the convergence problem of Eisenstein series associated to any automorphic form can be reduced to the case discussion in Step 1.
Step 1. Special case for Eisenstein series of constant function over minimal P_0.

Let $\mathbf{1}$ be constant function 1 over $M_0(\mathbb{A})$. It is L^2-automorphic. (Why?) Then for any $\lambda \in \mathfrak{a}_{M_0}^*$, we get the automorphic form $\lambda\mathbf{1}$ over $U_0(\mathbb{A})M_0(F)\backslash M_0(\mathbb{A})$, given by $g \mapsto m_P(g)^\lambda \cdot 1 = m_P(g)^\lambda$. Form the so-called *relative Eisenstein series*

$$E^{M/M_0}(\mathbf{1},\lambda;g) := \sum_{\delta\in\big(M(F)\cap P_0(F)\big)\backslash M(F)} m_P(\delta g)^\lambda \cdot 1$$

$$= \sum_{\delta\in\big(M(F)\cap P_0(F)\big)\backslash M(F)} m_P(\delta g)^\lambda.$$

Claim 1. *If $\mathrm{Re}\,\lambda \in \rho_0 + \mathcal{C}_0^+$, then E^{M/M_0} is well-defined.*

To understand the proof, let us recall the basic convergence proof in calculus for the Riemann zeta $\zeta(s) := \sum_{n=1}^\infty \frac{1}{n^s}$. It suffices to prove that $\int_1^\infty x^{-\mathrm{Re}s}dx$ converges. But this late integration may be essentially written as $x^{-(\mathrm{Re}s-1)}|_1^\infty$. Clearly, for convergence, it suffices that $\mathrm{Re}s - 1 > 0$.

Similarly, for the associated Eisenstein series, we first use properties related with U_0 to reduce the convergence problem to that for a certain integration. With a suitable manipulation, the final problem becomes the one for the integration

$$\int_c^\infty t_i^{-\langle \mathrm{Re}\,\mu+\rho_0-2\rho_0,\varpi_\alpha\rangle} dt,$$

where $\mathrm{Re}\,\mu+\rho_0$ comes from the exponent involved, while $-2\rho_0$ comes from the part $a^{-2\rho_0}da$ of the corresponding Haar measure. Therefore, for all $\alpha \in \Delta_0$, it suffices to have

$$0 < \langle \mathrm{Re}\,\mu + \rho_0 - 2\rho_0, \alpha\rangle = \langle \mathrm{Re}\,\mu - \rho_0, \alpha\rangle.$$

In other words, $\mathrm{Re}\,\mu \in \rho_0 + \mathcal{C}_0^+$.
Step 2. Reduce to Claim 1.
(a) Eisenstein series $E^{M/M_0}(\mathbf{1}, \lambda; g)$ provides a standard model for all automorphic forms.

Clearly, if ϕ is a P-level automorphic form, then so is $\lambda\phi$. Hence $\lambda\phi$ is of moderate growth. Consequently there exist a constant $c > 0$ and a sufficiently positive $\mu \in \mathrm{Re}\,\mathfrak{a}_{M_0}^*$ such that for all $k \in K$ and $m \in M(\mathbb{A})^1 \cap S^P$, we have

$$\big|\lambda\phi(mk)\big| \leq c \cdot m_{P_0}(mk)^{\mu+\rho_0}$$

which is then $\leq c \cdot E^{M/M_0}(\mathbf{1}, mk)$, since E^{M/M_0} is a summation of all positive terms $m_{P_0}(\delta mk)$ a special case of which, i.e., $\delta = 1$, gives $m_{P_0}(mk)$.
Claim 2. $|\lambda\phi_\pi(g)| \leq c \cdot m_P(g)^{\mathrm{Re}\pi+\mathrm{Re}\lambda} \cdot E^{M/M_0}(\mathbf{1}, g), \qquad \forall g \in G(\mathbb{A})$.

From the above discussion, we have the following inequality works only for $k \in K,\ m \in M(\mathbb{A})^1 \cap S^P$,

$$\big|\lambda\phi(mk)\big| \leq c \cdot E^{M/M_0}(\mathbf{1}, mk) \qquad (*).$$

Thus we have to see how it works when these special mk is changed to general $g \in G(\mathbb{A})$. By the Langlands decomposition $G(\mathbb{A}) = U(\mathbb{A}) \cdot A_{M(\mathbb{A})} \cdot M(\mathbb{A})^1 \cdot K$, and the fact that both sides of (*) is $M(F)$-invariant, it suffices to see how the part $A_{M(\mathbb{A})}$ palys a role. But $A_{M(\mathbb{A})} \subset Z_{M(\mathbb{A})}$. This then reduces the problem to the one for center characters involved. This is then done at the end of 5.1.

(b) As such, to justify the convergence of $E(\phi_\pi, \pi)$, it suffices to study

$$\sum_{\delta \in P(F)\backslash G(F)} m_P(\delta g)^{\pi+\lambda} \cdot E^{M/M_0}(\mathbf{1}, \delta g).$$

By writing out E^{M/M_0} in terms of $\sum_{\gamma \in \left(M(F)\cap P_0(F)\right)\backslash M(F)}$, it is sufficient to justify the convergence of the series

$$\sum_{\delta \in P_0(F)\backslash G(F)} m_P(\delta g)^{\pi+\lambda},$$

which is done in Step 1, or better by Claim 1. This completes the proof.

Remarks. (1) In Step I, if we replace the constant function $\mathbf{1}$ by a cusp form, a rapidly decreasing function, the same argument works.
Theorem B′. *Let $\phi \in A_0(U_0(\mathbb{A})M_0(F)\backslash G(\mathbb{A}))$ be a P_0-level cusp form. Then its associated relative Eisenstein series*

$$E^{M/M_0}(\phi, \lambda; g) := \sum_{\delta \in (M(F)\cap P_0(F))\backslash M(F)} m_P(\delta g)^\lambda \cdot \phi(\delta g)$$

converges absolutely if $\mathrm{Re}\lambda \in \rho_0 + \mathcal{C}_0^+$.
(2) There is a parallel theory of relative Eisenstein series E^{M_1/M_2} for $M_2 \subset M_1$. In particular, as indicated in Step 2(b), E^{M_1/M_3} is naturally related with the composition of E^{M_2/M_3} with E^{M_1/M_2} if $M_1 \supset M_2 \supset M_3$.

4 Constant Terms of Eisenstein Series

We will see the importance of constant terms of Eisenstein series later. Here let us give a nice formula for a special but most important case of them. That is, the one for Eisenstein series associated to cusp forms.
Theorem C. *Let $\phi_\pi \in A_0\Big(U(\mathbb{A})M(F)\backslash G(\mathbb{A})\Big)_\pi$ be a cusp form. Then for any standard parabolic subgroup $P' = U'M'$, we have*

$$E_{P'}(\phi_\pi, \pi) = \sum_{w \in W(M,M')} E^{M'/M}\Big(M(w,\pi)\phi_\pi, w\pi\Big),$$

where

$$W(M,M') := \Big\{ w \in W \,\big|\, w^{-1}(\alpha) > 0 \forall \alpha \in R^+(T_0, M'), \\ wMw^{-1} \hookrightarrow M' \text{ standard levi} \Big\}.$$

Remark. If $W(M,M') = \emptyset$, we understand that the above summation is zero.

Proof. By definiton,

$$\begin{aligned} E_{P'}(\phi_\pi, \pi) &= \int_{U'(F)\backslash U'(\mathbb{A})} \sum_{\gamma \in P(F)\backslash G(F)} \phi_\pi(\gamma u g)\, du \\ &= \sum_{\gamma \in P(F)\backslash G(F)} \int_{U'(F)\backslash U'(\mathbb{A})} \phi_\pi(\gamma u g)\, du \end{aligned}$$

due to the absolute convergence property.

Now using the Bruhat decomposition

$$G(F) = \cup_{w \in W_{M'}\backslash W/W_M} P(F) w^{-1} P'(F),$$

where W_M and $W_{M'}$ are the Weyl group of the Levi M and M' respectively, we can rewrite $G(F)$ accordingly to get

$$\sum_{w \in \left(W_{M'}\backslash W/W_M\right)} \sum_{\gamma \in \left(P(F)\backslash P(F) w^{-1} P'(F)\right)} \int_{U'(F)\backslash U'(\mathbb{A})} \phi_\pi(\gamma u g)\, du$$

Note then that $W_{M,M'} :=$

$$\Big\{ w \in W \,|\, w(\alpha) > 0 \ \forall \alpha \in R^+(T_0, M),\ w^{-1}(\alpha) > 0 \forall \alpha \in R^+(T_0, M') \Big\}$$

is a system of representatives for $W_{M'}\backslash W/W_M$, we can continue to get

$$\sum_{w \in W_{M,M'}} \sum_{\gamma \in \left(P(F)\backslash P(F) w^{-1} (M'(F)U'(F))\right)} \int_{U'(F)\backslash U'(\mathbb{A})} \phi_\pi(\gamma u g)\, du$$

$$= \sum_{w \in W_{M,M'}} \sum_{m' \in \left((M'(F) \cap w P(F) w^{-1})\backslash M'(F)\right)}$$

$$\int_{U'(F)\backslash U'(\mathbb{A})} \sum_{u' \in \left(\left(U'(F) \cap (m')^{-1} w P(F) w^{-1} m'\right)\backslash U'(F)\right)} \phi_\pi(w^{-1} m' u' u g)\, du$$

using the decompostion $P' = M'U'$ and rewrite the summation over P' to the double sum of M' and U'. Thus note that $m'u'\,ug = \Big(m'u'u(m')^{-1}\Big)\cdot m'g$ with $m'u'u(m')^{-1}$ still unipotent. So by using an unfolding argument to combine the integration and summmation over the unipotents into a single integration, we arrive at

$$\sum_{w\in W_{M,M'}}\ \sum_{m'\in\Big(M'(F)\cap wP(F)w^{-1}\backslash M'(F)\Big)} \int_{\Big(U'(F)\cap wP(F)w^{-1}\Big)\backslash U'(\mathbb{A})}\phi_\pi(w^{-1}um'g)\,du$$

$$=\sum_{w\in W_{M,M'}} E^{M'/M}\Bigg(\int_{\Big(U'(F)\cap wP(F)w^{-1}\Big)\backslash U'(\mathbb{A})}\phi_\pi(w^{-1}um'g)\,du;\pi\Bigg)(g)$$

which converges absolutely and uniformly if g remains in a compact subset.

So far, the cuspidal condition has not been used. Yet still we obtain a general formula for constant terms in terms of relative Eisenstein series. Next, we use the cuspidality to simplify the above general formula.

Note that if $w^{-1}U'w\cap M\neq\{1\}$, wMw^{-1} is not contained in M'. So M contains certain unipotent elements, and we obtain a proper parabolic subgroup in P. Thus, by the cuspidality of ϕ_π we see that for such P''s,

$$\int_{\Big(U'(F)\cap wP(F)w^{-1}\Big)\backslash U'(\mathbb{A})}\phi_\pi(w^{-1}um'g)\,du=0.$$

Consequently,

$$E_{P'}(\phi_\pi,\pi)=\sum_{w\in W_{M,M'},wMw^{-1}\subset M'}\ \sum_{m'\in\Big(M'(F)\cap wP(F)w^{-1}\backslash M'(F)\Big)} \int_{\Big(U'(F)\cap wP(F)w^{-1}\Big)\backslash U'(\mathbb{A})}\phi_\pi(w^{-1}um'g)\,du$$

$$= \sum_{w \in W(M,M')} \sum_{m' \in \left(M'(F) \cap wP(F)w^{-1} \backslash M'(F)\right)} \int_{\left(U'(F) \cap wP(F)w^{-1}\right) \backslash U'(\mathbb{A})} \phi_\pi(w^{-1}um'g)\, du,$$

since

$$\left\{ w \in W_{M,M'} \,|\, wMw^{-1} \subset M' \right\} = W(M, M').$$

Now for $w \in W(M, M')$,
(i) $M' \cap wPw^{-1}$ is a standard parabolic subgroup of M' whose Levi is wMw^{-1}; and
(ii) $U'(F) \cap wP(F)w^{-1} = U'(F) \cap wU(F)w^{-1}$ since M' contains no unipotent elements.

Therefore, $E_{P'}(\phi_\pi, \pi)(g)$

$$= \sum_{w \in W(M,M')} \sum_{m' \in \left(M'(F) \cap wP(F)w^{-1} \backslash M'(F)\right)} \int_{\left(U'(F) \cap wU(F)w^{-1}\right) \backslash U'(\mathbb{A})} \phi_\pi(w^{-1}um'g)\, du$$

$$= \sum_{w \in W(M,M')} \sum_{m' \in \left(M'(F) \cap wP(F)w^{-1} \backslash M'(F)\right)} \left(M(w,\pi)\phi_\pi\right)(m'g)$$

$$= \sum_{w \in W(M,M')} E^{M'/wMw^{-1}}\left(\left(M(w,\pi)\phi_\pi\right), \pi\right)(g),$$

as required.

As a direct consequence of the above calculation, we also established the convergence of the integral defining the intertwining operator.

A special case of the above theorem is of **particular importance**. That is, if $M' = wMw^{-1}$ is just conjugate to M, then $E^{M'/wMw^{-1}}$ becomes trivial and

$$W(M, M') = \left\{ w \in W(M) \,|\, wMw^{-1} = M' \right\}$$

where $W(M) :=$ the set of elements w of the Weyl group W of G whose length is minimal in the left coset of W_M and such that

wMw^{-1} is again a standard Levi. That is to say, we have obtained the following

Theorem C′. *Let* $\phi_\pi \in A_0\Big(U(\mathbb{A})M(F)\backslash G(\mathbb{A})\Big)_\pi$ *be a cusp form. Then for any standard parabolic subgroup* $P' = U'M'$ *whose standard Levi component* M' *is conjugate to* M,

$$E_{P'}(\phi_\pi, \pi) = \sum_{w \in W(M)} M(w, \pi)\phi_\pi.$$

We end this section by point out that in the convergence statement, for any automorphic forms, we assume that $\mathrm{Re}\,\pi$ is positive, and in particular if π is cuspidal, $\mathrm{Re}\,\pi \in \rho_P + \mathcal{C}_P^+$. For certain purposes, the following stronger technical result proves to be useful.

Proposition. *For convergence of* $E(\phi, \pi)$ *and* $M(w, \pi)$, *it suffice to assume that*

(i) $\langle \mathrm{Re}\,\pi, \alpha^\vee \rangle \gg 0\ \forall \alpha \in R(T_M, G)$ such that $\alpha > 0$ but $w\alpha < 0$, *when* ϕ *is a general automorphic form;*

(ii) $\langle \mathrm{Re}\pi - \rho_P, \alpha^\vee \rangle > 0\ \forall \alpha \in R(T_M, G)$ such that $\alpha > 0$ but $w\alpha < 0$, *when* ϕ *is a cusp form.*

5 Fundamental Properties of Eisenstein Series

All results are due to Langlands in this generality.

Theorem I. (**Meromorphic Continuation**) (a) *Eisenstein series* $E(\phi_\pi, \pi)$ *defined by*

$$E(\phi_\pi, \pi)(g) := \sum_{\gamma \in P(F)\backslash G(F)} \phi_\pi(\gamma g)$$

over a certain positive cone $\mathcal{C}$ *in* $\mathrm{Re}\,\mathfrak{a}_M^*$ *admits a unique meromorphic continuation to the whole complex space* $\mathfrak{a}_M^*$;

(b) *For any* $w \in W$, *intertwining operator* $M(w, \pi)$:

$$\begin{array}{ccc} A\Big(U(\mathbb{A})M(F)\backslash G(\mathbb{A})\Big)_\pi & \to & A\Big(U'(\mathbb{A})\cdot wMw^{-1}(F)\backslash G(\mathbb{A})\Big)_{w\pi} \\ \phi_\pi & \to & M(w, \pi)\phi_\pi \end{array}$$

defined by

$$g \mapsto \int_{U'(F)\cap wU(F)w^{-1}\backslash U'(\mathbb{A})} \phi_\pi(ug)\,du$$

over a certain positive cone $\mathcal{C}$ *of* $\mathrm{Re}\,\mathfrak{a}_M^*$ *admits a unique meromorphic continuation to the whole complex space* $\mathfrak{a}_M^*$.

Theorem II. (**Functional Equations**) *Meromorphically on* $\mathfrak{P} = \pi \otimes \mathfrak{a}_M^*$,
(a) $E\Big(M(w,\pi)\phi_\pi, w\pi\Big) = E(\phi_\pi, \pi)$ *for all* $w \in W(M)$; and
(b) $M(w', w\pi) \circ M(w,\pi) = M(w'w, \pi)$ *for all* $w \in W(M), w' \in W(wMw^{-1})$.

Theorem III. (**Holomorphicity and Singularities**) *Assume that* π_0 *is* **cuspidal.**
(a) (i) *If* $\mathrm{Re}\,\lambda$ *belongs to the* ρ_P*-shifted positive Weyl chamber* $\rho_P + \mathcal{C}_M^+$,

$$E\Big(\phi_{\pi_0\otimes\lambda}, \pi_0 \otimes \lambda\Big) \qquad \text{and} \qquad M\big(w, \pi_0 \otimes \lambda\big)$$

are holomorphic in λ*;*
(ii) *If* λ *belongs to* $\mathrm{Im}\,\mathfrak{a}_M^*$, $E\Big(\phi_{\pi_0\otimes\lambda}, \pi_0 \otimes \lambda\Big)$ *and* $M\big(w, \pi_0 \otimes \lambda\big)$ *are holomorphic in* λ*;*

(b) (i) *There exists a locally finite set* D *of root hyperplanes* D *such that the singularities of* $E\Big(\phi_{\pi_0\otimes\lambda}, \pi_0 \otimes \lambda\Big)$ *and* $M\big(w, \pi_0 \otimes \lambda\big)$ *lie over* D*;*
(ii) *The singularities of* $E\Big(\phi_{\pi_0\otimes\lambda}, \pi_0 \otimes \lambda\Big)$ *and* $M\big(w, \pi_0 \otimes \lambda\big)$ *are without multiplicities at* $\langle \mathrm{Re}\lambda, \alpha^\vee\rangle \geq 0\ \forall \alpha \in \Delta_M$*, i.e., the generic multiplicities are at most one;*
(iii) *There are only finitely many singular hyperplanes of*

$$E\Big(\phi_{\pi_0\otimes\lambda}, \pi_0 \otimes \lambda\Big) \qquad \text{and} \qquad M\big(w, \pi_0 \otimes \lambda\big)$$

which intersect $\Big\{\, \pi_0 \otimes \lambda \,|\, \langle \mathrm{Re}\lambda, \alpha^\vee\rangle \geq 0\ \forall \alpha \in \Delta_M \Big\}$.

Day Three: Pseudo-Eisenstein Series

6 Paley-Wiener Functions

6.1 Paley-Wiener Functions

Over the complex space $X_M^G := (\mathfrak{a}_M^G)^*$, denote by $P(X_M^G)$:= the space of $\mathbb{C}$-valued functions on X_M^G which are Paley-Wiener, i.e., the space consisting of $\mathbb{C}$-valued holomorphic functions on X_M^G such that

$$\mathrm{sup}_{\lambda\in X_M^G}\Big\{|f(\lambda)|\cdot e^{r\|\mathrm{Re}\lambda\|}(1+\|\lambda\|^n)\Big\}<\infty.$$

More generally, for a fixed $R>0$, set

$$D_R := \Big\{\lambda\in X_M^G\mid \|\mathrm{Re}\lambda\|<R\Big\}$$

and $P^R(X_M^G)$:= the space consisting of $\mathbb{C}$-valued holomorphic functions on D_R such that

$$\mathrm{sup}_{\lambda\in D_R}\Big\{|f(\lambda)|\cdot(1+\|\lambda\|^n)\Big\}<\infty.$$

6.2 Fourier Transforms

For any $f\in P^{(R)}(X_M^G)$, its Fourier transform is defined by

$$\hat{f}(m) := \int_{\lambda\in X_M^G,\mathrm{Re}\lambda=\lambda_0} f(\lambda)\cdot m^\lambda\, d\lambda \qquad \forall m\in Z_{G(\mathbb{A})}M(\mathbb{A})^1\backslash M(\mathbb{A}).$$

It is well known that
(i) $\hat{f}$ *becomes a* C^∞*-function with compact support*; and
(ii) (**Fourier Inversion**) *there exists a constant depending only on the Haar measures involved such that for all Paley-Wiener* f,

$$f(\lambda)=c\cdot\int_{Z_{G(\mathbb{A})}M(\mathbb{A})^1\backslash M(\mathbb{A})}\hat{f}(m)\cdot m^{-\lambda}dm.$$

Denote by $\widehat{P}^{(R)}(X_M^G)$:= the space consisting of all Fourier transforms of Paley-Wiener functions. We then also easily have

(iii) (**Estimations**) $\widehat{P}^R(X_M^G) \subset \widetilde{P}^R(X_M^G)$:= *the space consisting of* C^∞*-functions* $\widetilde{f}$ *on* $Z_{G(\mathbb{A})}M(\mathbb{A})^1\backslash M(\mathbb{A})$ *such that there exists* $c > 0$ *satisfying*

$$\left|\widetilde{f}(m)\right| \leq c \cdot e^{-R\|\log_M^G m\|} \; \forall m \in M(\mathbb{A}).$$

6.3 Paley-Wiener on $\mathfrak{P}$

Let $\mathfrak{P} = \pi_0 \otimes X_M^G$ be an equivalence class of automorphic representations of $G(\mathbb{A})$. We are interested in the functions on $\mathfrak{P}$ satisfying the following conditions:
(i) for any $\pi \in \mathfrak{P}$, $\Phi(\pi) \in A\Big(U(\mathbb{A})M(F)\backslash G(\mathbb{A})\Big)_\pi$; and
(ii) Φ is K-finite.

Such functions are in one-to-one correspondence to functions $\widetilde{\Phi}$ over X_M^G satisfying
(i′) for all $\lambda \in X_M^G$, $\widetilde{\Phi}(\lambda) \in A\Big(U(\mathbb{A})M(F)\backslash G(\mathbb{A})\Big)_{\pi_0}$; and
(ii′) the vector subspace of $A\Big(U(\mathbb{A})M(F)\backslash G(\mathbb{A})\Big)_{\pi_0}$ generated by the functions $\widetilde{\Phi}(\lambda)$ where λ runs over all X_M^G is finite dimensional.

More precisely, if $\pi = \pi_0 \otimes \mu_\pi$ then

$$\Phi(\pi) := \big(\mu_\pi \widetilde{\Phi}\big)(\mu_\pi) \quad \leftrightarrow \quad \widetilde{\Phi}(\mu_\pi) := \big(\lambda^{-1}\Phi\big)(\pi)$$

where as usual $(\lambda\phi)(g) := m_P(g)^\lambda \cdot \phi(g)$ for a P-level automorphic form ϕ.

With this, define the space of *Paley-Wiener functions* on $\mathfrak{P}$ to be
(a) $P_{(M,\mathfrak{P})}$:= the space of functions Φ over the total space $\mathfrak{P}$ such that $\widetilde{\Phi} \in A\Big(U(\mathbb{A})M(F)\backslash G(\mathbb{A})\Big)_{\pi_0} \otimes P(X_M^G)$; and
(b) $P_{(M,\mathfrak{P})}^R$:= the space of functions Φ over $\mathfrak{P}_R := \Big\{\pi \in \mathfrak{P} \big| \|\mathrm{Re}\pi\| < R\Big\}$ such that $\widetilde{\Phi} \in A\Big(U(\mathbb{A})M(F)\backslash G(\mathbb{A})\Big)_{\pi_0} \otimes P^R(X_M^G)$.

As such, for any function $\Phi \in P_{(M,\mathfrak{P})}^{(R)}$, its Fourier transform $F_\pi(\Phi)$ (associated to π) is a function over $Z_{M(\mathbb{A})}M(\mathbb{A})^1\backslash M(\mathbb{A})$ valued in

$A\Big(U(\mathbb{A})M(F)\backslash G(\mathbb{A})\Big)_\pi$ given by

$$F_\pi(\Phi)(m'): g \mapsto \int_{\lambda\in \mathrm{Im} X_M^G} (m')^{-(\lambda+\rho_P)}\cdot \Big(\Phi(\pi\otimes\lambda)\Big)(m'mk)\, d\lambda$$

for all $g=umk\in G(\mathbb{A})$ with $u\in U(\mathbb{A}), m\in M(\mathbb{A}), k\in K$. Clearly,
(i) $F_\pi(\Phi)(\bullet)(g)\in \widetilde{A}(M(F)\backslash M(\mathbb{A}))_\pi :=$ the space obtained as the left-translation under the adelic group $M(\mathbb{A})$ of $A\Big(M(F)\backslash M(\mathbb{A})\Big)_\pi$ (Recall that the difficulty as always is that $A\Big(M(F)\backslash M(\mathbb{A})\Big)_\pi$ is not closed under the action of adelic group as said earlier);
(ii) $F_{\pi\otimes\lambda}(\Phi)(\bullet)(g)=\lambda\, F_\pi(\Phi)(\bullet)(g)\ \forall\lambda\in X_M^G$; and
(iii) $F_\pi(\Phi)(\bullet)(umg) = \delta(m)\cdot F_\pi(\Phi)(\bullet)(g)\ \forall m\in Z_{G(\mathbb{A})}M(\mathbb{A})^1$ and $u\in U(\mathbb{A})$.

In particular, we conclude that
(a) At the unit element $\mathbf{e}$ of $M(\mathbb{A})$,

$$\begin{aligned}\Big(F_\pi(\Phi)(\mathbf{e})\Big)(g=umk) &= \int_{\lambda\in \mathrm{Im} X_M^G} \mathbf{e}^{-(\lambda+\rho_P)}\Phi(\pi\otimes\lambda)(emk)\, d\lambda\\ &= \int_{\lambda\in \mathrm{Im} X_M^G} \Phi(\pi\otimes\lambda)(mk)\, d\lambda\\ &= \int_{\lambda\in X_M^G, \mathrm{Re}\lambda=\mathrm{Re}\lambda_0} \Phi(\pi\otimes\lambda)(mk)\, d\lambda\end{aligned}$$

for any $\lambda_0\in X_M^G$ since Φ is holomorphic. It is independent of π;
(b) We have a commutative diagram

$$\begin{array}{ccc} A\Big(U(\mathbb{A})M(F)\backslash G(\mathbb{A})\Big)_{\pi_0}\otimes_{\mathbb{C}} P^{(R)}(X_M^G) & \stackrel{\tilde{\bullet}}{\to} & P^{(R)}_{(M,\mathfrak{P})}\\ \mathrm{Id}\otimes \mathrm{F}\downarrow & & \downarrow F_\pi\ ,\\ A\Big(U(\mathbb{A})M(F)\backslash G(\mathbb{A})\Big)_{\pi_0}\otimes_{\mathbb{C}} C_c^\infty\Big(Z_{G(\mathbb{A})}M(\mathbb{A})^1\backslash M(\mathbb{A})\Big) & \stackrel{\nu}{\to} & \widehat{P}^{(R)}_{(M,\mathfrak{P})}\end{array}$$

where the morphism μ is defined by $\nu\Big(\sum_i\varphi_i\otimes f_i\Big)(m'):$

$$g\mapsto \sum_i f_i(m_P(g))\cdot (m')^{-\rho_P}\varphi_i(m'g)\ \forall m'\in M(\mathbb{A}), g\in G(\mathbb{A});$$

In particular, we have the following result which controls the growth of the Fourier transform of elements in $P^R_{(M,\mathfrak{P})}$

(c) (Estimations) $\widehat{P}^R_{(M,\mathfrak{P})} \subset \mathrm{Im}\Big(A\big(U(\mathbb{A})M(F)\backslash G(\mathbb{A})\big)_{\pi_0} \otimes_{\mathbb{C}} \widetilde{P}^R(X^G_M)\Big)$.

7 Pseudo-Eisenstein Series

In this section we assume that $\mathfrak{P}$ is an equivalence class of cuspidal representations.

Definition. For $\Phi \in P^R_{(M,\mathfrak{P})}$, its associated *pseudo-Eisenstein series* is a function on $G(F)\backslash G(\mathbb{A})$ defined by

$$\begin{aligned}\theta_\Phi(g) &:= \sum_{\delta\in P(F)\backslash G(F)} \Big(F_\pi(\Phi)(e)\Big)(\delta g)\\ &= \sum_{\delta\in P(F)\backslash G(F)} \int_{\lambda\in X^G_M, \mathrm{Re}\lambda=\mathrm{Re}\lambda_0} \Phi(\pi\otimes\lambda)(\delta g)\,d\lambda\\ &= \sum_{\delta\in P(F)\backslash G(F)} \int_{\pi\in\mathfrak{P}, \mathrm{Re}\pi=\mathrm{Re}\pi_0} \Phi(\pi)(\delta g)\,d\pi.\end{aligned}$$

Clearly, in the convergence region $\mathrm{Re}\,\pi \in \mathcal{C}$ of $E\Big(\Phi(\pi),\pi\Big)(g)$ we have

$$\theta_\Phi(g) = \int_{\pi\in\mathfrak{P}, \mathrm{Re}\,\pi=\mathrm{Re}\,\pi_0} E\Big(\Phi(\pi),\pi\Big)(g)\,d\pi.$$

This explain why θ_Φ is called a pseudo-Eisenstein series.

Theorem A. (a) *θ_Φ is well-defined;*

(b) (i) *θ_Φ is square-integrable over $Z_{G(\mathbb{A})}G(F)\backslash G(\mathbb{A})$ if $\Phi \in P^R_{(M,\mathfrak{P})}$;*

(ii) *θ_Φ is rapidly decreasing over $Z_{G(\mathbb{A})}G(F)\backslash G(\mathbb{A})$ if $\Phi \in P_{(M,\mathfrak{P})}$;* and

(c) $\overline{\langle\theta_\Phi \,|\, \Phi \in P_{(M,\mathfrak{P})},\ \forall M,\mathfrak{P}\rangle} = L^2\Big(G(F)\backslash G(\mathbb{A})\Big)$.

The moral of this result is that it provides a natural bridge between general square-integrable functions and Eisenstein series: the decomposition of L^2 functions may be transformed to the one for pseudo-Eisenstein series.

Proof. Let us start with (c). It is a direct consequence of the philosophy of cusp forms. Indeed, for any $f \in L^2\Big(G(F)\backslash G(\mathbb{A})\Big)$, $(M,\mathfrak{P})$ and $\Phi \in P_{(M,\mathfrak{B})}$,

$$\begin{aligned}
&\int_{Z_{G(\mathbb{A})}G(F)\backslash G(\mathbb{A})} \overline{f(g)} \cdot \theta_\Phi(g)\, dg \\
&= \int_{Z_{G(\mathbb{A})}G(F)\backslash G(\mathbb{A})} \overline{f(g)} \cdot \sum_{\delta \in P(F)\backslash G(F)} \int_{\pi\in\mathfrak{P},\mathrm{Re}\pi=\mathrm{Re}\lambda_0} \Phi_\pi(\delta g)\, d\pi\, dg \\
&= \int_{Z_{G(\mathbb{A})}P(F)\backslash G(\mathbb{A})} \overline{f(g)} \cdot \int_{\pi\in\mathfrak{P},\mathrm{Re}\pi=\mathrm{Re}\lambda_0} \Phi_\pi(g)\, d\pi\, dg\, dg \\
&= \int_{Z_{G(\mathbb{A})}U(\mathbb{A})M(F)\backslash G(\mathbb{A})} \overline{f_P(g)} \cdot \int_{\pi\in\mathfrak{P},\mathrm{Re}\pi=\mathrm{Re}\lambda_0} \Phi_\pi(g)\, d\pi\, dg\, dg
\end{aligned}$$

(by an unfolding argument and the definition of constant terms f_P of f along P). Clearly, by the convergence property, this latest expression can be rewritten as

$$\int_{\pi\in\mathfrak{P},\mathrm{Re}\pi=\mathrm{Re}\lambda_0} \Big[\int_{Z_{G(\mathbb{A})}U(\mathbb{A})M(F)\backslash G(\mathbb{A})} \overline{f_P(g)} \cdot \Phi_\pi(g)\, dg\Big]\, dg.$$

Thus if f is orthogonal to all pseudo-Eisenstein series, then

$$\Big[\int_{Z_{G(\mathbb{A})}U(\mathbb{A})M(F)\backslash G(\mathbb{A})} \overline{f_P(g)} \cdot \bullet(g)\, dg\Big] = 0$$

for any $\bullet \in C_0\Big(U(\mathbb{A})M(F)\backslash G(\mathbb{A})\Big)$ since Φ can be taken to be any Paley-Wener functions. In other words, all the cuspidal P-forms of f are identically zero. Consequently, by the philosophy of cusp forms, $f = 0$ almost everywhere.

(a, b) By the **estimations** on Fourier transforms of elements in $P^R_{(M,\mathfrak{P})}$, we know that if $f \in \hat{P}^R(X_M^G)$, there exists μ very near ρ_P such that $|f(m)| \leq c \cdot m^{\mu_M}$ $\forall m \in M(\mathbb{A})$. Consequently

$$|F_\pi(\Phi)(\bullet)(g)| \leq c \cdot m_{P_0}^{\rho_P+\mu+\mathrm{Re}\xi} \qquad \forall g \in S^P.$$

Thus using the convergence proof for Eisenstein series, we conclude that θ_Φ is well-defined. This gives (a).

Moreover, we have

$$\big|\theta_\Phi(g)\big| \leq c \cdot E^{G/M_0}(\mathbf{1}, g).$$

Thus for (b.i), it suffices to show that

$$\int_{A_{G(\mathbb{A})}G(F)\backslash G(\mathbb{A})} \big|\theta_\Phi(g)\big| \cdot E^{G/M_0}(\mathbf{1}, g)\, dg < \infty.$$

By an unfolding argument, we need to show that

$$\int_{A_{G(\mathbb{A})}\backslash A_{M(\mathbb{A})}} \int_{M(F)\backslash M(\mathbb{A})^1} \int_K \Big|\Big(F_\pi(\Phi)(e)\Big)(amk)\Big| \cdot E_P(amk)\, dk\, (am)^{-2\rho_P} dmda$$

converges. Recall that we have a natural upper bound for $F_\pi(\Phi)(e)$. Using it, we are lead to show the convergence of the integration

$$\int_{A_{G(\mathbb{A})}\backslash A_{M(\mathbb{A})}} a^{\mu'} e^{-R\|\log_M a\|} da$$

for fixed μ'. Clearly, it is a comparison of two exponential functions: with R sufficiently large, the second term dominates the first. This gives b.i)

With the proof above, for (b.ii), what is left here is to apply the fact that cusp forms are rapidly decreasing.

8 First Decomposition of $L^2\Big(G(F)\backslash G(\mathbb{A})\Big)$

8.1 Inner Product Formula for P-ESes

The following inner product formula for pseudo-Eisenstein series plays a key role in the first level decomposition for the space of L^2 functions on $Z_{G(\mathbb{A})}G(F)\backslash G(\mathbb{A})$. It also gives us a first example to realize why the constant formula for Eisenstein series associated to cusp forms is fundamental to all the following works.

Main Theorem A. *For* $\Phi \in P_{(M,\mathfrak{P})}$, $\Phi' \in P_{(M',\mathfrak{P}')}$, $\Big\langle \theta_{\Phi'}, \theta_\Phi \Big\rangle =$

$$\int_{\pi\in\mathfrak{P},\mathrm{Re}\pi=\lambda_0} \sum_{w\in W\Big((M,\mathfrak{P}),(M',\mathfrak{P}')\Big)} \Big\langle M(w^{-1}, -w\bar{\pi})\Phi'(-w\bar{\pi}), \Phi(\pi)\Big\rangle_M d\pi$$

where

$$W\Big((M,\mathfrak{P}),(M',\mathfrak{P}')\Big) := \Big\{ w\in W \,|\, w\in W \,\big|\, w(M,\mathfrak{P}) = (M',\mathfrak{P}')\Big\}$$

and

$$\Big\langle \phi, \varphi \Big\rangle_M := \int_K \int_{Z_{M(\mathbb{A})}M(F)\backslash M(\mathbb{A})} \overline{\phi(mk)}\,\varphi(mk)\cdot m^{-2\rho_P}\, dm\cdot dk.$$

Remarks. (i) As usual, if $W\Big((M,\mathfrak{P}),(M',\mathfrak{P}')\Big) = \emptyset$, i.e., if $(M,\mathfrak{P})$ and $(M',\mathfrak{P}')$ are not equivalent to each other, then we understand the above summation as empty which gives the value 0.
(ii) We also would like to reminder the reader that by using Paley-Wiener functions, automorphic forms involved are all cuspidal. Accordingly, $\lambda_0 \in \rho_P + \mathcal{C}_M$ can be taken to be any vector in the shifted positive Weyl chamber in $\mathrm{Re}\,\mathfrak{a}_M^*$.
Proof. The proof of this formula is surprisingly easy. In fact, by a standard unfolding argument, we have

$$\begin{aligned}\Big\langle \theta_{\Phi'}, \theta_\Phi \Big\rangle &:= \int_{\big(Z_{G(\mathbb{A})}G(F)\backslash G(\mathbb{A})} \overline{\theta_{\Phi'}(g)}\cdot\theta_\Phi(g)\cdot dg \\ &= \int_{\big(Z_{G(\mathbb{A})}U(\mathbb{A})M(F)\backslash G(\mathbb{A})} \overline{\theta_{\Phi',P}(g)}\cdot\Big(F_\pi(\Phi)(e)\Big)(g)\cdot dg.\end{aligned}$$

This leads to a calculation on the constant term of pseudo-Eisenstein series.

Now λ_0 is in the absolutely and uniformly convergent region, so the problem may be transformed to that for constant terms of Eisenstein series. Also it is cusp form that shows up in Paley-Wiener functions. Hence, we only need to use the formula of constant forms for Eisenstein series associated to cusp forms. Recall then that by the formula of Day 2, for such a constant form to be non-zero, standard parabolic subgroups involved should satisfy the condition that one is conjugate to a standard sub-Levi of the other. Consequently when apply to our situation here, to get a non-zero inner product, by switching Φ' and Φ if necessary, we conclude that M and M' must be in the same association class, i.e., there should be a $w \in W$ such that $wMw^{-1} = M'$. But, constant terms for the associated Eisenstein series become extremely simple – it can be written in terms of the associated intertwining operator directly even without using relative Eisenstein series. All this, plus a simple application of Fourier inversion formula then give the result above, except the following easy point: if $\mathfrak{P}$ is not transformed to $\mathfrak{P}'$, then the corresponding cusp forms (appeared in Φ' and Φ) are orthogonal.

8.2 Decomposition of L^2-Spaces According to Cuspidal Data

A direct application of the above inner product formula (and the fact that P-ESes are dense) is the following first decomposition of the L^2-space.

As above, cuspidal data $(M, \mathfrak{P}), (M', \mathfrak{P}')$ are called equivalent if there exists w such that $w(M, \mathfrak{P}) = (M', \mathfrak{P}')$ or the same, $wMw^{-1} = M'$ and $w\mathfrak{P} = \mathfrak{P}'$. Denote by

$$\mathfrak{X} := \Big[(M, \mathfrak{P})\Big] =: \Big\{(M, \mathfrak{P})\Big\}/\sim$$

be an equivalence class of cuspidal data. Denote by ξ the unique central character associated to $\mathfrak{X}$. (How?) We call then $\mathfrak{X}$ as ξ-*admissible.* Denote by $\mathfrak{G}$ the collection of all ξ-admissible $\mathfrak{X}$'s and set

$$L^2\Big(G(F)\backslash G(\mathbb{A})\Big)_{\mathfrak{X}} := \Big\langle \theta_\Phi \,\big|\, \Phi \in P_{(M,\mathfrak{P})},\ (M, \mathfrak{P}) \in \mathfrak{X} \Big\rangle_{\mathbb{C}}.$$

Main Theorem A$'$. *There is a natural orthogonal decomposition*

$$L^2\Big(G(F)\backslash G(\mathbb{A})\Big)_{\xi} = \widehat{\oplus}_{\mathfrak{X}\in\mathfrak{G}} L^2\Big(G(F)\backslash G(\mathbb{A})\Big)_{\mathfrak{X}}.$$

8.3 Constant Terms of P-SEes

The description of a general formula for constant terms of pseudo-Eisenstein series in literature appears to be unnecessarily complicated: As stated above, it is a simple translation of that for Eisenstein series associated with cusp forms. To help the reader, and also for the purpose to introduce the so-called relative pseudo-Eisenstein series, we add this subsection. Recall that

$$\theta_\Phi(g) := \int_{\pi\in\mathfrak{P},\mathrm{Re}\pi=\lambda_0} E\Big(\Phi(\pi),\pi\Big)(g)\,d\pi.$$

Thus to write down the constant terms of θ_Φ along say P', the key is to apply the following result for Eisenstein series associated to cusp forms:

Theorem. $E_{P'}\Big(\Phi(\pi),\pi\Big) = \sum_{w\in W(M,M')} E^{M'/M}\Big(M(w,\pi),w\pi\Big).$

As such, relative pseudo-Eisenstein series should be introduced. We will not give the most original one, but their cousins. (Pay attention to the notation we use.) Fix an extension $\xi_{M'}$ of ξ to $Z_{M'(\mathbb{A})}$. View M as a subgroup of M' (instead of G), set

$$\mathfrak{P}_{M'} := \Big\{\pi\in\mathfrak{P}\,\Big|\,\pi \text{ is } \xi_{M'}-\text{admissible}\Big\},$$

which is an equivalence class of automorphic representations of $M(\mathbb{A})$. If $\mathfrak{P}_{M'}\neq\emptyset$, we can have the associated Paley-Wiener functions $P^{R'}_{(M,\mathfrak{P}_{M'})}$. In particular, for $w\in W(M,M')$ (so that $wMw^{-1}\hookrightarrow M'$), with a fixed $\pi'_w\in w\mathfrak{P}$, we have an equivalence class of automorphic representations $\pi'_w\otimes X^{M'}_{wMw^{-1}} =: [\pi'_w]_{M'}$. For a fixed $R'>0$, and a G-level Paley-Wiener function $\Phi\in P^R_{(M,\mathfrak{P})}$ we have then a subset

$$\Big\{\pi'\in[\pi'_w]_{M'}\,\Big|\,\|\mathrm{Re}\,\pi'-\mu_0\|<R'\Big\} =: [\pi'_w]_{M';R'}$$

and hence a Paley-Wiener function in $P^{R'}_{(wMw^{-1},[\pi'_w]_{M'})}$ given by

$$[\pi'_w]_{M';R'}\ni\pi'\mapsto M(w,w^{-1}\pi')\Phi(w^{-1}\pi').$$

(Here, to make sure that everything is well-defined, $\mu_0 \in \mathrm{Re}\, X_{M'}^G$ is a fixed vector satisfying

$$\|\mu_0\| + R' < R, \quad \text{and} \quad \langle -\mu_0, \alpha^\vee \rangle \gg 0 \quad \forall \alpha \in R^+(T_{M'}, G).)$$

Set also P^w be the standard parabolic associated to wMw^{-1} with unipotent radical U^w and $\nu_0 \in \mathrm{Re}\, X_{wMw^{-1}}^{M'}$ satisfying

$$\|\nu_0\| < R' \quad \text{and} \quad \langle \nu_0 - \rho_{P^w}^{P'}, \alpha^\vee \rangle \gg 0 \quad \forall \alpha \in R^+(T_{wMw^{-1}}, M').$$

Then define a *relative pseudo-Eisenstein series* by

$$\theta_{M(w,w^{-1}\bullet)\Phi(w^{-1}\bullet)}^{M'/wMw^{-1}}(\pi'_w)(g) := \int_{\pi' \in [\pi'_w]_{M'}, \mathrm{Re}\pi' = \mu_0 + \nu_0} \sum_{\left(\delta \in \left(M'(F) \cap P^w(F)\right) \backslash M'(F)\right)} \Big(M(w, w^{-1}\pi')\Phi(w^{-1}\pi')\Big)(\delta g)\, d\pi'.$$

Lemma. *If* $\pi'_w \in \pi''_w + \mathrm{Im} X_{wMw^{-1}}^{M'}$, *then*

$$\theta_{M(w,w^{-1}\bullet)\Phi(w^{-1}\bullet)}^{M'/wMw^{-1}}(\pi'_w) = \theta_{M(w,w^{-1}\bullet)\Phi(w^{-1}\bullet)}^{M'/wMw^{-1}}(\pi''_w).$$

Consequently, $\theta_{M(w,w^{-1}\bullet)\Phi(w^{-1}\bullet)}^{M'/wMw^{-1}}(\pi'_w)$ may be viewed as a function on the quotient space

$$D_{w,\mu_0}^\bullet := \Big\{\pi'_w \in w\mathfrak{P} \,\Big|\, \mathrm{Re}\pi'_w = \mu_0\Big\} / \mathrm{Im} X_{wMw^{-1}}^{M'}.$$

With this, we are now ready to state the following easy consequence of the formula on constant terms of Eisenstein series:

Proposition. (i) $\theta_{M(w,w^{-1}\bullet)\Phi(w^{-1}\bullet)}^{M'/wMw^{-1}}(\pi'_w)$ *is rapidly decreasing on* $D_{w,\mu_0}^\bullet$; and
(ii) $\theta_{\Phi,P'} = \sum_{w \in W(M,M')} \int_{D_{w,\mu_0}^\bullet} \theta_{M(w,w^{-1}\bullet)\Phi(w^{-1}\bullet)}^{M'/wMw^{-1}}(\pi'_w)\, d\pi'_w.$

9 Decomposition of Automorphic Forms According to Cuspidal Data

9.1 Main Result

From the discussion above, the key to the success of our first decomposition of L^2-space is that there is a natural pairing $\langle\,,\,\rangle$ on L^2-functions. On the other hand, general automorphic forms are only of moderate growth, so they are in general not integrable. This posts a huge difficulty for the study of (space of) automorphic forms.

Luckily, we have a way to overcome this, still using a certain pairing. The essential points then may be summarized as follows:
(i) Pseudo-Eisenstein series associated to Paley-Wiener functions (or better, to cusp forms) are rapildy decreasing;
(ii) There is a natural pairing between rapidly decreasing functions and moderate growth functions; and
(iii) There is a huge algebra of operators there ready to be used.
Simply put, by (i) and (ii) we can talk about the pairing between pseudo-Eisenstein series and general automorphic forms, which gives us a way to understand automorphic forms in terms of cusp forms; and by (iii), with the common eigen-space decomposition for ceratin operators we obtain a decomposition for the space of automorphic forms as well, since operators acting on P-ESes can be 'transformed' to that for automorphic forms via the 'adjoint' operation w.r.t. the pair in (ii).

Fix then a central character ξ. Two ξ-admissible cuspidal representations (M,π) and (M',π') are said to be *equivalent* if there exists a $w\in W$ such that $w(M,\pi)=(M',\pi')$. Let $\eta:=[(M,\pi)]$ be the equivalence class associated to (M,π) and Ξ be the collection of all ξ-admissible equivalence classes.

Recall that from Day 1, for any $\phi\in A\Big(G(F)\backslash G(\mathbb{A})\Big)_\xi$, and any standard parabolic $P=UM$, there exists a finite subset

$$D(M,\phi)\subset \mathbb{C}[\mathrm{Re}\,\mathfrak{a}_M^G]\times\Pi_0(M(\mathbb{A}))\times A_0\Big(U(\mathbb{A})M(F)\backslash G(\mathbb{A})\Big)$$

such that

(i) For all $(Q,\pi,\varphi)\in D(M,\phi)$, $\varphi\in A_0\Big(U(\mathbb{A})M(F)\backslash G(\mathbb{A})\Big)_\pi$; and
(ii) $\phi_P^{\rm cusp}(g)=\sum_{(Q,\pi,\varphi)\in D(M,\phi)} Q\Big(\log_M^G m_P(g)\Big)\cdot\varphi(g)$.
(We further assume that no (Q,π,φ) is wasted in an obvious sense.)
Set then $A\Big(G(F)\backslash G(\mathbb{A})\Big)_\eta :=$

$$\Big\{\phi\in A\Big(G(F)\backslash G(\mathbb{A})\Big)_\xi \,\Big|\, (M,\pi)\in\eta\ \forall (Q,\pi,\varphi)\in D(M,\phi)\Big\}.$$

Main Theorem B. *There is a direct sum decomposition*

$$A\Big(G(F)\backslash G(\mathbb{A})\Big)_\xi=\oplus_{\eta\in\Xi}A\Big(G(F)\backslash G(\mathbb{A})\Big)_\eta.$$

9.2 Langlands Operators

Fix $R>0$ and a ξ-admissible equivalence classes $\mathfrak{X}$ of cuspidal data say associated to $(M,\mathfrak{P})$. Set

$$\widetilde{\mathfrak{X}}:=\Big\{(M,\pi)\,\Big|\,\pi\in\mathfrak{P}\ \exists(M,\mathfrak{P})\in\mathfrak{X}\Big\}$$
$$\widetilde{\mathfrak{X}}^R:=\Big\{(M,\pi)\in\widetilde{\mathfrak{X}}\,\Big|\,\|\mathrm{Re}\,\pi-\mathrm{Re}\,\xi\|<R\Big\}.$$

Then a *Langlands operator* A is an assignment on $\widetilde{\mathfrak{X}}^R$ satisfying the following properties:
(0) For any $(M,\pi)\in\widetilde{\mathfrak{X}}^R$, $A(M,\pi)\in\mathrm{End}_{M(\mathbb{A})}\Big(A_0\big(M(F)\backslash M(\mathbb{A})\big)\Big)_\pi$;
(i) A is holomorphic on $\widetilde{\mathfrak{X}}^R$;
(ii) A is of moderate growth; and
(iii) A is admissible.

To understand this definition, let us first shift from moving target spaces to a fixed one by introducing the operator

$$\begin{aligned}X_M^G\ni\lambda\mapsto &A_{M,\pi}(\lambda):=\underline{-\lambda}\circ A(M,\pi\otimes\lambda)\circ\underline{\lambda}\\ &\in\mathrm{End}_{M(\mathbb{A})}\Big(A_0\big(M(F)\backslash M(\mathbb{A})\big)\Big)_\pi\end{aligned}$$

where $\Big(\underline{\lambda}\phi\Big)(m):=m^\lambda\cdot\phi(m)$. As such,
(i) means that operators $A_{M,\pi}(\lambda)$ depend holomorphically on λ;

(ii) means that there exist positive constants c and N such that (for one and hence all operator norms)

$$\left\|A_{M,\pi}(\lambda)\right\| \leq c \cdot \left(a + \|\lambda\|\right)^N;$$

(iii) For $w \in W$, set $M' := wMw^{-1}, \pi' := w\pi$. Then we have a natural induced morphism

$$\mathrm{ad}(w) : A_0\Big(M(F)\backslash M(\mathbb{A})\Big)_\pi \to A_0\Big(M'(F)\backslash M'(\mathbb{A})\Big)_{\pi'}.$$

The admissibility says that the following diagram is commutative:

$$\begin{array}{ccc} A_0\Big(M(F)\backslash M(\mathbb{A})\Big)_\pi & \stackrel{\mathrm{ad}(w)}{\to} & A_0\Big(M'(F)\backslash M'(\mathbb{A})\Big)_{\pi'} \\ A(M,\pi) \downarrow & & \downarrow A(M',\pi') \\ A_0\Big(M(F)\backslash M(\mathbb{A})\Big)_\pi & \stackrel{\mathrm{ad}(w)}{\to} & A_0\Big(M'(F)\backslash M'(\mathbb{A})\Big)_{\pi'}. \end{array}$$

Denote by $\mathcal{H}_{\mathfrak{X}}^R$ the collection of all Langlands operators. Clearly, under composition, $\mathcal{H}_{\mathfrak{X}}^R$ becomes a $\mathbb{C}$-algebra, and if $A \in \mathcal{H}_{\mathfrak{X}}^R$, A is determined by its value at a single $(M,\pi) \in \widetilde{\mathfrak{X}}^R$.

Ex. A special type of Langlands operators is of particular interests. They are Langlands operators such that for each fixed $(M,\pi) \in \widetilde{\mathfrak{X}}^R$, the endmorphism $A(M,\pi)$ on the space $A_0\Big(M(F)\backslash M(\mathbb{A})\Big)_\pi$ is induced by a scalar, say $f(M,\pi)$, multiplication. We will denote the collection of all such Langlands operators by $\mathcal{H}_{\mathfrak{X};\mathbb{C}}^R$, which is clearly a subalgebra of $\mathcal{H}_{\mathfrak{X}}^R$. For example, fixed a π_0, and write $\pi = \pi_0 \otimes \lambda_\pi$. Set, in particular, the scalar $f(M,\pi)$ above to be $\langle\lambda_\pi, \lambda_\pi\rangle$. We will denote the associated Langlands operator by A_0.

Langlands operators A naturally act on Paley-Wiener functions $\Phi \in P_{(M,\mathfrak{P})}^R$ via

$$\Phi \mapsto A\Phi \left(\pi \mapsto A(M,\pi)\Phi(\pi)\right).$$

This then also induces a natural action of Langlands operator on pseudo-Eisenstein series via

$$\Delta(A) : \theta_\Phi \mapsto \Big(\Delta(A)\Big)(\theta_\Phi) := \theta_{A\Phi}.$$

For later use set $\Delta := \Delta(A_0)$. Δ, which is essentially the standard Laplace operator in question, can be extended to the entire L^2-space over which it is self-adjoint. As such, the so-called spectral decomposition of the L^2-space is indeed a decomposition with respect to this Δ.

9.3 Key Bridge

If A is a Langlands operator, set A^* to be the Langlands operator defined by

$$A^*(M, -\bar{\pi}) := A(M, \pi).$$

Set also $\mathcal{H}_\xi^R = \oplus_{\mathfrak{X}} \mathcal{H}_{\mathfrak{X}}^R$.

The reason why Langlands operator plays a central role in the decomposition of space of automorphic forms is the following

Bridge Lemma. *Let $\phi \in A\Big(G(F)\backslash G(\mathbb{A})\Big)_\xi$ and $R \gg_\phi 0$. Then for any $A \in \mathcal{H}_\xi^R$, there exists a unique element $\Delta(A)\phi \in A\Big(G(F)\backslash G(\mathbb{A})\Big)_\xi$ such that for any pseudo-Eisenstein series θ_Φ, we have*

$$\Big\langle \Delta(A^*)\theta_\Phi, \phi \Big\rangle = \Big\langle \theta_\Phi, \Delta(A)\phi \Big\rangle.$$

(The notation $\Delta(A)\phi$ is copied from [MW] and reminds us of Chinese characters.)

Proof. Two main points:

Sublemma 1. *For a fixed $g \in G(\mathbb{A})$, and write $m_P(g)^\bullet$ for the function on X_M^G whose value at λ is given by $m_P(g)^\lambda$. Then for any polynomial Q in* $\mathrm{Re}\,\mathfrak{a}_M^G$,

$$\Big(\partial_Q m_P(g)^\bullet\Big)(\lambda) = Q\Big(\log_M^G(m_P(g)\Big) \cdot m_P(g)^\lambda.$$

Sublemma 2. *For $\Phi \in P_{(M,\mathfrak{P})}^R$ and $\varphi \in A_0\Big(U(\mathbb{A})M(F)\backslash G(\mathbb{A})\Big)_\pi$, set*

$$I\Big(\phi, \varphi; \lambda\Big) := \int_{Z_{G(\mathbb{A})}U(\mathbb{A})M(F)\backslash G(\mathbb{A})} \overline{\Big(F_\pi(\Phi)(e)\Big)(g)} \cdot m_P(g)^\lambda \varphi(g)\, dg.$$

Then

$$I\Big(\phi,\varphi;\lambda\Big)=\begin{cases}0, & \text{if } -\overline{\pi\otimes\lambda}\notin\mathfrak{P};\\ C\cdot\Big\langle\Phi\big(-\overline{\pi\otimes\lambda},\lambda\circ\varphi\Big\rangle, & \text{if } -\overline{\pi\otimes\lambda}\in\mathfrak{P}.\end{cases}$$

where $C:=\mathrm{Vol}\Big(Z_{G(\mathbb{A})}\big(Z_M(F)\cap Z_{M(\mathbb{A})}\big)\backslash Z_{G(\mathbb{A})}\big(Z_{M(\mathbb{A})}\cap M(\mathbb{A})^1\big)\Big).$

The reader should realize that $\mathcal{H}_{\mathfrak{X}}^R$ is a very powerful tool in studying automorphic forms. In Day 4, we will use it to get 'extra' flexibility which then helps us to obtain a spectral decomposition for L^2-spaces. We too expect that *an alternative proof of meromorphic continuation and functional equations for general Eisenstein series can be established by pushing the study here on the pairing between pseudo-Eisenstein series and automorphic forms.*

Day Four: Spectrum Decomposition: Residual Process

10 Why Residue?

10.1 Pseudo-Eisenstein Series and Residual Process

For $\Phi \in P^{R,\mathfrak{F}}_{(M,\mathfrak{P})}$, its associated pseudo-Eisenstein series $\theta_\Phi(g)$ is defined by

$$\theta_\Phi(g) := \sum_{\delta\in P(F)\backslash G(F)} \int_{\pi\in\mathfrak{P},\mathrm{Re}\pi=\lambda} \Phi_\pi(g)\,d\pi$$

for any $\lambda \in \rho_P + \mathcal{C}_M^+$. As such, the series converges absolutely and even uniformly for g in a compact subset of $G(\mathbb{A})$. Consequently,

$$\begin{aligned}\theta_\Phi(g) &= \int_{\pi\in\mathfrak{P},\mathrm{Re}\pi=\lambda} \sum_{\delta\in P(F)\backslash G(F)} \Phi_\pi(g)\,d\pi\\ &= \int_{\pi\in\mathfrak{P},\mathrm{Re}\pi=\lambda} E\Big(\Phi_\pi,\pi;g\Big)\,d\pi,\end{aligned}$$

which belongs to $L^2\Big(G(F)\backslash G(\mathbb{A})\Big)$.

Moreover, if $\Phi \in P^{\mathfrak{F}}_{(M,B)}$, then $E\Big(\Phi_\pi,\pi;g\Big)$ is holomorphic on $\Big\{\pi \in \mathfrak{P} : \mathrm{Re}\pi = 0\Big\}$. So it too makes sense to talk about the integration $\int_{\pi\in\mathfrak{P},\mathrm{Re}\pi=0} E\Big(\Phi_\pi,\pi;g\Big)\,d\pi$, which can be easily seen to be L^2 as well. As they stand, the difference we should notice here is that this latest integration is canonical since now for the integral domain, we replace an arbitrary $\lambda_0 \in \rho_P + \mathcal{C}_M^+$ above by 0, the zero element of $\Big(\mathfrak{a}_M^G\Big)^*$.

It is quite natural then to use these latest integrations to generate the L^2-space. On the other hand, by shifting the contour from $\int_{\pi\in\mathfrak{P},\mathrm{Re}\pi=\lambda}$ to $\int_{\pi\in\mathfrak{P},\mathrm{Re}\pi=0}$, singularities of Eisenstein series contribute additional data via the residues, as one may imagine with the experience from Cauchy's residue theorem. This is exactly where the residue theory naturally gets into the picture.

Clearly, a far-reaching generalization of the classical residue theorem is needed in order to cover the generality encountered here. As one may imagine, while complicated in stating it, the main idea is rather simple. In literature, this is achieved via the so-called residue theorem. We are not going to give all the details here, instead let us concentrate our attention to the outcome.

Assume then $\mathfrak{H}$ is a singular root hyperplane of $E\Big(\Phi(\pi),\pi\Big)$ which sits between $\Big\{\pi\in\mathfrak{P}\,\Big|\,\mathrm{Re}\pi=\lambda\Big\}$ to $\Big\{\pi\in\mathfrak{P}\,\Big|\,\mathrm{Re}\pi=0\Big\}$. Then by shifting the integration from $\int_{\pi\in\mathfrak{P},\mathrm{Re}\pi=\lambda}$ to $\int_{\pi\in\mathfrak{P},\mathrm{Re}\pi=0}$ along a 'nice' path, joining 0 and λ, we need to add up the contribution from the residue of E along $\mathfrak{H}$. Thus, if $z(\mathfrak{H})$ is the intersection point of $\mathfrak{H}$ and the nice path above, then the contribution is supposed to be given by

$$\int_{\pi\in\mathfrak{H},\mathrm{Re}\pi=z(\mathfrak{H})}\mathrm{Res}_{\mathfrak{H}}^{\mathfrak{P}}E(\Phi_\pi,\pi,g)\,d\pi.$$

Similarly, as above, this time at $\mathfrak{H}$-level (instead of the $\mathfrak{P}$-level at the very beginning,) the integration is again not canonical. To make it canonical, we need to shift the contour from $\int_{\pi\in\mathfrak{H},\mathrm{Re}\pi=z(\mathfrak{H})}$ to $\int_{\pi\in\mathfrak{H},\mathrm{Re}\pi=o(\mathfrak{H})}$, where $o(\mathfrak{H})$ represents the so-called 'origin' of $\mathfrak{H}$, provided that $\mathrm{Res}_{\mathfrak{H}}^{\mathfrak{P}}E(\Phi_\pi,\pi,g)$ is holomorphic at every point of $\Big\{\pi\in\mathfrak{H}\,\Big|\,\mathrm{Re}\pi=o(\mathfrak{H})\Big\}$. (See the precise definition below.) Accordingly, the residue of $\mathrm{Res}_{\mathfrak{H}}^{\mathfrak{P}}E(\Phi_\pi,\pi,g)$ along singular hyperplanes, which is of codimension 2 in $\mathfrak{P}$, of $\mathfrak{H}$ have to be added.

With this informal discussion, we are now ready to follow Arthur [Ar4] to give the general picture as follows (but we reminder the reader here that not all singularities are of interests to us – we only care about these 'sitting' between λ and 0):

(**a**) There are a pair of standard parabolic subgroups $P'\subset P$ and a filtration of affine spaces

$$\mathfrak{t}_0\subset\mathfrak{t}_1\subset\cdots\subset\mathfrak{t}_r=\mathrm{Re}\,\mathfrak{a}_{P'}^*$$

with

$$\mathfrak{t}_i=\Lambda_i+\widetilde{\mathfrak{t}}_i$$

where $\widetilde{\mathfrak{t}}_i$ is a linear subspace of $\mathrm{Re}\,\mathfrak{a}_{P'}^*$ and Λ_i is a vector in $\mathrm{Re}\,\mathfrak{a}_{P'}^*$ which is orthogonal to $\widetilde{t}_i$. Assume that $\widetilde{t}_0 = \mathfrak{a}_P^*$ and that for any i,

$$\widetilde{t}_{i-1} = \left\{\lambda \in \widetilde{t}_i : \lambda(\beta^\vee) = 0\right\}$$

for some root β in $R(T_{P'}, P')$ such that $\beta^\vee$ does not vanish on $\widetilde{t}_i$. Suppose that for each i we have also chosen a unit vector ν_i in $\widetilde{t}_i$ which is orthogonal to $\widetilde{t}_{i-1}$. It is uniquely determined up to a sign. We will denote this sequence of affine subspaces together with the choices of unit normals by the letter S and denote the smallest space $\mathfrak{t}_0$ by

$$\mathfrak{t}_S :=: \Lambda_S + \mathrm{Re}\,\mathfrak{a}_P^*.$$

(**b**) Suppose now that φ is a meromorphic function on $\mathfrak{a}_{P'}^*$ whose singularities lie along hyperplanes. Then we can obtain a meromorphic function on $\mathfrak{t}_{S,\mathbb{C}} = \Lambda_S + \mathfrak{a}_P^*$ by taking successive residues with respect to S. More precisely, let Λ_0 be a point in $\mathfrak{t}_{S,\mathbb{C}}$ with the properties that any singular hyperplane of φ which contains Λ_0 also contains $\mathfrak{t}_{S,\mathbb{C}}$. Set

$$\Lambda_0(u) := \Lambda_0 + z_1\nu_1 + \cdots + z_r\nu_r$$

for $\mathbf{z} = (z_1, \ldots, z_r) \in \mathbb{C}^r$. Let $\Gamma_1, \ldots, \Gamma_r$ be small positively oriented circles about the origin in the complex plane $\mathbb{C}$ such that for each i, the radius of Γ_{i+1} is much smaller than that of Γ_i. Then

$$\frac{1}{\left(2\pi\sqrt{-1}\right)^r}\int_{\Gamma_1}\cdots\int_{\Gamma_r}\varphi\Big(\Lambda_0(\mathbf{z})\Big)\,dz_r\,\cdots\,dz_1$$

is a meromorphic function of Λ_0. We will denote it by $\mathrm{Res}_{S,\Lambda\to\Lambda_0}\Phi(\Lambda)$ or simply $\mathrm{Res}_{\mathfrak{S}}^{P/B}$ if $\mathfrak{S}$ is the subspace in $\mathfrak{P}$ corresponding to $\mathfrak{t}_S$. In particular, if $P' = P_0$ and $P = G$, we write it as $\mathrm{Res}_{\mathfrak{S}}^G$ or even simply $\mathrm{Res}_{\mathfrak{S}}$.

10.2 What do we have?

(a) **Singularities** The above discussion on residue is clearly based on the fact that meromorphically extended Eisenstein series (associated to cusp forms) admit singularities only along root hyperplanes. Accordingly, let $S_{(M,\mathfrak{P})} :=$ be the collection of all hyperplanes which are singular for at least one such Eisenstein series (associated to M-level cusp forms of type $\mathfrak{P}$), their intersections and the Weyl-conjugations of these intersections. $S_{(M,\mathfrak{P})}$ is not *a priori* locally finite, so we need to introduce $S_{(M,\mathfrak{P})}^{\mathfrak{F}}$ to get the locally finiteness by assuming that cusp forms are of fixed finite K-type $\mathfrak{F}$. (Here, locally finiteness means that for any compact subset C, there are only finitely many hyperplanes which intersect C.) For later use, for $\mathfrak{S} \in S_{(M,\mathfrak{P})}$, set

$$\mathrm{Norm}\,\mathfrak{S} := \Big\{ w \in \mathrm{Stab}(M,\mathfrak{P}) | w\mathfrak{S} = \mathfrak{S} \Big\}.$$

(b) **Pairing between Pseudo-ESes**

Instead of using directly the pseudo-Eisenstein series θ_Φ, one finds that it is more effective to take the pairings $\langle \theta_{\Phi'}, \theta_\Phi \rangle$. As we have already seen, by working with such pairings, no essentially information is lost, yet we get a maximal freedom in manipulating the situation at hands: after all, there is a huge Hecke algebra $\mathfrak{H}_{\mathfrak{X}}^R$ there ready to be used (as we experienced in the decomposition for space of automorphic forms in Day 3).

As such, we, by introducing such a new flexibility, may then only need to restrict ourselves to a certain 'natural but easy' subclass of Paley-Wiener functions Φ's with the hope that the general and wild Φ's can be studied using approximation.

The choice of these special Paley-Wiener is surprisingly easy: We will assume that Φ are such that they vanish very high order along singular hyperplanes of Eisenstein series.

To end this discussion, let us recall the following starting point:
Main Theorem A of Day 3. We have

$$\Big\langle \theta_{\Phi'}, \theta_\Phi \Big\rangle = \int_{\pi \in \mathfrak{P}, \mathrm{Re}\,\pi = \lambda_0} A(\Phi', \Phi)(\pi)\, d\pi$$

where $A(\Phi', \Phi)(\pi)$

$$:= \Big\langle \sum_{w \in W(M)} M(w^{-1}, -w\bar{\pi}) \Phi'(-w\bar{\pi}), \Phi(\pi) \Big\rangle$$

$$:= \sum_{(M',\mathfrak{P}') \in \mathfrak{X}} \sum_{w \in W((M,\mathfrak{P}),(M',\mathfrak{P}'))} \Big\langle M(w^{-1}, -w\bar{\pi}) \Phi_{(M',\mathfrak{P}')}(-w\bar{\pi}), \Phi(\pi) \Big\rangle.$$

10.3 Difficulties

(a) **Compactness**. Note that integration domains $\{\pi \in \mathfrak{P} | \mathrm{Re}\pi = \lambda\} \simeq X_M^G$ and so are elements in $S_{(M,\mathfrak{P})}$ are in general unbounded spaces. However our integrands appears only conditionally convergent (since we are talking not just ESes but also their residues). It is (to solve this problem that we'd) better to consider compact subsets. Thus, for any $\mathfrak{S} \in S_{(M,\mathfrak{P})}^{\mathfrak{F}}$ and $T > 0$, set

$$\mathfrak{S}_{\leq T} := \Big\{ \pi \in \mathfrak{S} \,\big|\, \|\mathrm{Im}\,\lambda_\pi\|^2 \leq T + \|\mathrm{Re}\,\lambda_\pi\|^2 \Big\}$$

where $\pi = \pi_0 \otimes \lambda_\pi$. Clearly this is a compact subset at least for a fixed real part of π.

(b) **Singularities**. For certain constructions, singularities cause serious problems. To avoid this, in our initial discussion, we will in fact only allow ourselves work over a nice subset of $\mathfrak{S}_{\leq T}$, i.e.,

$$\mathfrak{S}_{\leq T}^{o} := \mathfrak{S}_{\leq T} - \cup_{\mathfrak{S}' \in S_{(M,\mathfrak{P})}^{\mathfrak{F}}, \mathfrak{S}' \subsetneq \mathfrak{S}} \mathfrak{S}'.$$

That is to say, by first removing any possible singularities which are 'trouble makers', while wishing that somehow at a certain later stage, approximation argument with allow us to deal general cases as well.

11 Main Results

11.1 Functional Analysis

For ξ a unitary central character and consider the Hilbert space $H := L^2\Big(G(F)\backslash G(\mathbb{A})\Big)_\xi$.

Identify $\mathfrak{P}$ with X_M^G and define an operator A_0 on Paley-Wiener functions by

$$\Phi \mapsto A_0\Phi : \pi \mapsto \langle \lambda_\pi, \lambda_\pi \rangle \Phi_\pi.$$

A_0 then induces a natural operator Δ on pseudo-Eisenstein series

$$\Delta(\theta_\Phi) = \theta_{A_0(\Phi)}.$$

Δ is self-adjoint and can be extended to $L^2\Big(G(F)\backslash G(\mathbb{A})\Big)_\xi$. (Essentially, it is the standard Laplace operator.) As such, if $\mathcal{Q}$ denotes the set of $z \in \mathbb{C}$ such that $\Delta - z \cdot \mathrm{Id}$ is bijective and admits a bounded inverse, then $\mathcal{Q}$ is open and contains $\mathbb{C} - \mathbb{R}$. The set $\mathbb{C} - \mathcal{Q}$ is called the *spectrum* of Δ and $\Big(\Delta - z \cdot \mathrm{Id}\Big)^{-1}$ for $z \in \mathcal{Q}$ is called the *resolvent* of Δ.

It is well known that for every $x \in \mathbb{R}$, there exist a closed subspace $H_x \subset H$ together with an orthogonal projection $p_x : H \to H_x$ such that
(i) $H_x \subset H_{x'}$ *if* $x \leq x'$;
(ii) $H_x = \cap_{x':x'>x} H_{x'}$;
(iii) $\cap_x H_x = \{0\}, \cup_x H_x$ *is dense in* H; and
(iv) *There exists a real number* $R > 0$ *such that the spectrum of* Δ *is contained in* $(-\infty, R]$;

Set also $H_{x^-} := \overline{\cup_{x':x'<x} H_{x'}}$ and $p_{x^-} : H \to H_{x^-}$ the associated orthogonal projection. Then $x \in \mathbb{R}$ is called a point of *continuity* of (H_x) or simply of Δ if $H_x = H_{x^-}$.
(v) *For* $T \gg 0$ *with* $-T$ *a point of continuity of* (H_x),

$$\langle h', q_T h\rangle = \lim_{\varepsilon \to 0} -\frac{1}{2\pi\sqrt{-1}} \int_{\Gamma(-T,R';\varepsilon)} \langle h', (\Delta - z\mathrm{Id})^{-1} h\rangle \, dz,$$

where $q_T := \mathrm{Id} - p_{-T}$ *and* $\Gamma(-T, R'; \varepsilon)$ *is the contour consisting of sides* $\{(x,y) \in \mathbb{C} : x = -T, \varepsilon \leq |y| \leq c\} \cup \{(x,y) \in \mathbb{C} : x = R', \varepsilon \leq |y| \leq c\} \cup \{(x,y) \in \mathbb{C} : -T \leq x \leq R, |y| \pm c\}$ *with counter-clockwise orientation.* (Do draw a picture for $\Gamma(-T, R'; \varepsilon)$.)

11.2 Main Theorem: Rough Version

Two elements $\mathfrak{S}$ and $\mathfrak{S}'$ are called *equivalent* if there exists $w \in W$ such that $w\mathfrak{S} = \mathfrak{S}'$ and $w(M, \mathfrak{P}) = (M', \mathfrak{P}')$ where $(M, \mathfrak{P})$ and $(M', \mathfrak{P}')$ are associated cuspidal data for $\mathfrak{S}$ and $\mathfrak{S}'$ respectively. Denote by

$$S_{\mathfrak{X}}^{\mathfrak{F}} = \cup_{(M,\mathfrak{P}):(M,\mathfrak{P})\in\mathfrak{X}} S_{(M,\mathfrak{P})}^{\mathfrak{F}}$$

and let $[S_{\mathfrak{X}}^{\mathfrak{F}}] := S_{\mathfrak{X}}^{\mathfrak{F}} / \sim$ be the collection of equivalence classes.

Fix $T, R \gg 0$ and $T' \gg_{T,R} 0$ be such that $-T$ is a point of continuity of Δ.

Main Theorem. *For any $\mathfrak{C} \in [S_{\mathfrak{X}}^{\mathfrak{F}}]$ there exists an orthogonal projection $\mathrm{Proj}_{\mathfrak{C},T}^{\mathfrak{F}}$ onto a closed subspace of*

$$L^2\Big(G(F)\backslash G(\mathbb{A})\Big)_{\mathfrak{X},\mathfrak{F}} := \overline{\langle \theta_\Phi | \Phi \in P_{\mathfrak{X}}^{\mathfrak{F}} \rangle}$$

satisfying the follows:

(0) $\mathrm{Proj}_{\mathfrak{C},T}^{\mathfrak{F}} = 0$ *if for all* $\mathfrak{S} \in \mathfrak{C}$, $\Big\{\pi \in \mathfrak{S}_{\leq T} \,\Big|\, \mathrm{Re}\pi = o(\mathfrak{S})\Big\} = \emptyset$;

(1) $\mathrm{Proj}_{\mathfrak{C},T}^{\mathfrak{F}} \perp \mathrm{Proj}_{\mathfrak{C}',T}^{\mathfrak{F}}$ *if* $\mathfrak{C} \neq \mathfrak{C}' \in [S_{\mathfrak{X}}^{\mathfrak{F}}]$;

(2) $q_T \theta_\Phi = \sum_{\mathfrak{C} \in [S_{\mathfrak{X}}^{\mathfrak{F}}]} \mathrm{Proj}_{\mathfrak{C},T}^{\mathfrak{F}} \theta_\Phi$ *for all* $\Phi \in P_{\mathfrak{X}}^{\mathfrak{F}}$.

Before we give a refined version of this theorem in which we describe the projection $\mathrm{Proj}_{\mathfrak{C},T}^{\mathfrak{F}}$ in details, let us give an application of this result. For this, set then

$$L_{\mathfrak{C},T}^{2,\mathfrak{F}} := \mathrm{Proj}_{\mathfrak{C},T}^{\mathfrak{F}} L^2\Big(G(F)\backslash G(\mathbb{A})\Big)_{\mathfrak{X},\mathfrak{F}}.$$

Lemma. $L_{\mathfrak{C},T_1}^{2,\mathfrak{F}} \subset L_{\mathfrak{C},T_2}^{2,\mathfrak{F}}$ *if* $T_1 < T_2$.

Consequently, let

$$L^2\Big(G(F)\backslash G(\mathbb{A})\Big)_{\mathfrak{C},\mathfrak{F}} := \overline{\cup_{T:T\in\mathbb{R}_{>0}} L_{\mathfrak{C},T}^{2,\mathfrak{F}}}.$$

Spectrum Decomposition. (**Rough Version**) *There exists an orthogonal decomposition*

$$L^2\Big(G(F)\backslash G(\mathbb{A})\Big)_{\mathfrak{X},\mathfrak{F}} = \hat{\oplus}_{\mathfrak{C}\in[S_{\mathfrak{X}}^{\mathfrak{F}}]} L^2\Big(G(F)\backslash G(\mathbb{A})\Big)_{\mathfrak{C},\mathfrak{F}}.$$

11.3 Main Theorem: Refined Version

The projection $\mathrm{Proj}_{\mathfrak{C},T}^{\mathfrak{F}}$ and hence the spaces $L_{\mathfrak{C},T}^{2,\mathfrak{F}}$ in the decomposition theorem may be understood in great detail. There are two stages of this. Here is the first one using residues. (The second, using Levi subgroups, will be one of the main topics of Day 5.)

Recall that here $A(\Phi',\Phi)(\pi)$

$$:=\Big\langle \sum_{w\in W(M)} M(w^{-1},-w\bar\pi)\Phi'(-w\bar\pi),\Phi(\pi)\Big\rangle:$$

$$=\sum_{(M',\mathfrak{P}')\in\mathfrak{X}}\ \sum_{w\in W((M,\mathfrak{P}),(M',\mathfrak{P}'))}\Big\langle M(w^{-1},-w\bar\pi)\Phi_{(M',\mathfrak{P}')}(-w\bar\pi),\Phi(\pi)\Big\rangle.$$

Theorem A. *For $\mathfrak{C}\in[S_{\mathfrak{X}}^{\mathfrak{F}}]$, $\Phi',\Phi\in P_{\mathfrak{X}}^{\mathfrak{F}}$, and $\mathfrak{S}_{\mathfrak{C}}\in\mathfrak{C}$,*
(i) *The function*

$$r_{\mathfrak{C}}(\Phi',\Phi)=\frac{1}{|\mathrm{Norm}\,\mathfrak{S}_{\mathfrak{C}}|}\sum_{w\in W(M)}\Big(\mathrm{Res}^G_{w\mathfrak{S}_{\mathfrak{C}}}A(\Phi',\Phi)\Big)(w\pi)$$

is holomorphic at every point of $\Big\{\pi\in\mathfrak{S}_{\mathfrak{C},\leq T}\,\Big|\,\mathrm{Re}\pi=o(\mathfrak{S}_{\mathfrak{C}})\Big\}$;
(ii) $\Big\langle\theta_{\Phi'},\mathrm{Proj}^{\mathfrak{F}}_{\mathfrak{C},T}\theta_\Phi\Big\rangle$

$$=m_{\mathfrak{C}}(\Phi',\Phi):=\int_{\pi\in\mathfrak{S}_{\mathfrak{C},\leq T},\mathrm{Re}\pi=o(\mathfrak{S}_{\mathfrak{C}})}r_{\mathfrak{C}}(\Phi',\Phi)(\pi)\,d_{\mathfrak{S}_{\mathfrak{C}}}\pi.$$

Back to pseudo-Eisenstein series and hence Eisenstein series, set

$$e_{\mathfrak{C}}(\Phi,\pi):=\frac{1}{|\mathrm{Norm}\,\mathfrak{S}_{\mathfrak{C}}|}\sum_{w\in W(M)}\mathrm{Res}^G_{w\mathfrak{S}_{\mathfrak{C}}}E(\Phi,w\pi).$$

It is a meromorphic function with polynomial singularities over $S_{\mathfrak{X}}^{\mathfrak{F}}$.

Theorem B. *For $\mathfrak{C}\in[S_{\mathfrak{X}}^{\mathfrak{F}}]$, $\Phi\in P_{\mathfrak{X}}^{R,\mathfrak{F}}$, and $\mathfrak{S}_{\mathfrak{C}}\in\mathfrak{C}$,*
(i) *The function $e_{\mathfrak{C}}(\Phi,\pi)$ is holomorphic at every point of* $\Big\{\pi\in\mathfrak{S}_{\mathfrak{C},\leq T}\,\Big|\,\mathrm{Re}\pi=o(\mathfrak{S}_{\mathfrak{C}})\Big\}$;
(ii) $\mathrm{Proj}^{\mathfrak{F}}_{\mathfrak{C},T}\theta_\Phi=\displaystyle\int_{\pi\in\mathfrak{S}_{\mathfrak{C},\leq T},\mathrm{Re}\pi=o(\mathfrak{S}_{\mathfrak{C}})}e_{\mathfrak{C}}(\Phi,\pi)\,d_{\mathfrak{S}_{\mathfrak{C}}}\pi.$

With this, to see the structure

$$L^{2,\mathfrak{F}}_{\mathfrak{C},T}:=\mathrm{Proj}^{\mathfrak{F}}_{\mathfrak{C},T}L^2\Big(G(F)\backslash G(\mathbb{A})\Big)_{\mathfrak{X},\mathfrak{F}}$$

more closely, for every $\pi \in \left\{\pi \in \mathfrak{S}_{\mathfrak{C},\leq T} \,\middle|\, \mathrm{Re}\pi = o(\mathfrak{S}_{\mathfrak{C}})\right\}$, introduce the associated $A^{\mathfrak{F}}_{\mathfrak{C},\pi}$:= the space of automorphic forms on $G(F)\backslash G(\mathbb{A})$ generated by (holomorphic) $e_{\mathfrak{C}}(\Phi,\pi)$ above.

Theorem C. (0) $A^{\mathfrak{F}}_{\mathfrak{C},\pi} \simeq A^{\mathfrak{F}}_{\mathfrak{C},w\pi}$ *for all* $w \in \mathrm{Norm}\,\mathfrak{S}_{\mathfrak{C}}$;
(i) *For all* $\pi \in \mathfrak{S}_{\mathfrak{C},\leq T}$ *with* $\mathrm{Re}\pi = o(\mathfrak{S}_{\mathfrak{C}})$, $A^{\mathfrak{F}}_{\mathfrak{C},\pi}$ *is equipped with a positive definite scalar product* $\langle\cdot,\cdot\rangle$ *such that for all* $\Phi',\Phi \in P^{\mathfrak{F}}_{\mathfrak{X}}$,

$$\Big\langle e_{\mathfrak{C}}(\Phi',\pi), e_{\mathfrak{C}}(\Phi,\pi)\Big\rangle = r_{\mathfrak{C}}(\Phi',\Phi)(\pi);$$

(ii) *There is a natural isometry*

$$L^{2,\mathfrak{F}}_{\mathfrak{C},T} \simeq \mathrm{Hilb}^{\mathfrak{F}}_{\mathfrak{C},T}$$

where $\mathrm{Hilb}^{\mathfrak{F}}_{\mathfrak{C},T}$:= *the Hilbert space consisting of all measurable functions* F *on* $\left\{\pi \in \mathfrak{S}_{\mathfrak{C},\leq T}, \mathrm{Re}\pi = o(\mathfrak{S}_{\mathfrak{C}})\right\}$ *such that*

- $F(\pi) \in A^{\mathfrak{F}}_{\mathfrak{C},\pi}$ *almost everywhere;*
- $F(w\pi) = F(\pi)$ *for all* $w \in \mathrm{Norm}\,\mathfrak{S}_{\mathfrak{C}}$; and
- $\displaystyle\int_{\pi\in\mathfrak{S}_{\mathfrak{C},\leq T},\mathrm{Re}\pi=o(\mathfrak{S}_{\mathfrak{C}})} \|F'\pi)\|^2 d_{\mathfrak{S}_{\mathfrak{C}}}\pi < \infty.$

In particular, let $T \to \infty$, we get the spectrum decomposition for $L^{2,\mathfrak{F}}_{\mathfrak{C}}$ stated above.

11.4 How to Prove?

It is based on an induction on the semi-simple rank of G. As stated above, we first deal with a narrow class of Paley-Wiener functions Φ, i.e., those Φ which vanish to sufficiently high order along singular hyperplanes; then work over general Φ via an approximation argument. Indeed, if we let $\left(\mathrm{Hilb}^{\mathfrak{F}}_{\mathfrak{C},T}\right)^o$:= be similarly defined as that for $\mathrm{Hilb}^{\mathfrak{F}}_{\mathfrak{C},T}$ above but with the additional condition that F vanishes to a sufficiently high order along singular hyperplanes, then we have the following

Key Approximation Lemma. $\overline{\left(\mathrm{Hilb}_{\mathfrak{C},T}^{\mathfrak{F}}\right)^o} = \mathrm{Hilb}_{\mathfrak{C},T}^{\mathfrak{F}}$. *That is to say,* $\left(\mathrm{Hilb}_{\mathfrak{C},T}^{\mathfrak{F}}\right)^o$ *is dense in* $\mathrm{Hilb}_{\mathfrak{C},T}^{\mathfrak{F}}$.

The reader can see at least that with this vanishing condition to high order along singular hyperplanes, the statement in (i) of Theorems A and B on the holomorphicity then becomes trivial. As for the rest of statements, we need the following

Main Theorem′. *There exists* $z(\mathfrak{S}) \in \mathrm{Re}(\mathfrak{S})$ *and residue datum* $\mathrm{Res}_{\mathfrak{S}}^G$ *such that for all* $\Phi' \in P_{\mathfrak{X}}^{\mathfrak{F}}, \Phi \in P_{(M,\mathfrak{P})}^{\mathfrak{F}}$,

$$\langle \theta_{\Phi'}, q_T\theta_\Phi\rangle = \sum_{\mathfrak{S}\in S_{(M,\mathfrak{P})}^{\mathfrak{F}}} \frac{1}{|\mathrm{Norm}\mathfrak{S}|} \times \int_{\pi\in\mathfrak{S}_{\leq T},\mathrm{Re}\pi=z(\mathfrak{S})} \left(\sum_{w\in\mathrm{Norm}\mathfrak{S}} \mathrm{Res}_{\mathfrak{S}}^G A(\Phi',\Phi)(w\pi)\right) d_{\mathfrak{S}}\pi.$$

This theorem is indeed the heart of all theorems stated above. The start point is of course the following inner product formula for pseudo-Eisenstein series

$$\left\langle \theta_{\Phi'}, \theta_\Phi\right\rangle = \int_{\pi\in\mathfrak{P},\mathrm{Re}\pi=\lambda_0} A(\Phi',\Phi)(\pi)\, d\pi$$

where $\lambda_0 \in \rho_P + \mathcal{C}_M^+$. From here, to shift the integral domain to that with $\mathrm{Re}\pi = o(\mathfrak{S})$, we need to apply the residue theorem. With all this done, then we use functional analysis to have the formula

$$\left\langle \theta_{\Phi'}, q_T\theta_\Phi\right\rangle = \lim_{\varepsilon\to 0} -\frac{1}{2\pi\sqrt{-1}} \int_{\Gamma(-T,R';\varepsilon)} \langle\theta_{\Phi'}, \theta_{\Phi_z}\rangle\, dz$$

where

$$\Phi_z(\pi) := \frac{1}{z-\langle\lambda,\lambda\rangle}\cdot\Phi(\pi).$$

Since

$$\Big(z\mathrm{Id}-\Delta\Big)\theta_{\Phi_z} = \theta_\Phi,$$

we are then in a position to apply the following

Key Technical Lemma. *Let $T' \gg_{T,R} 0$ with $T \gg 0$ such that $-T$ is a point of continuity of Δ. Suppose that for all $\mathfrak{S} \in S^{\mathfrak{F}}_{(M,\mathfrak{P})}$,*

(1) $\sum_{w\in\mathrm{Stab}(M,\mathfrak{P})} \Big(\mathrm{Res}^G_{w\mathfrak{S}} A(\Phi',\Phi)\Big)(w\pi)$ *is holomorphic at every point of the imaginary space* $\Big\{\pi \in \mathfrak{S}_{\leq T} \,|\, \mathrm{Re}\pi = o(\mathfrak{S})\Big\}$; and

(2) *For all* $F \in H^R_{\mathfrak{X},\mathbb{C}}$, $\mathrm{Res}^G_{\mathfrak{S}} A(\Phi', F\Phi) = F|_{\mathfrak{S}} \cdot \mathrm{Res}^G_{\mathfrak{S}} A(\Phi',\Phi)$.

Then
$$\Big\langle \theta_{\Phi'}, q_T\theta_\Phi \Big\rangle = \sum_{\mathfrak{S}\in S^{\mathfrak{F}}_{(M,\mathfrak{P})}} \frac{1}{|\mathrm{Norm}\mathfrak{S}|} \times \int_{\pi\in\mathfrak{S}_{\leq T},\mathrm{Re}\pi=o(\mathfrak{S})} \Bigg(\sum_{w\in\mathrm{Norm}\mathfrak{S}} \mathrm{Res}^G_{\mathfrak{S}} A(\Phi',\Phi)(w\pi)\Bigg)\, d_{\mathfrak{S}}\pi$$

where the origin $o(\mathfrak{S})$ is defined by $o(\mathfrak{S}) = \mathrm{Re}\mathfrak{S} \cap (\mathrm{Re}\mathfrak{S}^0)^\perp$ with $\mathfrak{S}^0 := \{\lambda \in X^G_M | \mathfrak{S}\otimes\lambda = \mathfrak{S}\}$.

Clearly, to verify the theorem, the point is to verify the two conditions listed in this lemma. The first is about holomorphicity while the other is about why taking residue process is compactible with the action of 'scalar' Hecke operators.

As such, to say the least, by direct applying the residue theorem, which is long and hence tedious, we have
$$\Big\langle \theta_{\Phi'}, q_T\theta_\Phi \Big\rangle =_T \sum_{\mathfrak{S}\in S^{\mathfrak{F}}_{(M,\mathfrak{P})}} \frac{1}{\mathrm{Norm}\mathfrak{S}} \times \int_{\pi\in\mathfrak{S}_{\leq T},\mathrm{Re}\pi=z(\mathfrak{S})} \Bigg(\sum_{w\in\mathrm{Norm}\mathfrak{S}} \mathrm{Res}^G_{\mathfrak{S}} A(\Phi',\Phi)(w\pi)\Bigg)\, d_{\mathfrak{S}}\pi$$

where $=_T$ means that up to certain terms outside the compact subset $\mathfrak{S}_{\leq T}$, we have the equality. But then with the use of Φ_z, we can get a holomorphicity out from this $=_T$ relation, which by the functional analysis relation togeher with condition (2) then gives a real equality for $\langle\theta_{\Phi'}, q_T\theta_\Phi\rangle$, since $\frac{1}{z-\langle\lambda_\pi,\lambda_\pi\rangle}$ may be taken out from $\mathrm{Res}A(\Phi',\Phi_z)$ to get $\mathrm{Res}A(\Phi',\Phi)$ which is what we want in the lemma and the so-called horizontal lines are well-behaved. In this sense the key is to establish the relation $\mathrm{Res}^G_{\mathfrak{S}} A(\Phi', F\Phi) = F|_{\mathfrak{S}} \cdot \mathrm{Res}^G_{\mathfrak{S}} A(\Phi',\Phi)$ for the narrow class of Paley-Wiener functions used.

Day Five: Eisenstein Systems and Spectral Decomposition (II)

12 Eisenstein Systems

12.1 Relative Theory

Let $P_L = L \cdot U_L$ be a standard parabolic subgroup of G. For any unitary character ξ_L of $Z_{L(\mathbb{A})}$, set

$$L^2\Big(U_L(\mathbb{A})L(F)\backslash G(\mathbb{A})\Big)_{\xi_L} := \Big\{f \in L^2\Big(U_L(\mathbb{A})L(F)\backslash G(\mathbb{A})\Big) \\ \,\Big|\, f(zg) = z^{\rho_P} \cdot z^{\xi_L} \cdot f(g), \forall z \in Z_{L(\mathbb{A})}, g \in G(\mathbb{A})\Big\}.$$

A parallel discussion with relative pseudo-Eisenstein series $\theta_\Phi^{L/M}$ for Paley-Wiener functions $\Phi \in P_{(M,\mathfrak{P}_L)}^{R,\mathfrak{F}}$ gives a spectral decomposition for $L^2\Big(U_L(\mathbb{A})L(F)\backslash G(\mathbb{A})\Big)_{\xi_L}$ in terms of residues

$$\begin{aligned}\theta_\Phi^{L/M} &= \lim_{T\to\infty} \sum_{\mathfrak{C}_L\in[S_{\mathfrak{X}_L}^{\mathfrak{F}}]} \mathrm{Proj}_{\mathfrak{C}_L,T}^{L/M,\mathfrak{F}}\Big(\theta_\Phi^{L/M}\Big) \\ &= \lim_{T\to\infty} \sum_{\mathfrak{C}_L\in[S_{\mathfrak{X}_L}^{\mathfrak{F}}]} \int_{\pi\in\mathfrak{S}_{\mathfrak{C}_L;\leq T}, \mathrm{Re}\,\pi=o(\mathfrak{S}_{\mathfrak{C}_L})} \frac{1}{|\mathrm{Norm}_{W_L}\mathfrak{S}_{\mathfrak{C}_L}|} \\ &\qquad \times \sum_{w\in W_L(M)} \mathrm{Res}_{w\mathfrak{S}_{\mathfrak{C}_L}}^{L} E^{L/M}(\Phi, w\pi)\, d_{\mathfrak{S}_{\mathfrak{C}_L}}\pi\end{aligned}$$

with the parallel notation index by L (instead of often omitted G),
Ex. For every $(M,\mathfrak{P}) \in \mathfrak{X}$ with central character $\xi : Z_{G(\mathbb{A})} \to \mathbb{C}^*$, let ξ_L be an extension of ξ to $Z_{L(\mathbb{A})}$. Set

$$\mathfrak{P}_L := \Big\{\pi \in \mathfrak{P} \,\Big|\, \pi|_{Z_{L(\mathbb{A})}} = \xi_L\Big\}.$$

Assume that $\mathfrak{P}_L \neq \emptyset$. Then we have
(i) (**Compatibility**) *For $w \in W(M)$ such that wLw^{-1} is a standard Levi, $w\mathfrak{P}_L = (w\mathfrak{P})_{wLw^{-1}}$*; and

(ii) (**Admissibility**) $\mathfrak{P} = \mathfrak{P}_L + X_L^G$.
Accordingly let $\mathfrak{X}_L := [(M, \mathfrak{P}_L)]$ be the equivalence class of cuspidal data at the level of L/M associated to $(M, \mathfrak{P}_L)$ (instead of that for G/M).

12.2 Discrete Spectrum

Note that if $\dim\mathfrak{S}_L \neq 0$, then

$$\int_{\pi\in\mathfrak{S}_{\mathfrak{C}_L;\leq T}, \mathrm{Re}\,\pi = o(\mathfrak{S}_{\mathfrak{C}_L})} \frac{1}{|\mathrm{Norm}_{W_L}\mathfrak{S}_{\mathfrak{C}_L}|} \sum_{w\in W_L(M)} \mathrm{Res}^L_{w\mathfrak{S}_{\mathfrak{C}_L}} E^{L/M}(\Phi, w\pi)\, d_{\mathfrak{S}_{\mathfrak{C}_L}}\pi$$

is not an automorphic form (since the $\mathfrak{z}_L$-finiteness is not satisfied). We conclude that $\dim\mathfrak{S}_L$ has to be zero if the resulting residue is indeed an automorphic form.

Viewing cusp forms as trivial Eisenstein series for which we take the residue process to be the identity map, then we have the following:

Theorem A. *Automorphic forms in* $L^2\Big(U_L(\mathbb{A})L(F)\backslash G(\mathbb{A})\Big)_{\xi_L}$ *are the residues of relative Eisenstein series associated to cusp forms.*

Equivalently, $A^2\Big(U_L(\mathbb{A})L(F)\backslash G(\mathbb{A})\Big)_{\xi_L}$

$$:= L^2\Big(U_L(\mathbb{A})L(F)\backslash G(\mathbb{A})\Big)_{\xi_L} \cap A\Big(U_L(\mathbb{A})L(F)\backslash G(\mathbb{A})\Big)$$

is spanned by cusp forms and $\mathrm{Res}^L_{w\pi_L} E^{L/M}(\Phi, w\pi)$ for all $M \subset L$ and all Paley-Wiener functions Φ which we remind the reader are induced from cusp forms. This explains why in literature

$$A^2 = A_0 \oplus \mathrm{Res}\Big(\mathrm{ES}(A_0)\Big)$$

whose corresponding part in L^2-space is often called *discrete*.

12.3 Eisenstein Systems

Usually we call $\mathrm{Res}^L_{w\mathfrak{S}_{\mathfrak{C}_L}} E^{L/M}(\Phi, w\pi)$ an Eisenstein system or better but more narrowly $\mathrm{Res}^L_{w\pi_L} E^{L/M}(\Phi, w\pi_\mu)$ an *Eisenstein system* where $\pi_\mu := \pi_L \otimes \mu$ for $\mu \in X^G_L$.

Eisenstein system is the essential part in understanding the residue spectrum. Osborne and Warner devote a whole volume [OW] on the beautiful theory involved. Key properties of Eisenstein systems may be summarized as the following

Theorem B. (i) (**Compatibility**) *For* $\pi_\mu := \pi_L \otimes \mu,\ \ \mu \in X^G_L$ *and* $\mathfrak{S} := \pi_L \otimes X^G_L, w \in W_L(M)$,

$$E^{G/L}\Big(\mathrm{Res}^L_{w\pi_L} E^{L/M}(\Phi, w\pi_\mu); u\Big) = \mathrm{Res}^G_{w\mathfrak{S}} E^{G/M}(\Phi, w\pi_\mu);$$

(ii) (**Functional Equations**) *For* $t \in W(L), w \in W_L(M)$, *and* $\Phi \in P_{\mathfrak{X}}$,

$$\begin{aligned} &M(t,\mu)\Big(\mathrm{Res}^L_{w\pi_L} E^{L/M}(\Phi, \pi')\Big)_{\pi'=w\pi_\mu} \\ =&\Big(\mathrm{Res}^{tLt^{-1}}_{tw\pi_L} E^{tLt^{-1}/tMt^{-1}}(M(t,\mu)\Phi, \pi')\Big)_{\pi'=tw\pi_\mu}. \end{aligned}$$

Consequently, we have the following

Theorem C. *Eisenstein series associated to L^2-automorphic forms admit unique meromorphic continuation and satisfy the functional equations* (as stated in Theorems I and II of Day 2).

Proof. As said, every L^2 automorphic form can be written as a summation of a cusp form and a residue form obtained from an Eisenstein series associated to a cusp form of lower level. For the cusp form part, it will be proved in Day 6, while for the residue part, it is a direct consequence of Theorem B based on that for cusp forms just mentioned.

13 Why or Better How to Get Theorem B?!

13.1 Bridge

Let $L \subset M$. For a center character ξ and a ξ-admissible cuspidal datum $(M, \mathfrak{P}) \in \mathfrak{X}$, fix a unitary character ξ_L of $Z_{L(\mathbb{A})}$ extending ξ

and trivial on $A_{L(\mathbb{A})}^G$. Fix a nontrivial $\mathfrak{P}_L$. Also for any $\mathfrak{S}_L \in S_{(M,\mathfrak{P}_L)}$, set $\mathfrak{S} := \mathfrak{S}_L + \mathrm{Re}\, X_L^G \in S_{(M,\mathfrak{P})}$. Then it is known that
(i) $o(\mathfrak{S}) = o(\mathfrak{S}_L)$;
(ii) $\mathfrak{S}_L = \mathfrak{S} \cap \mathfrak{P}_L$; and
(iii) $\Big\{\pi \in \mathfrak{S}_{L,\leq T} \,\Big|\, \mathrm{Re}\pi = o(\mathfrak{S}_L)\Big\} \neq \emptyset$

$$\Leftrightarrow \Big\{\pi \in \mathfrak{S}_{\leq T} \,\Big|\, \mathrm{Re}\pi = o(\mathfrak{S})\Big\} \neq \emptyset.$$

(Condition (i) says that the origin keeps unchanged; (ii) says that $\mathfrak{S}_L$ can be recovered from $\mathfrak{S}$; while (iii) claims that the key non-empty condition are equivalent for both G- and L-levels.)

Recall that for $\Phi', \Phi \in P_{\mathfrak{X}}^{\mathfrak{F}}$,

$$A(\Phi', \Phi) := \sum_{\tau \in W(M)} \Big\langle M(\tau^{-1}, -\tau\bar{\pi})\Phi'(-\tau\bar{\pi}), \Phi(\pi)\Big\rangle.$$

Key Bridge. *For all $\mathfrak{S}_L \in S_{\mathfrak{X}_L}^{\mathfrak{F}}$, we have*

$$\mathrm{Res}_{\mathfrak{S}}^G A(\Phi', \Phi) = \mathrm{Res}_{\mathfrak{S}_L}^L A(\Phi', \Phi).$$

That is to say, for functions $A(\Phi', \Phi)$, the residue taken with respect to $(\mathfrak{P}, \mathfrak{S})$ at the G-level is the same as that with respect to $(\mathfrak{P}_L, \mathfrak{S}_L)$ at the L-level. Clearly, this then essentially gives Theorem B(i).

13.2 Basic Facts

Let $\mathrm{Sing}_T^{G,\mathfrak{F}}$:= the set of elements $\mathfrak{S}$ of $S_{\mathfrak{X}}^{\mathfrak{F}}$ such that
(i) $\Big\{\pi \in \mathfrak{S}_{\leq T} \,\Big|\, \mathrm{Re}\pi = o(\mathfrak{S})\Big\} \neq \emptyset$; and
(ii) There exist $\Phi', \Phi \in P_{\mathfrak{X}}^{R,\mathfrak{F}}$ such that $\mathrm{Res}_{\mathfrak{S}}^G A(\Phi', \Phi) \neq 0$.
In particular, (with residue non-zero), $A(\Phi', \Phi)$ is singular along $\mathfrak{S}$. One checks that
Lemma. $\mathrm{Sing}_{T_1}^{G,\mathfrak{F}} \subset \mathrm{Sing}_{T_2}^{G,\mathfrak{F}}$ *if* $T_1 \leq T_2$.

Set accordingly

$$\mathrm{Sing}^{G,\mathfrak{F}} := \cup_{T \gg 0} \mathrm{Sing}_T^{G,\mathfrak{F}} \qquad \text{and} \qquad \mathrm{Sing}^G := \cup_{\mathfrak{F}} \mathrm{Sing}^{G,\mathfrak{F}}.$$

Now for any $\mathfrak{S}$, denote by $(M, \mathfrak{P})$ its associated cuspidal datum, set

$M_{\mathfrak{S}}$:= the smallest standard Levi $\supseteq M$ such that $\mathfrak{S}$ is $\mathrm{Re}X_{M_{\mathfrak{S}}}^G$-stable.

Also as usual, set

$$\mathfrak{S}^0 := \Big\{\lambda \in X_M^G \,|\, \mathfrak{S}\otimes\lambda = \mathfrak{S}\Big\}.$$

Then $\mathrm{Re}\mathfrak{S}^0 := \Big\{\mathrm{Re}\lambda \,|\, \lambda \in \mathfrak{S}^0\Big\}$ is a linear subspace of $\mathrm{Re}X_M^G$, and $\mathrm{Re}\mathfrak{S} := \Big\{\mathrm{Re}\pi \,|\, \pi \in \mathfrak{S}\Big\}$ is an affine subspace of $\mathrm{Re}X_M^G$. In particular, $o(\mathfrak{S})$, the so-called *origin* of $\mathfrak{S}$, is simply an element of $\mathrm{Re}\mathfrak{S}$ defined by

$$o(\mathfrak{S}) := \mathrm{Re}\mathfrak{S} \cap \Big(\mathrm{Re}\mathfrak{S}^0\Big)^{\perp}.$$

(We here suggest the reader to draw a picture in $\mathbb{R}^2$ for the subspace $\mathrm{Re}\mathfrak{S}^0 = \{x+y=0\}$ and the affine line $\mathrm{Re}\mathfrak{S} = \{x+y=1\}$ to find what is $o(\mathfrak{S})$.)

Clearly, for any $L \supset M_{\mathfrak{S}}$, we have

$$\begin{aligned}\mathrm{Re}\mathfrak{S}^0 &= \Big(\mathrm{Re}\mathfrak{S}^0 \cap \mathrm{Re}X_M^L\Big) + \mathrm{Re}X_L^G\\ &= \Big(\mathrm{Re}\mathfrak{S}^0 \cap \mathrm{Re}X_M^{M_{\mathfrak{S}}}\Big) + \mathrm{Re}X_{M_{\mathfrak{S}}}^G\end{aligned}$$

in particular. We will call $\mathrm{Re}X_{M_{\mathfrak{S}}}^G$ the *fixed part* of $\mathfrak{S}$, due to the following

Key Lemma. *Let* $\mathfrak{T} \subset \mathfrak{S}$ *be a singular hyperplane of a certain* $\mathrm{Res}_{\mathfrak{S}}^G E(\Phi,\pi), \Phi \in P_{(M,\mathfrak{P})}$ *such that* $\mathfrak{T} \cap \{\pi \in \mathfrak{S}_{\leq T} | \mathrm{Re}\pi = o(\mathfrak{S})\} \neq \emptyset$. *Then*

(i) $o(\mathfrak{T}) = o(\mathfrak{S})$.

(ii) $M_{\mathfrak{T}} = M_{\mathfrak{S}}$; and

(iii) *There exists* $\tau \in W_{M_{\mathfrak{S}}}$ *such that*

(a) $\tau\pi = -\bar{\pi}$ *for any* $\pi \in \mathfrak{S}$ *with* $\mathrm{Re}\pi = o(\mathfrak{S})$; and

(b) $\mathfrak{T}$ *is a singular hyperplane for at least one of the functions*

$$\pi \in \mathfrak{S} \mapsto \mathrm{Res}_{\mathfrak{S}}^G\Big\langle M(\tau, -\tau\bar{\pi})\Phi'(-\tau\bar{\pi}), \Phi(\pi)\Big\rangle.$$

Indeed, for the singular plane $\mathfrak{T}$, we have

$$\mathrm{Re}\,\mathfrak{T} =: \mathrm{Re}\mathfrak{S}_1^0 = \Big(\mathrm{Re}\mathfrak{S}_1^0 \cap \mathrm{Re}X_M^{M_{\mathfrak{S}}}\Big) + \mathrm{Re}X_{M_{\mathfrak{S}}}^G.$$

That is to say, the part $\mathrm{Re}X^G_{M_{\mathfrak{S}}}$ kept unchanged from the level $\mathfrak{S}$ to the level $\mathfrak{T}$, while $\left(\mathrm{Re}\mathfrak{S}^0 \cap \mathrm{Re}X^{M_{\mathfrak{S}}}_M\right)$ is changed to $\left(\mathrm{Re}\mathfrak{S}^0_1 \cap \mathrm{Re}X^{M_{\mathfrak{S}}}_M\right)$, which is one dimension lower. Hence ideally, we are finally supposed to get, after say N steps,

$$\begin{aligned}\mathrm{Re}\mathfrak{S}^0_N &= \left(\left(\mathrm{Re}\mathfrak{S}^0_N \cap \mathrm{Re}X^{M_{\mathfrak{S}}}_M\right) + \mathrm{Re}X^G_{M_{\mathfrak{S}}}\right) \\ &= \{0\} + \mathrm{Re}X^G_{M_{\mathfrak{S}}}.\end{aligned}$$

Thus 0 corresponding to $o(\mathfrak{S})$ which gives a discrete spectrum, while $\mathrm{Re}X^G_{M_{\mathfrak{S}}}$ kept unchanged even after N steps. All in all, the upshot here is the following fundamental

Theorem D. *If* $\mathfrak{S} \in \mathrm{Sing}^G$. *Then*
(i) $\mathfrak{S} \ni \pi_0 \otimes o(\mathfrak{S})$ *where* π_0 *is* **the** *unique element of* $\mathfrak{P}$ *which is trivial on* $A^G_{M(\mathbb{A})}$; and
(ii) $o(\mathfrak{S})$ *is in the interior of the cone generated by* $R^T(T_M, M_{\mathfrak{S}})$.

Reasoning. We start with (ii). There are two main reasons.
(ii.a) We have the following
Geometric Consequences of the Residual Argument. *For any* $\mathfrak{S} \in \mathfrak{C} \cap \mathrm{Sing}^G$, *we have*
(α) $\mathfrak{S}^* := -\overline{\mathfrak{S}} \in \mathfrak{C} \cap \mathrm{Sing}^G$;
(β) *There exists* $w_{\mathfrak{S}} \in \mathrm{Stab}(M, \mathfrak{P})$ *such that*

$$w_{\mathfrak{S}}\, o(\mathfrak{S}) = -o(\mathfrak{S}), o(\mathfrak{S})|_{\mathrm{Im}\mathfrak{S}} = \mathrm{Id}, o(\mathfrak{S})|_{\mathrm{Re}\mathfrak{S}^0} = \mathrm{Id};$$

(γ) *Let* $\Phi \in P^{R,\mathfrak{F}}_{\mathfrak{X}}, \pi_1 \in \mathfrak{S}$ *such that* $\mathrm{Res}^G_{\mathfrak{S}} E(\Phi, \pi) \not\equiv 0$ *and is holomorphic at* π_1.
Then the set of cuspidal supports of $\mathrm{Res}^G_{\mathfrak{S}} E(\Phi, \pi)$ *is contained in*

$$\left\{w\pi_1 \,\middle|\, w \in W(M), -w\overline{\mathfrak{S}} \in \mathrm{Sing}^{G,\mathfrak{F}}\right\}.$$

(ii.b) With (a), being L^2, π has to be 'real' and the L^2 criterion of automorphic forms, i.e., the negativity of the cuspidal supports transforms to a positivity statement here.
As for (i), we have the following
Theorem D′ $L^2_{\mathfrak{C}} \subset$ Discrete Spectrum $\Leftrightarrow M_{\mathfrak{S}} = G\ \forall \mathfrak{S} \in \mathfrak{C} \Leftrightarrow \mathfrak{S} = \{\pi_L := \pi_0 \otimes o(\mathfrak{S})\}\ \forall \mathfrak{S} \in \mathfrak{C}$.

14 Spectrum Decomposition: Levi Interpretation

14.1 Spaces $A_{\mathfrak{C},\pi_\mu}$

Let $\mathfrak{C} \in [S_{\mathfrak{X}}]$. We want to attach it an association class of Levi subgroups. (Recall that two standard Levis are called *associated* if they are W-conjugated.) This is given by

Existence Lemma. *If $\mathfrak{C} \cap \mathrm{Sing}^{G,\mathfrak{F}} \neq \emptyset$, then there exists a unique association class $L(\mathfrak{C})$ of Levi subgroups of G such that for any $L \in L(\mathfrak{C})$, there exists an $\mathfrak{S} \in \mathfrak{C} \cap \mathrm{Sing}^{G,\mathfrak{F}}$ satisfying* $\mathrm{Re}\mathfrak{S}^0 = \mathrm{Re}X_L^G$. *In particular, $M_{\mathfrak{S}} = L$.*

Set then $\pi_L := \pi_0 \otimes o(\mathfrak{S})$ which we now know is always in $\mathfrak{S}$ by Theorem D(i). In particular, $\{\pi_L\} \in \mathrm{Sing}^L$. Consequently, let $\mathfrak{C}_L$ be the singular class attached to $\{\pi_L\}$. (By Theorem D$'$, then π_L belongs to the discrete spectrum of $L(\mathbb{A})$.) Moreover, for $\mu \in X_L^G$ (with $L \neq G$), set $\pi_\mu := \pi_L \otimes \mu \in \mathfrak{S}$, and μ is called *generic* if
(i) none of conjugates of π_μ is in the cuspidal supports of any L^2-automorphic forms;
(ii) π_μ does not belong to any hyperplane of $\mathfrak{S}$ contained in $S_{\mathfrak{X}}^{\mathfrak{F}}$.

Set

$$A\Big(L(F)\backslash L(\mathbb{A})\Big)_{\mathfrak{C}_L} := A\Big(L(F)\backslash L(\mathbb{A})\Big) \cap L^2\Big(L(F)\backslash L(\mathbb{A})\Big)_{\mathfrak{C}_L}.$$

Then as usual, we get the spaces $A\Big(U_L(\mathbb{A})L(F)\backslash G(\mathbb{A})\Big)_{\mathfrak{C}_L}$ and it cousin

$$A\Big(U_L(\mathbb{A})L(F)\backslash G(\mathbb{A})\Big)_{\mathfrak{C}_L;\mu} := (\mu \circ m_{P_L}) \cdot A\Big(U_L(\mathbb{A})L(F)\backslash G(\mathbb{A})\Big)_{\mathfrak{C}_L}.$$

For $\Phi \in P_{\mathfrak{X}}$, $\mu \in \mathrm{Im}\, X_L^G$ generic, set

$$e^L(\Phi, \pi_\mu) := \frac{1}{|\mathrm{Norm}_{W_L(M)}\{\pi_L\}|} \sum_{w \in W(M)} \mathrm{Res}^L_{w\pi_L} E^{L/M}(\Phi, w\pi_\mu).$$

Recall that

$$A_{\mathfrak{C},\pi_\mu} := \Big\langle \mathrm{Res}^L_{w\mathfrak{S}} E(\Phi, w\pi_\mu) | w \in W(M), \mathfrak{S} \in \mathfrak{C}, \Phi \in P_{\mathfrak{X}} \Big\rangle_{\mathbb{C}}.$$

Proposition. (i) $e^L(\Phi,\pi_\mu)\in A\Big(U_L(\mathbb{A})L(F)\backslash G(\mathbb{A})\Big)_{\mathfrak{C}_L;\mu}$; and
(ii) $A\Big(U_L(\mathbb{A})L(F)\backslash G(\mathbb{A})\Big)_{\mathfrak{C}_L;\mu}=\Big\langle e^L(\Phi,\pi_\mu)|\Phi\in P_{\mathfrak{X}}\Big\rangle_{\mathbb{C}}$.

With this, we are ready to state the following
Main Theorem. (i) *For* $\Phi\in P_{\mathfrak{X}}$, $E^{G/L}(e^L(\Phi,\pi_\mu);\mu)\in A_{\mathfrak{C},\pi_\mu}$; and
(ii) *Up to a constant multiple, we have an isometry*

$$E^{G/L}(\bullet;\mu):A\Big(U_L(\mathbb{A})L(F)\backslash G(\mathbb{A})\Big)_{\mathfrak{C}_L;\mu}\simeq A_{\mathfrak{C},\pi_\mu}.$$

Proof. (i) Direct consequence of Thm B, or better of the Key Bridge; and
(ii) The injectivity comes from the fact that $E^{G/L}(\bullet;\mu)$ preserves the scalar product; while the surjectivity is a bit difficult, which may be established using an induction on the semi-simple rank. It breaks into the following
Sublemma 1. *Let* $w\in W(M),\Phi\in P_{\mathfrak{X}}$. *Then* $\mathrm{Res}^L_{w\mathfrak{S}}E(\Phi,w\pi_\mu)$

$$\in\langle\mathrm{Res}^L_{\sigma\mathfrak{S}}E(\Phi',\sigma\pi_\mu)\,|\,\sigma\in W(M)\text{ s.t. }M_{\sigma\mathfrak{S}}\neq G,\Phi'\in P_{\mathfrak{X}}\rangle_{\mathbb{C}}.$$

Sublemma 2.(Major One) *Let* $\sigma\in W(M)$ *be such that* $M_{\sigma\mathfrak{S}}\neq G$ *and* $\Phi'\in P_{\mathfrak{X}}$. *Then* $\mathrm{Res}^L_{\sigma\mathfrak{S}}E(\Phi',\sigma\pi_\mu)$

$$\in\langle\mathrm{Res}^L_{\tau\mathfrak{S}}E(\Phi'',\tau\pi_\mu)\,|\,\tau\in W(M)\text{ s.t. }M_{\sigma\mathfrak{S}}\in L(\mathfrak{C}),\Phi''\in P_{\mathfrak{X}}\rangle_{\mathbb{C}}.$$

Sublemma 3. *Let* $\tau\in W(M)$ *be such that* $M_{\tau\mathfrak{S}}\in L(\mathfrak{C})$ *and* $\Phi''\in P_{\mathfrak{X}}$. *Then there exists* $\Phi'''\in P_{\mathfrak{X}},\tau_L\in W_L(M)$ *such that*

$$\mathrm{Res}^L_{\tau\mathfrak{S}}E(\Phi'',\tau\pi_\mu)=\mathrm{Res}^L_{\tau_L\mathfrak{S}}E(\Phi''',\tau_L\pi_\mu).$$

15 Spectral Decomposition: Arthur's Version of Langlands

15.1 In Terms of Levi

Let L be a standard Levi of G. Denote by $L^2\Big(L(F)\backslash L(\mathbb{A})\Big)_{d,\xi}:=$ the part of the discrete spectrum of L (modulo $Z_{L(\mathbb{A})}$) whose central character is trivial on $A_{L(\mathbb{A})}^G$ and whose restriction to $Z_{G(\mathbb{A})}$ is ξ. Denote by

$$A\Big(L(F)\backslash L(\mathbb{A})\Big)_{d,\xi} := L^2\Big(L(F)\backslash L(\mathbb{A})\Big)_{d,\xi} \cap A\Big(L(F)\backslash L(\mathbb{A})\Big).$$

As before, set the associated P-level space by

$$A\Big(U_L(\mathbb{A})L(F)\backslash G(\mathbb{A})\Big)_{d,\xi}$$
$$:= \Big\{\phi \in A\Big(U_L(\mathbb{A})L(F)\backslash G(\mathbb{A})\Big) \mid \phi_k \in A(L(F)\backslash L(\mathbb{A})), \forall k \in K\Big\}.$$

Denote by $P_{L,\mathrm{dis}}^R :=$ the space of holomorphic functions on $\Big\{\mu \in X_L^G \mid \|\mathrm{Re}\mu\| < R\Big\}$ with values in $A\Big(U_L(\mathbb{A})L(F)\backslash G(\mathbb{A})\Big)_{d,\xi}$ satisfying the growth condition (required in the definition of Paley-Wiener functions). Then we have

Lemma. *For all* $\Phi \in P_{L,\mathrm{dis}}^R$,

$$\lim_{T\to\infty} \int_{\mu\in\mathrm{Im}X_L^G, \|\mu\|<T} E(\Phi,\mu)\, du$$

defines an element of $L^2\Big(G(F)\backslash G(\mathbb{A})\Big)_\xi$.

Let $L^2\Big(G(F)\backslash G(\mathbb{A})\Big)_{\xi;L}$ be the closed subspace of $L^2\Big(G(F)\backslash G(\mathbb{A})\Big)_\xi$ generated by the elements of the lemma above.

Lemma and Definition. *Suppose that* $L \sim L'$ *are associated, then*

$$L^2\Big(G(F)\backslash G(\mathbb{A})\Big)_{\xi;L} = L^2\Big(G(F)\backslash G(\mathbb{A})\Big)_{\xi;L'},$$

which we denoted by $L^2\Big(G(F)\backslash G(\mathbb{A})\Big)_{\xi;[L]}$.

Let also $W^L :=$ the set of elements of W of minimal length modulo W_L and introduce its subset $\mathrm{Stab}_{W^L}L := \{w \in W^L | wLw^{-1} = L\}$. With this, we are ready to state the following

Main Theorem$'$. (i) $L^2\Big(G(F)\backslash G(\mathbb{A})\Big)_{\xi;[L]}$ *is isomorphic to the Hilbert space* $\mathrm{Hilb}_{L,\xi}^{\mathfrak{F}}$ *of measurable functions* F *on* $\mathrm{Im}X_L^G$ *with values in the space* $L^2\Big(U_L(\mathbb{A})L(F)\backslash G(\mathbb{A})\Big)_{d,\xi}$ *satisfying the following conditions*

- $M(t^{-1}, t\mu)F(t\mu) = F(\mu)$ *almost everywhere* $\forall t \in \mathrm{Stab}_{W^L} L$;
- $\lim_{T\to\infty} \int_{\mu \in \mathrm{Im} X_L^G, \|\mu\| < T} \frac{1}{|\mathrm{Stab}_{W^L} L|} \|F(\mu)\|^2 \, d\mu < +\infty$;

(ii) *There exists an orthogonal decomposition*

$$L^2\Big(G(F)\backslash G(\mathbb{A})\Big)_\xi = \oplus_{[L]} L^2\Big(G(F)\backslash G(\mathbb{A})\Big)_{\xi;[L]}.$$

15.2 Relation with Residual Approach: The Proof

Let $\mathfrak{X}$ be an equivalence class of cuspidal data and $\mathfrak{C}$ be an equivalence class in $S_{\mathfrak{X}}$ such that $\mathfrak{C} \cap \mathrm{Sing}^G \neq \emptyset$. Then there exists an associated subspace $L^2\Big(G(F)\backslash G(\mathbb{A})\Big)_{\mathfrak{C}}$ of $L^2\Big(G(F)\backslash G(\mathbb{A})\Big)_\xi$ by the residual approach of the spectral decomposition.

Now let (L, δ) be a discrete parameter for $\mathfrak{C}$, that is to say, $L \in L(\mathfrak{C})$ with $L = M_{\mathfrak{S}}$ for some $\mathfrak{S} \in \mathfrak{C}$ and

$$\delta = L^2\Big(L(F)\backslash L(\mathbb{A})\Big)_{\{\pi_0 \otimes o(\mathfrak{S})\} =: \delta}$$

so that $\mathrm{Re}\mathfrak{S} = \Big(\pi_0 \otimes o(\mathfrak{S})\Big) \otimes \mathrm{Re} X_L^G$. (This notation is well-defined as δ is uniquely determined by $\pi_0 \otimes o(\mathfrak{S})$.) Denote also by $\mathfrak{C}_L$ the singular class attached to δ, i.e., that attached to $\{\pi_0 \otimes o(\mathfrak{S}) =: \pi_L\}$. Define

$$\mathrm{Stab}_{W^L}(L, \delta) := \{w \in W(L) \,|\, wLw^{-1} = L, w\mathfrak{C} = \mathfrak{C}\}$$

which is known to be the same as

$$\Big\{w \in \mathrm{Stab}_{W^L} L \,|\, \exists w_L \in W_L(M), t\pi_L = w_L \pi_L\Big\}.$$

Proposition$'$. (i) *Let* Φ *be a function on* X_L^G *with values in the space* $A\Big(U_L(\mathbb{A})L(F)\backslash G(\mathbb{A})\Big)_\delta$ *holomorphic in a neighborhood of every points of* $\mathrm{Im} X_L^G$ *and is rapidly decreasing on this set. Then*

$$e(\Phi) := \lim_{T\to\infty} \int_{\mu \in \mathrm{Im} X_L^G, \|\mu\| \leq T} E(\Phi, \mu) \, d\mu$$

lies in $L^2\Big(G(F)\backslash G(\mathbb{A})\Big)_{\mathfrak{C}}$; and

(ii) $L^2\Big(G(F)\backslash G(\mathbb{A})\Big)_{\mathfrak{C}}$ *is the closure of the subspace generated by these* $e(\Phi)$*'s.*

Main Theorem″. $L^2\Big(G(F)\backslash G(\mathbb{A})\Big)_{\mathfrak{C}}$ *is isomorphic to the Hilbert space* $H_{\mathfrak{C}}$ *of measurable functions* F *on* $\mathrm{Im}X_L^G$ *with values in the space* $L^2(U_L(\mathbb{A})L(F)\backslash G(\mathbb{A}))_\delta$ *satisfying the following conditions*

- $M(t^{-1},t\mu)F(t\mu)=F(\mu)$ *almost everywhere* $\forall t\in \mathrm{Stab}_{W^L}(L,\delta)$;
- $\lim_{T\to\infty}\int_{\mu\in \mathrm{Im}X_L^G,\|\mu\|<T}\frac{1}{|\mathrm{Stab}_{W^L}L|}\|F(\mu)\|^2\,d\mu<+\infty.$

Proof. The point here is not really about the proof, rather is about **how to rephrase** the statement as essentials have already been given in Day 4 and the Main Theorem.

Day Six: Arthur's Truncation and Meromorphic Continuation

16 Arthur's Analytic Truncation

16.1 Positive Chamber and Positive Cone

16.1.1 Dual Basis

Let $P_1 = U_1M_1 \supset P_0 = U_0M_0$ be a standard parabolic subgroup. For simplicity, set $\mathfrak{a}_i := \mathfrak{a}_{M_i}^G$ for $i = 0, 1$ and $\mathfrak{a}_0^1 := \mathfrak{a}_{M_0}^{M_1}$.

(**a**) Level $\left(\mathfrak{a}_0\right)^*$. Δ_0 gives a natural basis for $(\mathfrak{a}_0)^*$. For any $\alpha \in \Delta$, denote by $\alpha^\vee \in \mathfrak{a}_0$ the corresponding coroot. That is to say, $\alpha^\vee$ is the element of $\mathfrak{a}_0$ such that as a linear form on $\left(\mathfrak{a}_0\right)^*$, it is characterized by $\alpha^\vee(\alpha) = 2$ and $\mathrm{Ker}(\alpha^\vee) = S_\alpha$, the hyperplane fixed by $w_\alpha \in W$, the element of the Weyl group corresponding to α. (Recall that for simple root α, w_α is the unique reflection of $\left(\mathfrak{a}_0\right)^*$ determined by the property that it maps α to $-\alpha$ while keeps all points on a unique hyperplane $S_\alpha := \{\alpha\}^\perp$ unchanged.) Denote the collection of all coroots by

$$\Delta_0^\vee := \left\{\alpha^\vee \,\middle|\, \alpha \in \Delta_0\right\},$$

which gives a basis for $\mathfrak{a}_0$. Set

$$\widehat{\Delta}_0 := \left\{\varpi_\alpha \,\middle|\, \alpha \in \Delta_0\right\}$$

the basis of $\left(\mathfrak{a}_0\right)^*$ dual to $\Delta_0^\vee$.

(**b**) Level $\left(\mathfrak{a}_1\right)^*$. Then we have the decomposition

$$\begin{array}{ccccc} \left(\mathfrak{a}_0\right)^* & = & \left(\mathfrak{a}_0^1\right)^* & \oplus & \left(\mathfrak{a}_1\right)^* \\ \cup & & \cup & & \cup \\ \Delta_0 & \supset & \Delta_0^1 & & \Delta_1. \end{array}$$

Here Δ's below are bases of the corresponding spaces above. Recall that Δ_1 may be obtained from non-trivial restrictions of elements of Δ_0 to $\mathfrak{a}_1$. Consequently, there is a natural bijection

$$\Delta_0 \backslash \Delta_0^1 \leftrightarrow \Delta_1.$$

Set then

$$\Delta_1^\vee := \Big\{ \alpha^\vee \,\big|\, \alpha \in \Delta_0 \backslash \Delta_0^1 \Big\}$$

be the collection of coroots corresponding to elements in $\Delta_0 \backslash \Delta_0^1$. View $\Delta_1^\vee$ as a basis of $\mathfrak{a}_1$, then

$$\Big\{ \varpi_\alpha \,\big|\, \alpha \in \Delta_0 \backslash \Delta_0^1 \Big\},$$

i.e., the collection of ϖ_α's for $\alpha \in \Delta_0 \backslash \Delta_0^1$ gives a basis of $\Big(\mathfrak{a}_1\Big)^*$ dual to $\Delta_1^\vee$. (Surely, then elements ϖ_α's are viewed as linear forms on $\mathfrak{a}_1$ via the restriction to $\mathfrak{a}_1$.)

Equivalently, compactible with the correspondence $\Delta_0 \backslash \Delta_0^1 \leftrightarrow \Delta_1$, $\widehat{\Delta}_1$ may also be constructed as follows. For any $\beta \in \Delta_1$, let $\alpha \in \Delta_0$ be the unique element such that $\alpha\big|_{\mathfrak{a}_1} = \beta$. This β then gives the coroot $\alpha^\vee \in \mathfrak{a}_0$ and hence also element ϖ_α in $\Big(\mathfrak{a}_0\Big)^*$ by (a). Restricting it to $\mathfrak{a}_1$, we get then an element $\varpi_\beta := \varpi_\alpha\big|_{\mathfrak{a}_1}$. Clearly, we have a natural correspondence

$$\Big\{ \varpi_\alpha \,\big|\, \alpha \in \Delta_0 \backslash \Delta_0^1 \Big\} \leftrightarrow \Big\{ \varpi_\beta \,\big|\, \beta \in \Delta_1 \Big\}$$

which is compactible with the correspondence $\Delta_0 \backslash \Delta_0^1 \leftrightarrow \Delta_1$. For this reason, we also denote the set $\Big\{ \varpi_\beta \,\big|\, \beta \in \Delta_1 \Big\}$ by $\widehat{\Delta}_1$. Moreover, for late use, write $\beta^\vee$ for the projection of $\alpha^\vee$ to the space $\mathfrak{a}_1$. It is clear that

$$\Delta_1^\vee := \Big\{ \beta^\vee \,\big|\, \beta \in \Delta_1 \Big\}$$

is the basis of $\mathfrak{a}_1$ dual to the basis $\widehat{\Delta}_1$ (of $\Big(\mathfrak{a}_1\Big)^*$).

One of the basic fact here is the following

Lemma. $\widehat{\Delta_1}$ *is in the positive* $\mathbb{Q}$*-span of* Δ_1.

16.1.2 Positive Chamber and Positive Cone

In the space $\mathrm{Re}\,\mathfrak{a}_1$, define the so-called *positive chamber* $\mathcal{C}_1$ and *positive cone* $\mathfrak{C}_1$ respectively by

$$\mathcal{C}_1 := \Big\{t \in \mathrm{Re}\,\mathfrak{a}_1 \,\Big|\, \alpha(t) > 0,\ \forall \alpha \in \Delta_1\Big\}$$

$$\mathfrak{C}_1 := \Big\{t \in \mathrm{Re}\,\mathfrak{a}_1 \,\Big|\, \varpi(t) > 0,\ \forall \varpi \in \widehat{\Delta_1}\Big\}.$$

Denote by τ_1 and $\widehat{\tau}_1$ the characteristic functions for the subset $\mathcal{C}_1$ and for $\mathfrak{C}_1$ in $\mathfrak{a}_0$ respectively. For example, if $t = t_0^1 + t_1 \in \mathfrak{a}_0$ is a decomposition of t according to the decomposition $\mathfrak{a}_0 = \mathfrak{a}_0^1 \oplus \mathfrak{a}_1$, then $\hat{\tau}_1(t) = 0$ if and only if $t_1 \notin \mathfrak{C}_1$.
Lemma$'$. $\mathcal{C}_1 \subset \mathfrak{C}_1$.
Ex. $SL(3)$ as in [Ar4].

16.2 Arthur's Analytic Truncation

As seen earlier, automrphic forms are only of moderate growth, hence not integrable over $Z_{G(\mathbb{A})}G(F)\backslash G(\mathbb{A})$. Arthur's analytic truncation is a universal construction which help us to obtain integrable functions out of these only with moderate growth.

An element $T \in \mathfrak{a}_0$ is said to be *sufficiently positive* and denoted by $T \gg 0$ if for all $\alpha \in \Delta_0$, $\langle \alpha, T\rangle \gg 0$ are large enough. Fix such a T. Let $f : G(F)\backslash G(\mathbb{A}) \to \mathbb{C}$ be a smooth function. We define *Arthur's analytic truncation* $\Lambda^T f$ (for f with respect to the parameter T) to be the function on $G(F)\backslash G(\mathbb{A})$ given by $\Big(\Lambda^T f\Big)(g)$

$$:= \sum_{P:\,\mathrm{standard}} (-1)^{\dim\big(Z_P/Z_G\big)} \sum_{\delta \in P(F)\backslash G(F)} f_P(\delta g) \cdot \hat{\tau}_P\Big(H_P(\delta g) - T\Big),$$

where f_P denotes the constant term of f along with the standard parabolic subgroup P and $H_P(g) := \log_M m_P(g)$ is an element in $\mathfrak{a}_P$. (As noticed above, since $\hat{\tau}_P(\lambda)$ depends only on the projection λ_P of λ in $\mathfrak{a}_P$, it does not matter whether we view H_P as a map onto $\mathfrak{a}_P$ or a map to $\mathfrak{a}_0$.)

Remarks. (1) If $P = G$, the corresponding term on the RHS is simply $\phi(g)$. So

$$\Lambda^T f(g) = f(g) - \sum_{P:\text{maximal}} f_P(g) \cdot \hat{\tau}_P\Big(H(g) - T\Big) + \cdots .$$

That is to say, $\Lambda^T f$ is obtained from f by removing its constant terms along proper parabolic subgroups;
(2) Since every parabolic subgroup is conjugated with a standard one, the second summation $\sum_{\delta \in P(F)\backslash G(F)}$ simply says that in the definition, constant terms along all parabolic, not just standard parabolic, have to be properly removed. In other words, we have

$$\Big(\Lambda^T f\Big)(g) := \sum_P (-1)^{\dim\left(Z_P/Z_G\right)} f_P(g) \cdot \hat{\tau}_P\Big(H_P(g) - T\Big).$$

Fundamental properties of Arthur's truncation may be summarized as

Theorem. (Arthur) *For $T \gg 0$, i.e., T sufficiently positive, in $\mathfrak{a}_0$, and f an automorphic form on $G(F)\backslash G(\mathbb{A})$,*
(i) *$\Lambda^T f$ is rapidly decreasing*;
(ii) $\Lambda^T \circ \Lambda^T = \Lambda^T$; and
(iii) *Λ^T is self-adjoint.*

The proof of these properties is by no means simple: a refined reduction theory and certain fine estimations, of which Theorem A of Day 1 is a starting point, must be worked out. We will omit it entirely due to two reasons: Arthur explains basic ideals behind the truncation in [Ar5] beautifully; and at least for this brief note we only need such a truncation for relative rank one which is indeed a toy model (so very simple).

17 Meromorphic Continuation of Eisenstein Series: Deduction

This part is beautifully presented in [MW], following Jacquet.

17.1 Preparation

Let ξ be a central character of $G(\mathbb{A})$, and $(M, \mathfrak{P})$ a ξ-admissible cuspidal datum. Set

$$A_\xi^{\mathfrak{F}} := A\Big(U(\mathbb{A})M(F)\backslash G(\mathbb{A})\Big)_\xi^{\mathfrak{F}}, \quad L_\xi^{2,\mathfrak{F}} := L^2\Big(U(\mathbb{A})M(F)\backslash G(\mathbb{A})\Big)_\xi^{\mathfrak{F}},$$

and $L_{\xi,\mathrm{loc}}^{2,\mathfrak{F}}$:= the space of $\mathbb{C}$-valued functions on $G(F)\backslash G(\mathbb{A})$ of central character ξ which are locally square integrable modulo $Z_{G(\mathbb{A})}$.

Set

$$\mathfrak{c}_{\mathfrak{P}} := \Big\{\pi \in \mathfrak{P} \,\big|\, \langle \mathrm{Re}\,\pi - \rho_P, \alpha^\vee\rangle > 0\ \forall \alpha \in \Delta_P\Big\}$$

be the shifted positive chamber. Then for $\phi \in A\Big(U(\mathbb{A})M(F)\backslash G(\mathbb{A})\Big)_\pi$ with $\pi \in \mathfrak{P}$, the Eisenstein series $E(\phi, \pi)$ and the intertwining operator $M(w, \pi)$ are holomorphic at π whenever $\mathrm{Re}\,\pi \in \mathfrak{c}_{\mathfrak{P}}$. As such, the only central problem left is to meromorphically continue $E(\phi, \pi)$ and $M(w, \pi)$ to the whole $\mathbb{C}$-space $\left(\mathfrak{a}_M^G\right)^* \simeq: X_M^G$.

Denote by $\mathcal{H}^{\mathfrak{F}}$:= the space of smooth functions $h : G(\mathbb{A}) \to \mathbb{C}$ with compact support such that the subspace generated by the translations of h under $K \times K$ acting on both right and left decomposes into a sum of irreducible subspaces under the action of $K \times K$ each of whose isomorphism classes belongs to $\mathfrak{F} \times \mathfrak{F}$. It is an algebra which acts on $A_\xi^{\mathfrak{F}}$, $L_\xi^{2,\mathfrak{F}}$, $L_{\xi,\mathrm{loc}}^{2,\mathfrak{F}}$ and so on via

$$\varphi \mapsto \delta(h)\varphi : \ x \mapsto \int_{G(\mathbb{A})} h(y)\varphi(xy)\, dy.$$

Accordingly, for $\pi \in \mathfrak{P}$, denote by $i(\pi, \bullet)$ the associated representation of $\mathcal{H}^{\mathfrak{F}}$ on

$$A(M, \pi)^{\mathfrak{F}} := A\Big(U(\mathbb{A})M(F)\backslash G(\mathbb{A})\Big)_\pi^{\mathfrak{F}}.$$

Recall that $W(M)$:= the collection of $w \in W$ such that (i) wMw^{-1} is a standard Levi of G; and (ii) w is of minimal length in its right coset modulo W_M. For any $\varphi \in A(M,\pi)^{\mathfrak{F}}$, $\pi \in \mathfrak{P}$ with $\mathrm{Re}\,\pi \in \mathfrak{c}_{\mathfrak{P}}$, $E(\varphi,\pi) \in L^{2,\mathfrak{F}}_{\xi,\mathrm{loc}}$. This defines a function on

$$E : \pi \mapsto E(\pi) \in \mathrm{Hom}_{\mathbb{C}}\Big(A(M,\pi)^{\mathfrak{F}}, L^{2,\mathfrak{F}}_{\xi,\mathrm{loc}}\Big).$$

Similarly, for $w \in W(M)$, we have the operator

$$M(w,\bullet) : \pi \mapsto M(w,\pi) \in \mathrm{Hom}_{\mathbb{C}}\Big(A(M,\pi)^{\mathfrak{F}}, A(wMw^{-1}, w\pi)^{\mathfrak{F}}\Big).$$

By definition, easily, we obtain the following
Admissible Lemma. *For any $\pi \in \mathfrak{c}_{\mathfrak{P}}$, $h \in \mathcal{H}^{\mathfrak{F}}$ and $w \in W(M)$,*
(i) $E(\pi) \circ i(\pi,h) = \delta(h) \circ E(\pi)$; and
(ii) $M(w,\pi) \circ i(\pi,h) = i(w\pi,h) \circ M(w,\pi)$.
Moreover, by the convergence proof, we have the following
Proposition. *The function E and the operator M are holomorphic over $\mathfrak{c}_{\mathfrak{P}}$.*

With this preparation, now what we want to prove may be stated as the following:
Theorem A. (**Meromorphic Extension**) *E and M admit unique meromoprphic continuations to the whole space $\mathfrak{P}$.*
Theorem B. (**Functional Equations**) Meromorphically on $\mathfrak{P}$;
(i) $E_M(\pi) = E_{wMw^{-1}}(w\pi) \circ M(w,\pi)$; and
(ii) $M(w',w\pi) \circ M(w,\pi) = M(w'w,\pi)$, *where $w \in W(M)$ and $w' \in W(wMw^{-1})$.*

17.2 Deduction to Relative Rank 1

Basic idea for proving Theorems A and B is to reduce them to the case when P is maximal, i.e, the case of relative rank one. This is based on the following elementary consideration.

For $w \in W(M)$, let $l = l(w)$ be the length of w. Then we can find a decomposition

$$w = w_{\alpha_l} \cdots w_{\alpha_1}$$

of w into product of reflections w_{α_i} such that
(0) l is the minimal length among all such decompositions;
(i) If we set $w_i = w_{\alpha_i} \cdots w_{\alpha_1}$, $M_i = w_i M w_i^{-1}$, then α_{i+1} is a simple root of $R(T_{M_i}, G)$ satisfying $w_i^{-1}(\alpha_{i+1}) > 0$; and

(ii) The set of positive multiple roots of an element of $\Big\{w_i^{-1}(\alpha_{i+1})\,\Big|\,i = 0,1,\ldots,l-1\Big\}$ coincides with $\Big\{\alpha \in R(T_M,G)\,\Big|\,\alpha > 0, w\alpha < 0\Big\}$.
Denote by U_i the unipotent radical of the parabolic subgroup P_i of Levi M_i. Note that in particular, $P^w = wMw^{-1}\cdot U^w$ coincides with $P_l = M_lU_l$. Accordingly, we can decompose integration over $(U_l(\mathbb{A})\cap wU(\mathbb{A})w^{-1})\backslash U_l(\mathbb{A})$ into l integrals over

$$\Big(U_l(\mathbb{A})\cap ww_{i-1}^{-1}U_{i-1}(\mathbb{A})w_{i-1}w^{-1}\Big)\Big\backslash\Big(U_l(\mathbb{A})\cap ww_i^{-1}U_i(\mathbb{A})w_iw^{-1}\Big)$$

which by $u\mapsto w_iwuww_i^{-1}$ is isomorphic to

$$\Big(U_i(\mathbb{A})\cap w_{\alpha_i}U_{i-1}(\mathbb{A})w_{\alpha_i}^{-1}\Big)\backslash U_i(\mathbb{A}).$$

This decomposition and some variable changes then lead to the following

Bridge Decomposition. *With the same notation as above,*

$$M(w,\pi) = M(w_{\alpha_l}, w_{l-1}\pi)\cdot M(w_{\alpha_{l-1}}, w_{l-2}\pi)\cdots M(w_{\alpha_1},\pi).$$

Consequently, in order to meromorphically continued $M(w,\pi)$, it suffices to do it for the operators $M\Big(w_{\alpha_i}, w_{i-1}\pi\Big)$, which comes from the case of relative rank 1 corresponding to $\{\alpha_i\}$'s.

18 The Situation of Relative Rank 1

18.1 Working Site

The advantage in this case is that now everything can be written down quite precisely. Indeed, being relative rank one, $\dim X_M^G = 1$, and $W(M)$ only contains of two elements: 1 for which $M(1,\lambda) = \mathrm{Id}$ for every λ; and a non-trivial element which we denote here by w. Accordingly, denote by ${}^wP = {}^wM\cdot{}^wU$ the standard parabolic subgroup associated to ${}^wM := wMw^{-1}$. (It can happen that ${}^wP = P$. e.g., $G = SL(2)$ and $P = P_0 = B$.) Denote by α the unique element of Δ_M. It is known that $-w\alpha$ is the unique element of $\Delta_{wMw^{-1}}$.

18.2 Constant Terms

For $\lambda \in \mathfrak{c} := \mathfrak{c}_{\mathfrak{P}}$, $\varphi \in A(M,\pi)^{\mathfrak{F}}$, the associated Eisenstein series $E(\varphi,\lambda) := E(\lambda\varphi, \pi\otimes\lambda)$ is well-defined and gives an automorphic form in $A_\xi \subset L^{2,\mathfrak{F}}_{\xi,\mathrm{loc}}$. In particular, for every standard parabolic subgroup P' of G, its associated constant term is precisely given by

Lemma. (i) $E(\varphi,\lambda)_{P'} = 0$ *if* $P' \neq G, P, {}^wP$

(ii.a) *If* $P \neq {}^wP$,

$$\begin{aligned} E(\varphi,\lambda)_P =& \lambda\varphi \\ E(\varphi,\lambda)_{^wP} =& (w\lambda)\Big(M(w,\lambda)\varphi\Big); \end{aligned}$$

(ii.b) *If* $P = {}^wP$, $\quad E(\varphi,\lambda)_P = \lambda\varphi + (w\lambda)\Big(M(w,\lambda)\varphi\Big)$.

Consequently, for $0 \ll T \in \mathfrak{a}_{M_0}$ sufficiently positive, Arthur's analytic truncation $\Lambda^T E(\varphi,\lambda)$ becomes

$$\Lambda^T \circ E(\lambda) = E(\lambda) - E^T(\lambda) - E^T_w(w\lambda)\circ M(w,\lambda),$$

or equivalently

$$\Lambda^T \circ E(\lambda) - E(\lambda) = -E^T(\lambda) - E^T_w(w\lambda)\circ M(w,\lambda),$$

where

$$E^T : X^G_M \to \mathrm{Hom}_{\mathbb{C}}\Big(A(M,\pi)^{\mathfrak{F}}, L^{2,\mathfrak{F}}_{\xi,\mathrm{loc}}\Big)$$

is given by

$$E^T(\varphi,\lambda) := \sum_{\gamma\in P(F)\backslash G(F)} (\lambda\varphi)(\gamma g)\cdot\hat{\tau}\big(H_P(\gamma g)-T\big)$$

while its counterpart

$$E^T_w : X^G_{wMw^{-1}} \to \mathrm{Hom}_{\mathbb{C}}\Big(A(wMw^{-1}, w\pi)^{\mathfrak{F}}, L^{2,\mathfrak{F}}_{\xi,\mathrm{loc}}\Big)$$

is given by

$$E^T_w(w\lambda)\phi, w\lambda) := \sum_{\gamma\in P(F)\backslash G(F)} \Big((w\lambda)\phi\Big)(\gamma g)\cdot\hat{\tau}\Big(H_{^wP}(\gamma g)-T\Big).$$

18.3 Basic Properties

Lemma. (i) *The map* E^T *(on* X_M^G *not just on* $\mathfrak{c}$*) is holomorphic*;
(ii) *For every* $\lambda \in X_M^G$, $\Lambda^T \circ E^T(\lambda) = 0$; and
(iii) *There exists* $c > 0$ *such that if* $\lambda \in X_M^G$ *satisfying* $\langle \mathrm{Re}\,\lambda, \alpha^\vee \rangle < c$, *then* $E^T(\varphi, \lambda) \in L_\xi^{2,\mathfrak{F}}$ *for all* $\varphi \in A(M, \pi)^{\mathfrak{F}}$.
All are from basic properties of Arthur's truncation. Indeed, (i) can even be deduced from a much weak version, Theorem A of Day 1.

Similar statements hold for E_w^T.

Thus the difficulty is single out: it concentrate on the intertwining operator $M(w, \lambda)$.

18.4 Functional Equation for Truncated Eisenstein Series

Recall that for $\pi \in \mathfrak{P}$, we have the natural action $i(\pi, \bullet)$ of the algebra $\mathcal{H}^{\mathfrak{F}}$ on $A(M, \pi)^{\mathfrak{F}}$ (to make an artificial distinguish of the general action δ of $\mathcal{H}^{\mathfrak{F}}$).

Set

$$F^T(\phi, \lambda) := \Lambda^T\Big(\delta(h)E^T(\varphi, \lambda)\Big).$$

Since $\delta(h)E^T(\varphi, \lambda)$ is of uniformly moderate growth, $F^T(\phi, \lambda)$ is rapidly decreasing. Consequently we get a map

$$F^T : X_M^G \to \mathrm{Hom}_{\mathbb{C}}\Big(A(M, \pi)^{\mathfrak{F}}, L_\xi^{2,\mathfrak{F}}\Big)$$

which is clearly holomorphic. Similarly, we can have a holomorphic

$$F_w^T : X_{wMw^{-1}}^G \to \mathrm{Hom}_{\mathbb{C}}\Big(A(wMw^{-1}, w\pi)^{\mathfrak{F}}, L_\xi^{2,\mathfrak{F}}\Big).$$

Lemma. *For* $\lambda \in \mathfrak{c}, h \in \mathcal{H}^{\mathfrak{F}}$,

$$\begin{aligned}\Big(\Lambda^T \circ \delta(h) \circ \Lambda^T\Big)\Big(\Lambda^T \circ E(\lambda)\Big) - \Big(\Lambda^T \circ E(\lambda)\Big) \circ i(\lambda, h)\\ = -\,F^T(\lambda) - F_w^T(w\lambda) \circ M(w, \lambda).\end{aligned}$$

Proof. A simple manipulation based on the fact that Λ^T is idempotent and Admissible Lemma of 17.1.

18.5 Application of Resolvent Theory

The advantage of using the operator $\Lambda^T \circ \delta(h) \circ \Lambda^T$ is that it is a compact operator. In particular, if $h = h^* : g \mapsto \overline{h(g^{-1})}$, $\Lambda^T \circ \delta(h) \circ \Lambda^T$ is self-adjoint. So the resolvent theory can be applied.

Indeed, with a fixed K-type, the finite dimensional representation $i(\pi, \bullet)$ of $\mathcal{H}^{\mathfrak{F}}$ on $A(M,\pi)^{\mathfrak{F}}$, it makes sense to introduce the determinant $\det i(\pi, h)$ for $h \in \mathcal{H}^{\mathfrak{F}}$. Set

$$V_h := \Big\{\lambda \in X_M^G : \det i(\pi, h) = 0\Big\}.$$

Note that now X_M^G is just a complex plane, so V_h, consisting of only zeros of $\det i(\pi,h)$, is a discrete subset of X_M^G.

Key Lemma. *There exists a unique meromorphic function*

$$e^T : X_M^G - V_h \to \mathrm{Hom}_{\mathbb{C}}\Big(A(M,\pi)^{\mathfrak{F}}, L_\xi^{2,\mathfrak{F}}\Big)$$

such that for every $\lambda \in X_M^G - V_h$,
(1) $\Lambda^T \circ e^T(\lambda) = e^T(\lambda)$; and
(2) $\Big(\Lambda^T \circ \delta(h) \circ \Lambda^T\Big) e^T(\lambda) - e^T(\lambda) \circ i(\lambda, h) = -F^T(\lambda)$.

Proof. (1) is a direct consequence of the uniqueness in (2) since $\Lambda^T \circ \Lambda^T = \Lambda^T$.
As for (2), first, using the self-adjoint compact operator $\Lambda^T \circ \delta(h) \circ \Lambda^T$ to decompose the space $L_\xi^{2,\mathfrak{F}}$ into its eigenspaces

$$L_\xi^{2,\mathfrak{F}} = \oplus_{\gamma\in\Gamma} H_\gamma.$$

Hence

$$\mathrm{Hom}_{\mathbb{C}}\Big(A(M,\pi)^{\mathfrak{F}}, L_\xi^{2,\mathfrak{F}}\Big) = \oplus_{\gamma\in\Gamma}\mathrm{Hom}_{\mathbb{C}}\Big(A(M,\pi)^{\mathfrak{F}}, H_\gamma\Big).$$

Accordingly decompose $x \in \mathrm{Hom}_{\mathbb{C}}\Big(A(M,\pi)^{\mathfrak{F}}, L_\xi^{2,\mathfrak{F}}\Big)$ as $x = \sum_{\gamma\in\Gamma} x_\gamma$, and set

$$V_h' := \Big\{\lambda \in X_M^G : \det\big(\gamma - i(\lambda,h)\big) = 0\ \exists \gamma \in \Gamma \cup \{0\}\Big\}.$$

Then $X_M^G - V_h'$ is open and dense in X_M^G.

Now let U be a relatively compact open subset of X_M^G whose closure $\overline{U}$ does not intersect V_h'. Then for any $\lambda \in U$ and $\gamma \in \Gamma$, the equation

$$\Big(\Lambda^T \circ \delta(h) \circ \Lambda^T\Big) e^T(\lambda)_\gamma - e^T(\lambda)_\gamma \circ i(\lambda, h) = -F^T(\lambda)_\gamma$$

has a unique solution in $\mathrm{Hom}_{\mathbb{C}}\Big(A(M,\pi)^{\mathfrak{F}}, H_\gamma\Big)$, namely

$$e^T(\lambda)_\gamma := -F^T(\lambda)_\gamma \big(\gamma - i(\lambda, h)\big)^{-1}$$

using the resolvent $\big(\gamma - i(\lambda, h)\big)^{-1}$. With this, it is now standard to show that *the series* $e^T(\lambda) := \sum_{\gamma\in\Gamma} e^T(\lambda)_\gamma$ *converges and gives the solution of the lemma.*

Now define

$$\widetilde{E} : X_M^G - V_h \to \mathrm{Hom}_{\mathbb{C}}\Big(A(M,\pi)^{\mathfrak{F}}, L^{2,\mathfrak{F}}_{\xi,\mathrm{loc}}\Big)$$

by setting

$$\widetilde{E}(\lambda) := E^T(\lambda) + e^T(\lambda).$$

Similarly, we obtain a meromorphic function e_w^T using the set

$$V_{w,h} := \Big\{\lambda \in X^G_{wMw^{-1}} : \det i_w(\lambda, h) = 0\Big\},$$

and hence the w-variation $\widetilde{E}_w$.

Corollary. (i) *Functions* $\widetilde{E}$ *on* $X_M^G - V_h$ *and* $\widetilde{E}_w$ *on* $X^G_{wMw^{-1}} - V_{w,h}$ *are meromorphic;*
(ii) *For every* $\lambda \in \mathfrak{c} - \Big(\mathfrak{c} \cap (V_h \cup w^{-1}V_{w,h})\Big)$,

$$E(\lambda) = \widetilde{E}(\lambda) + \widetilde{E}_w(w\lambda) \circ M(w, \lambda).$$

As such, using the relation

$$E(\pi) \circ i(\pi, h) = \delta(h) \circ E(\pi),$$

we find that for every $\lambda \in \mathfrak{c} - \Big(\mathfrak{c} \cap (V_h \cup w^{-1}V_{w,h})\Big)$,

$$\delta(h)\Big(\widetilde{E}(\lambda)+\widetilde{E}_w(w\lambda)\circ M(w,\lambda)\Big) = \Big(\widetilde{E}(\lambda)+\widetilde{E}_w(w\lambda)\circ M(w,\lambda)\Big)\circ i(\lambda, h)$$

So with the relation

$$M(w,\pi)\circ i(\pi,h)=i(w\pi,h)\circ M(w,\pi),$$

we finally arrive at the following crucial expression involving the intertwining operator $M(w,\lambda)$.

Proposition. *For every* $\lambda\in\mathfrak{c}-\Big(\mathfrak{c}\cap(V_h\cup w^{-1}V_{w,h})\Big)$,

$$\begin{aligned}&\Big(\delta(h)\circ\widetilde{E}(\lambda)-\widetilde{E}(\lambda)\circ i(\lambda,h)\Big)\\ &=-\Big(\delta(h)\circ\widetilde{E}_w(w\lambda)-\widetilde{E}_w(w\lambda)\circ i_w(w\lambda,h)\Big)\circ M(w,\lambda).\end{aligned}$$

18.6 Injectivity

The point of the above proposition is that first of all besides $M(\pi,\lambda)$, all other terms, in particular,

$$\Big(\delta(h)\circ\widetilde{E}(\lambda)-\widetilde{E}(\lambda)\circ i(\lambda,h)\Big)$$

and

$$\Big(\delta(h)\circ\widetilde{E}_w(w\lambda)-\widetilde{E}_w(w\lambda)\circ i_w(w\lambda,h)\Big)$$

are defined over $X_M^G-\big(V_h\cup w^{-1}V_{w,h}\big)$; and secondly, we have the following

Injectivity Lemma. *There exists* $\mu\in X_M^G-\big(V_h\cup w^{-1}V_{w,h}\big)$ *such that the operator*

$$\Big(\delta(h)\circ\widetilde{E}_w(w\lambda)-\widetilde{E}_w(w\lambda)\circ i_w(w\lambda,h)\Big)$$

in $\mathrm{Hom}_{\mathbb{C}}\Big(A(M,\pi)^{\mathfrak{F}},L^{2,\mathfrak{F}}_{\xi,\mathrm{loc}}\Big)$ *is injective.*

Assume this, being injective, a left inverse exists. That is to say, we can choose a continuous map

$$p\in\mathrm{Hom}_{\mathbb{C}}\Big(L^{2,\mathfrak{F}}_{\xi,\mathrm{loc}},A(wMw^{-1},w\pi)^{\mathfrak{F}}\Big)$$

such that

$$p\circ\Big(\delta(h)\circ\widetilde{E}_w(w\lambda)-\widetilde{E}_w(w\lambda)\circ i_w(w\lambda,h)\Big)$$

is an isomorphism. Define then a meromorphic map

$$R : X_M^G - V_h \to \mathrm{Hom}_{\mathbb{C}}\Big(A(M,\pi)^{\mathfrak{F}}, A(wMw^{-1}, w\pi)^{\mathfrak{F}}\Big)$$

by

$$R(\lambda) := p \circ \Big(\delta(h) \circ \widetilde{E}(\lambda) - \widetilde{E}(\lambda) \circ i(\lambda, h)\Big)$$

and similarly

$$R_w : X_{wMw^{-1}}^G - V_{w,h} \to \mathrm{End}_{\mathbb{C}}\Big(A(wMw^{-1}, w\pi)^{\mathfrak{F}}\Big).$$

Then we have for $\lambda \in \mathfrak{c} - \Big(\mathfrak{c} \cap \big(V_h \cup w^{-1}V_{w,h}\big)\Big)$,

$$-R(\lambda) = R_w(w\lambda) \circ M(w, \lambda).$$

Now since $R_w(w\mu)$ is invertible, the function $\det R_w(\bullet)$ is not identically zero, consequently the operator

$$\begin{array}{ccc} X_{wMw^{-1}}^G & \to & \mathrm{End}_{\mathbb{C}}\Big(A(wMw^{-1}, w\pi)^{\mathfrak{F}}\Big) \\ \lambda & \mapsto & R_w(\lambda)^{-1} \end{array}$$

is still meromorphic. Thus, we can continue the function $M(w, \bullet)$ meromorphically to $X_M^G - \big(V_h \cup w^{-1}V_{w,h}\big)$ by setting

$$M(w, \lambda) = -R_w(w\lambda)^{-1} \circ R(\lambda),$$

and hence continue E to the same domain by letting

$$E(\lambda) := \widetilde{E}(\lambda) + \widetilde{E}_w(w\lambda) \circ M(w, \lambda).$$

Thus to complete the proof, it suffices to show that by choosing different h, $X_M^G - \big(V_h \cup w^{-1}V_{w,h}\big)$ cover the whole X_M^G, since the continuations obtained by changing h's can be glued together by the uniqueness of the meromorphic extension.

18.7 Meromorphic Continuation

So the only essential problem left is to establish the injectivity for the operator

$$\delta(h)\circ\widetilde{E}_w(w\lambda)-\widetilde{E}_w(w\lambda)\circ i_w(w\lambda,h)\in \mathrm{Hom}_{\mathbb{C}}\Big(A(M,\pi)^{\mathfrak{F}},L^{2,\mathfrak{F}}_{\xi,\mathrm{loc}}\Big).$$

This clearly depends on the choice of h in the Hecke algebra. For this, we do it in general setting (i.e., without assuming that X_M^G is of dimension one).

Lemma. *Let $\lambda\in X_M^G$. There exists $h\in\mathcal{H}^{\mathfrak{F}}$ such that*
(i) $h=h^*$, *where* $h^*(g):=\overline{h(g^{-1})}$;
(ii) $i(\lambda,h)=1$ *for all* $w\in W(M)$;
(iii) *There exists* $\mu\in X_M^G$ *such that*
(a) $\langle \mathrm{Re}\mu,\hat{\alpha}\rangle\gg 0$; and
(b) $i_w(w\lambda,h)=\sqrt{-1}$ *for all* $w\in W(M)$.
Proof. $\mathcal{H}^{\mathfrak{F}}$ is large enough to allow such a choice [MW].
Injectivity Lemma′. *There exists* $\mu\in X_M^G-\big(V_h\cup w^{-1}V_{w,h}\big)$ *such that the operator*

$$\Big(\delta(h)\circ\widetilde{E}_w(w\lambda)-\widetilde{E}_w(w\lambda)\circ i_w(w\lambda,h)\Big)$$

in $\mathrm{Hom}_{\mathbb{C}}\big(A(M,\pi)^{\mathfrak{F}},L^{2,\mathfrak{F}}_{\xi,\mathrm{loc}}\big)$ *is injective.*
Proof. Let μ be as in the previous lemma, then by definition, $\mu\notin V_h\cup w^{-1}V_{w,h}$. Moreover, by the choice of h, $i_w(w\mu,h)=\sqrt{-1}$. Hence the operator

$$\delta(h)\circ\widetilde{E}_w(w\lambda)-\widetilde{E}_w(w\lambda)\circ i_w(w\lambda,h)=\Big(\delta(h)-\sqrt{-1}\cdot\mathrm{Id}\Big)\circ\widetilde{E}_w(w\mu),$$

Sublemma. (i) *For every $\lambda\in X_M^G$, $E^T(\lambda)$ and $E_w^T(w\lambda)$ are injective;*
(ii) $\Big(1-\Lambda^T\Big)\circ\widetilde{E}_w(w\mu)=E_w^T(w\mu)$.
Proof. (ii) is simple and comes from the proof of $\Lambda^T\circ\Lambda^T=\Lambda^T$. As for (i), even injectivity of $E^T(\lambda)$ is crucial to the proof, it is still a rather 'weak' statement: After all, it suffices to check that if ϕ is not identically zero, then *the function* $E^T(\phi)$ is not identically zero. See pp.150-151 of [MW].

Clearly, then the injectivity of $E_w^T(w\mu)$ implies the injectivity for $\widetilde{E}_w(w\mu)$, since the formal is the composition of the latter from the left by $1 - \Lambda^T$. (A simple fact about maps.) Moreover, by definition, $\widetilde{E}_w(w\mu) = e_w^T(w\mu) + E_w^T(w\mu)$. Thus from the fact that $e_w^T(w\mu)$ is in $L_\xi^{2,\mathfrak{F}}$, by 18.3, which applies here since we assume that $\langle \mathrm{Re} w\mu, -w\hat{\alpha}\rangle \ll 0$, $\widetilde{E}_w(w\mu) \in L_\xi^{2,\mathfrak{F}}$ as well. As such note that $\delta(h)$ is self-adjoint, so the restriction to the space $L_\xi^{2,\mathfrak{F}}$ of $\delta(h) - \sqrt{-1}\cdot \mathrm{Id}$ is injective. Consequently, the composition $\Big(\delta(h) - \sqrt{-1}\cdot\mathrm{Id}\Big)\circ \widetilde{E}_w(w\mu)$ is injective. This establishes the meromorphic continuation.

18.8 Functional Equation

Consider the meromorphic function

$$\begin{array}{rccl} e: & X_M^G & \to & \mathrm{Hom}_{\mathbb{C}}\Big(A(M,\pi)^{\mathfrak{F}}, L^{2,\mathfrak{F}}_{\xi,\mathrm{loc}}\Big) \\ & \lambda & \mapsto & E(\lambda) - E_w(w\lambda)\circ M(w,\lambda). \end{array}$$

Let $\varphi \in A(M,\pi)^{\mathfrak{F}}$ such that $e(\varphi,\lambda) \neq 0$. Then essentially by the calculation of constant terms of the associated Eisenstein series, we conclude that for every standard parabolic subgroup P' of G,

$$\begin{aligned} e(\varphi,\lambda)_{P'}^{\mathrm{cusp}} &= 0 \qquad \text{if } P' \neq P \\ e(\varphi,\lambda)_{P'}^{\mathrm{cusp}} &= \lambda\big(\varphi - M_w(w^{-1}, w\lambda)M(w,\lambda)\varphi\big). \end{aligned}$$

Thus

$$\pi\otimes\lambda \in \Pi_0\Big(M, e(\varphi,\lambda)\Big).$$

That is, $\pi\otimes\lambda$ is in the cuspidal support of $e(\varphi,\lambda)$. We now claim that this will result a contradiction.

In fact, according to Thm E of Day 1 on exponents of L^2 automorphic forms, if we put certain negative conditions on the real part of λ, $e(\varphi,\lambda)$ becomes square integrable. This then implies that λ is essential real, by Prop 3.1 [MW], which contradicts with the fact that for this λ, $\mathrm{Im}\,\lambda$ is chosen to be quite general. Therefore, $e = 0$ on X_M^G. That is to say,

$$E(\lambda) = E_w(w\lambda)\circ M(w,\lambda).$$

Consequently, $e(\varphi,\lambda)_P^{\text{cusp}}=0$ for all λ,φ which implies

$$M_w(w^{-1},w\lambda)M(w,\lambda)=1.$$

This gives (all) the functional equation(s) in relative rank one since $W(M)=\{1,w\}$ only consists of two elements.

References

[Ar1] Arthur, J. Eisenstein series and the trace formula, Proc. Sympos. Pure Math., XXXIII, 253-274, AMS, Providence, R.I., 1979.

[Ar2] Arthur, J. A trace formula for reductive groups. I. Terms associated to classes in $G(\mathbb{Q})$. Duke Math. J. **45** (1978), no. 4, 911–952

[Ar3] Arthur, J. A trace formula for reductive groups. II. Applications of a truncation operator. Compositio Math. **40** (1980), no. 1, 87–121.

[Ar4] Arthur, J. On the inner product of truncated Eisenstein series, Duke Math. J. **49** (1982), 35-70

[Ar5] Arthur, J. *An introduction to the trace formula*, to appear

[La] Langlands, R. *On the functional equations satisfied by Eisenstein series*, Springer LNM **544**, 1976

[MW] Moeglin, C. & Waldspurger, J.-L. *Spectral decomposition and Eisenstein series.* Cambridge Tracts in Math, **113**. Cambridge University Press, Cambridge, 1995.

[OW] Osborne, M.S. & Warner, G. *Theory of Eisenstein Systems*, Academic Press 1981

Lin WENG
Graduate School of Mathematics
Kyushu University
Fukuoka 812-8581, Japan
Email: weng@math.kyushu-u.ac.jp

Geometric Arithmetic: A Program

Lin WENG

In this article, we originate a program for what we call Geometric Arithmetic. Such a program would consist of four parts, if we were able to properly understand the essentials now. Namely, (1) Non-Abelian Class Field Theory; (2) Geo-Ari Cohomology Theory; (3) Non-Abelian Zeta and L Functions; and (4) Riemann Hypothesis. However, here we could only provide the reader with $\frac{1+1(=\frac{1}{2}+\frac{1}{2})+1}{4}$ of them. To be more precise, discussed in this article are the following particulars;

(A) Representation of Galois Group, Stability and Tannakian Category;
(B) Cohomology, Truncations and Non-Abelian Zeta and L-Functions; and
(C) Explicit Formula, Functional Equation and Geo-Ari Intersection.

So what are these ABC of the Geometric Arithmetic?!

As stated above, (A) is aimed at establishing a Non-Abelian Class Field Theory. The starting point here is the following classical result: Over a compact Riemann surface, a line bundle is of degree zero if and only if it is flat, i.e., induced from a representation of fundamental group of the Riemann surface. Clearly, being a bridge connecting divisor classes and fundamental groups, this result may be viewed as and is indeed a central piece of the classical (abelian) class field theory. (See e.g., [H] and [W1].) Thus it is then only natural to give a non-abelian generalization of it in order to offer a non-abelian class field theory. This was first done by Weil. In his fundamental paper on generalization of abelian functions [W1], Weil showed that over a compact Riemann surface, a vector bundle is of degree zero if and only if it is induced from a representation of fundamental group of the surface.

Thus far, two new aspects naturally emerge. That is, unitary representations and non-compact Riemann surfaces, reflecting finite quotients of Galois groups and ramifications in Class Fields Theory, CFT for short, respectively: In a (complex) representation class of a finite group, there always exists a unitary one, while a discussion for compact Riemann surfaces results only unramified CFT. Thus mathematics demands new results to couple with them. It is well known that to this end we then have (i) Mumford's stability of vector bundles in terms of intersection; (ii) Narasimhan-Seshadri's correspondence; and (iii) Seshadri's parabolic analog of (i) and (ii). That is to say, now the above result of Weil is further refined to the follows: Over (punctured) Riemann surfaces, (Seshadri) equivalence classes of semi-stable parabolic bundles of parabolic degree zero correspond naturally in one-to-one to isomorphism classes of unitary representations of fundamental groups.

On the other hand, the results above, while central, do only parts of the CFT – at its best, the Weil-Narasimhan-Seshadri correspondence reflects a micro reciprocity law. What CFT really stands should not be a relation between a single representation and an isolated bundle, instead, CFT should expose Galois groups intrinsically in terms of bundles globally. Thus an integration process aiming at constructing a global theory becomes a great necessity.

It is at this point where the theory of Tannakian category enters into the picture. Recall that a typical theory of Tannakian category takes the following forms: (i) groups may be reconstructed from their associated categories of representations; (ii) fiber functors equipped Tannakian categories are clone categories of (i), i.e., are equivalent to the categories of representations; and (iii) original groups may be recovered from the automorphism groups of fiber functors.

At it turns out, with this strongest form of the standard theory of Tannakian category, we have little hope to match it perfectly with the CFT we are looking for. Fortunately, there are still room to manoeuvre, since in CFT we only care about finite quotients of the associated groups, and in terms of representations finite quotients correspond to what we call finitely completed Tannakian subcategories according to Tannaka duality and van Kampen completeness theorem. In this way, we finally establish a non-abelian CFT for Riemann surfaces, or better, for function fields over complex numbers successfully. Main results include the Existence Theorem, the Conductor Theorem and the Reciprocity Law. See e.g. the Main Theorem in A.2.

By establishing a CFT for Riemann surfaces as above, possibly, we give the reader an impression that everything works smoothly. No, practically, it is not the case. For example, we do not need all unitary representations. Or put this in another way, all semi-stable parabolic bundles of parabolic degree zero lead us to nowhere. Consequently, we must carefully select among these semi-stable objects a handful portion so that (i) the theory of Tannakian category can be applied; and (ii) there are still enough ingredients to apply Tannaka duality and van Kampen completeness theorem. This then leads to what we call geo-ari representations and geo-ari bundles, for which the Narasimhan-Seshadri correspondence still holds. (For curves over finite fields, besides deformation theory, we need to introduce a new principle, called the Harder-Narasimhan correspondence.)

The experienced reader here naturally would ask how we overcome the difficulty about tensor products of geo-ari bundles, since, generally speaking, to show the closeness under tensor operation is the only point to check when applying the theory of Tannakian category. Here for Riemann surfaces, two approaches are available. For one, we use the Narasimhan-Seshadri correspondence, as easily tensors of unitary representations are again unitary. But this analytic approach is not a genuine one, since a micro reciprocity law, i.e., the Weil-Narasimhan-Seshadri correspondence, is used. Thus a purely algebraic proof should be pursued. This then leads to the works of Kempf and Ramanan-Ramanthan on instability flags, which we call the KR^2-trick. Moreover, as the original KR^2-trick only works for bundles without parabolic structures, so to stylize the non-abelian CFT (for Riemann surfaces), we ask for a parabolic version of the

RK^2-trick.

Motivated by such a success in non-abelian CFT for function fields over $\mathbb{C}$, we anticipate that in principle, the non-abelian CFT for local and global fields works similarly. So the building blocks of our program for a non-abelian CFT then are the follows:

(1) There should be a suitable type of representations of Galois groups, which we call geometric representations and a suitable type of intersection stability for bundles which we call geometric parabolic bundles such that an analog of the Weil-Narasimhan-Seshadri Correspondence holds;

(2) There should exist subclasses of geometric representations and geometric parabolic bundles, which we call geo-ari representations and geo-ari bundles, respectively, such that (i) these classes form naturally two abelian categories, (ii) an analog of Hader-Narasimhan Correspondence holds; and (iii) an analog of KR^2-trick works. Thus in particular, by (1) and (2), we obtain two equivalent (generalized) Tannakian categories together with natural fiber functors;

(3) The (generalized) Tannakian categories contain systems of the so-called finitely completed Tannakian subcategories, so that via an analog of Tannaka Duality and van Kampen completeness theorem, we obtain the so-called fundamental theorem of non-abelian CFT consisting of existence theorem, conductor theorem and the reciprocity law.

To end this brief discussion on Part (A), we would like to point out that to realize the above mentioned 123 for our non-abelian CFT, standard theories on GIT, Tannakian category and representations of Galois groups are far from being enough. For examples, to achieve (1), we require (i) a Geometric Invariant Theory over integral bases in the spirit of Arakelov; (ii) a deformation theory for geometric representations of Galois groups; and (iii) a suitable completeness for representation and stability along the line of Fountaine and Langton, respectively; and to achieve (3), we require a theory of Tannakian category over integral bases.

Our next main scheme is devoted to non-abelian zeta functions.

As stated at the very beginning, Part (B) is a combination of our partial understanding of our new non-abelian zeta functions and what we call geo-ari cohomologies. This part to a large extent is practical rather than theoretical, due to the fact that not only all studies here are based on practical constructions, but we have not yet understood the mathematics involved theoretically.

Unlike for classical Weil zeta functions, instead of working on general algebraic varieties and counting their rational points (over finite fields) in a very primitive way, for our non-abelian zeta functions, we concentrate our attentions to moduli spaces of semi-stable bundles and count their rational points from moduli point of view, in a similar way as what Shimura does for Shimura varieties.

To be more precise, consider function fields over finite fields first. Then, for each fixed natural number r, we, by using a work of Mumford-Narasimhan-Seshadri, obtain the associated moduli spaces of rank r semi-stable bundles. In particular, with the so-called Harder-Narasimhan correspondence, which claims that the rationalities of bundles and moduli points coincide, we could then introduce a new type of zeta functions by considering rational points of the moduli space as moduli points associated with rational semi-stable bundles.

This approach, while different from that of Weil, is indeed a natural generalization of that of Artin: When $r = 1$, our construction recovers classical Artin zeta functions. Moreover, just like classical abelian zeta functions, our non-abelian version satisfies rationality and function equation as well. Since we even can give uniform bounds for the coefficients of these (local) zeta functions, so via an Euler product, we further introduce a more global non-abelian zeta functions for curves defined over number fields. Needless to say, when $r = 1$, these global zeta functions are nothing but the classical Hasse-Weil zeta functions. So non-abelian arithmetic aspect of curves is supposed to be reflected by these new zeta functions.

Well, while this latest general statement should finally lead us to a mathematics wonderland, we have not yet found our theoretical feet: the non-abelian reciprocity law. For this purpose, we then turn our attention to some concrete examples. This directs us to the study of what we call refined Brill-Noether loci and their intersections: Beyond traditional considerations, refined Brill-Noether locus measures how automorphisms of the associated bundles change too. Consequently, we obtain a concrete reciprocity law for elliptic curves in ranks 2 and 3. All this is done in B.1.

It now becomes quite apparent that to introduce non-abelian zeta functions for number fields, we need to introduce
(i) stablity for lattices with which the associated moduli is compact; and
(ii) a canonical cohomology theory which satisfies the Serre Duality and Riemann-Roch.

All this is done in B.2. Indeed, with Arakelov's approach towards arithmetic, intersection stability for lattices becomes more or less trivial. (This explains why many mathematicians, including Stuhler and the author, introduce stability for lattices totally independently.) As for a cohomology theory associated to number fields, based on Tate's Thesis [T], we are able to offer a practical definition in terms of Chevalley adelic language. More precisely, our cohomology groups are locally compact groups with H^0 discrete and H^1 compact and their counts are given with the help of Fourier analysis. In particular, the Serre duality comes from Pontragin duality and the Riemann-Roch comes from the Poisson summation formula and Serre duality. (As such, it would be extremely interesting to compare our approach with Deninger's approach using infinite dimensional spaces.)

Consequently, non-abelian zeta functions for number fields can be introduced. Being a natural generalization of classical Dedekind zeta functions, these new zetas are canonical as well. That is, they admit natural meromorphic continuations, satisfy 'the' functional equation, and the residues of them at simple poles are nothing but the Tamagawa type volumes of the associated moduli spaces of semi-stable bundles (over number fields). In particular, when rank is one, our work essentially recovers Iwasawa's ICM talk at MIT about Dedekind zeta functions. For details, please refer to B.2.

By saying this, we have no intention to claim that we are satisfied with what we have achieved. Far from being it – we have little

understanding of these new zeta functions. For examples,
(1) we have no idea now on how the non-abelian reciprocity law, which, by (A), are supposed to hold naturally, can be read from our non-abelian zeta functions;
(2) generally speaking, we are less sure about the meaning of special values of our non-abelian zeta functions, except their residues at $s = 1$.

On the other hand, as to (i) and (ii) above for number fields, what is accomplished here is the introduction of intersection stability and hence moduli spaces of semi-stable lattices, and a practical formation of one-dimensional geo-ari cohomology for which the duality and Riemann-Roch hold.

Not only our non-abelian zetas satisfy stereotype properties, they can also be written as integrations of Epstein type of zeta functions. This then leads us to introduce more general non-abelian L-functions using Langlands' fundamental theory of Eisenstein series. Roughly speaking, non-abelian L-functions are integrations of Eisenstein series associated to L^2-automorphic forms over certain geometrically truncated domains. For more details, please go to B.3.

On the other hand, in the theory of automorphic forms, or better, in Arthur's study of trace formula, there is another way to get integrations from Eisenstein series. Recall that since constant terms are usually only of moderate growth, Eisenstein series are not integrable over entire fundamental domains. To overcome this, by suitably truncating constant terms, Arthur then is able to create new yet rapidly decreasing functions, the so-called Arthur's truncation (at least for sufficiently regular parameters).

A basic observation here is that integrations over fundamental domains of Arthur's truncated Eisenstein series can also be expressed as integrations of the original Eisenstein series over certain truncated domains. Moreover, following Lafforgue [Laf], we are able to build up a global bridge between algebraic truncation obtained using stability and analytic truncation of Arthur associated to the constant function $\mathbf{1}$. All this then transforms our non-abelian L-functions to a special kind of Arthur's periods, or better Eisenstein periods (at least in the case where parameters are sufficiently regular).

As such, we are in a position to apply an advanced version of Rankin-Selberg method ([JLR]). This then helps us to see a much clearer structure of (a special kind of) these non-abelian L-functions. For more details, please go to B.4 and B.5. As a by-product, we are able to expose a fundamental relation among special values of abelian and non-abelian zetas (see e.g., B.5.6).

As said, our practical cohomology works only in dimension one. However, based on linear compacity of Chevalley, as given in Iwasawa's Princeton lectures notes, and Parshin's approach to duality and residue in dimension two, we in Part (C) offer a program for what we call a half-theoretical geo-ari cohomology in lower dimensions.

Part (C) of the program is designed to give a geometric justification of the formal summation $\sum_{\rho:\xi(\rho)=0} t^\rho$ for $t \in \mathbf{R}$. For this purpose, we propose a model for two-dimensional geo-ari intersection theory. This geo-ari intersection turns out to be extremely interesting, since the Riemann Hypothesis may be naturally studied within its framework. Since the Cramer formula is behind the above summation for zeros of Riemann zeta and since the Explicit Formula of Weil is behind the above proposed geo-ari intersection, our approach to the Riemann Hypothesis is in appearance different from but in essence related to that of Deninger and Connes-Haran.

More precisely, this model of two-dimensional geo-ari intersection is based on six elegant axioms for intersections of what we call micro divisors, which themselves are motivated by standard properties of intersections in dimension two, the Riemann-Roch in dimension one, adjunction formula for curves in surfaces, global functional equation, and Weil's explicit formula. They are parametrized by $\mathbb{R}$ and can be used to build up global divisors using an integration process. (This is a huge step: in algebraic geometry, divisors are defined as *finite* $\mathbb{Q}$-combination of prime divisors, and in Arakelov theory, only finite $\mathbb{R}$-combination of certain prime divisors are allowed.)

The advantage of having this mathematics model on geo-ari intersection is that, as in geometry, then the Riemann Hypothesis may be deduced from an analog of the Hodge index theorem. Accordingly, with our experiences in algebraic geometry and in Arakelov theory, we are led to search for the corresponding geo-ari cohomology in dimension two for which the Riemann-Roch holds, a challenging yet reasonable task. (However, it is left as everyone's guess that whether geo-ari ampleness and hence a natural positivity can be obtained.)

While the Riemann Hypothesis for abelian zeta functions, the main driving force for the above consideration, is still far from being understood, surprisingly its analogue for rank two zetas can be established. This is treated in C.4, based on certain algebraic, analytic and geometric consideration.

This article is a revised version of 'A Program for Geometric Arithmetic' written around 1999-2001: We rewrote sections B.2.2 and B.2.4 and added chapters B.3-5 and C.4, based on important discoveries described in (1) Non-Abelian L-Functions for Number Fields; (2) Rank Two Non-Abelian Zeta and Its Zeros, and (3) Arthur's Periods, Regularized Integrals and Refined Structures of Non-Abelian L-Functions[1]. So as Weil once said:

Number Theory Is Not Standing Still.

[1]Since essentials are copied in Chapters B.4-5, this paper will not be published elsewhere.

Contents

A. Representations of Galois Groups, Stability and Tannakian Categories

A.1. Summary

Non-Abelian Class Field Theory: A Program		
Galois Aspect	**Principle**	**Bundle Aspect**
Geometric Reps	Narasimhan-Seshadri	S.S. Para. Bdles
$\Downarrow$ Rationality	Vanishing of Brauer $\Downarrow$	Rationality$\Downarrow$
Geo-Ari Reps	Harder-Narasimhan	Geo-Ari Bundles
$\Downarrow \otimes$	KR^2 Trick	$\otimes \Downarrow$
Tannakian Cat	Tannaka Duality	Clone Category
Galois Group	Reciprocity Map	$\mathrm{Aut}^{\otimes}$
Finite Quotient	$\exists$ Thm, Reciprocity Law	F. Completed Mod

A.2. Non-Abelian CFT for Function Fields over C

A.2.1. Weil-Narasimhan-Seshadri Correspondence

A.2.1.1. Unitary Representations of Fundamental Groups

Let M^0 be a punctured Riemann surface of signature (g, N) with M the smooth compactification. Then, $M^0 = M\backslash\{P_1, \ldots, P_N\}$, M is of genus g with $P_1, \ldots, P_N$ pairwise distinct points on M. Suppose that $2g - 2 + N > 0$. From the uniformization theorem, M^0 can be represented as a quotient $\Gamma\backslash\mathcal{H}$ of the upper half plane $\mathcal{H} = \{z \in \mathbf{C} : \operatorname{Im} z > 0\}$ modulo an action of a torsion-free Fuchsian group $\Gamma \in PSL_2(\mathbf{R})$, generated by $2g$ hyperbolic transformations $A_1, B_1, \ldots, A_g, B_g$ and N parabolic transformations $S_1, \ldots, S_N$ satisfying a single relation

$$A_1 B_1 A_1^{-1} B_1^{-1} \cdots A_g B_g A_g^{-1} B_g^{-1} S_1 \cdots S_N = 1.$$

Denote the fixed points, the so-called cusps, of the parabolic elements $S_1, \ldots, S_N$ by $z_1, \ldots, z_N$ respectively. Then, images of the cusps

$z_1, \ldots, z_N \in \mathbf{R} \cup \{\infty\}$ under the projection $p : \mathcal{H}^* := \mathcal{H} \cup \mathbf{R} \cup \{\infty\} \to \Gamma\backslash\mathcal{H}^* = M$ result the punctures $P_1, \ldots, P_N \in M$. For each $i = 1, \ldots, N$, denote by $\Gamma_i = \Gamma_{z_i}$ the stablizer of z_i in Γ. Then Γ_i is a cyclic subgroup in Γ generated by S_i. Moreover, for an element $\sigma_i \in PSL_2(\mathbf{R})$ such that $\sigma_i\infty = z_i$, we have $\sigma_i^{-1} S_i \sigma_i = \begin{pmatrix} 1 & \pm 1 \\ 0 & 1 \end{pmatrix}$, and hence $\langle \sigma_i^{-1} S_i \sigma_i \rangle = \Gamma_\infty$. (For simplicity, from now on, when only a local discussion is involved, we always assume that $z_i = \infty$.)

For a representation $\rho : \pi_1(M^0) \simeq \Gamma \to GL(n, \mathbf{C})$ of Γ into a complex vector space V, the vector bundle $\mathbf{V} := \mathcal{H} \times V$ on $\mathcal{H}$ admits a natural Γ-vector bundle structure via $\gamma(z, v) = (\gamma(z), \rho(\gamma)v)$ for $\gamma \in \Gamma, z \in \mathcal{H}$ and $v \in V$. The quotient of $\mathcal{H} \times V$ modulo the action of Γ is then a vector bundle of rank n over M^0. Moreover, since the same representation ρ defines also a Γ-vector bundle structure on $\mathcal{H}^* \times V$, we obtain a vector bundle V_ρ on $M = \Gamma\backslash\mathcal{H}^*$ as well.

Next assume that ρ is unitary. Then with respect to a suitable basis of V,

$$\rho(S_i) = \mathrm{diag}\Big(\exp(2\pi i \alpha_{i1}), \ldots, \exp(2\pi i \alpha_{i,n})\Big)$$

where $\alpha_{ij} \in [0, 1)$ for all $i = 1, \ldots, N, j = 1, \ldots, n$. Hence, V_ρ, or better, the associated sheaf $p_*^\Gamma(\mathbf{V})$ of sections may be interpreted as follows: On M^0, it corresponds to the Γ-invariant sections of $\mathbf{V}$, while near parabolic punctures $P = P_i \in M$, over a neighborhood U of P of the form $\mathcal{H}_\delta/\Gamma_\infty$ where $\mathcal{H}_\delta := \{z = x + iy : y > \delta > 0\}$, the sections are all bounded Γ_∞-invariant sections of $\mathbf{V}$ on $\mathcal{H}_\delta$. Thus, as an $\mathcal{O}_{M,p}$-module, a basis of $p_*^\Gamma(\mathbf{V})$ at P is given by the Γ_∞ sections $\theta_j : z \mapsto \exp(2\pi i \alpha_j z) \cdot e_j$ where $\{e_1, \ldots, e_n\}$ is a basis of V such that $S_i(e_j) = \exp(2\pi i \alpha_{ij}) \cdot e_j$.

As a direct consequence, in addition to the associated bundles V_ρ on M, there exist as well the following structures on the fibers of V_ρ at punctures $P_1, \ldots, P_N$: Over $P = P_i$, we obtain real numbers

$$\alpha_{i1} = \alpha_{i2} = \cdots = \alpha_{ik_1} < \alpha_{i,k_1+1} = \alpha_{i,k_1+2} = \cdots = \alpha_{ik_2} < \cdots = \alpha_{i,k_{r_i}}$$

and a decreasing flag of $V_\rho|_P$ defined by

(i) $F_1(V_\rho|_P) := V_\rho|_P$;

(ii) $F_2(V_\rho|_P)$ the subspace spanned by $\theta_{k_1+1}, \dots, \theta_n$;
(iii) $F_3(V_\rho|_P)$ the subspace spanned by $\theta_{k_2+1}, \dots, \theta_n$, etc.
Clearly $k_1 = \dim F_1(V_\rho|_P) - \dim F_2(V_\rho|_P), \dots, k_r = \dim F_r(V_\rho|_P)$, and these additional structures are indeed determined by $\alpha'_{ij} := \alpha_{ik_j}, j = 1, \dots, r_i$, and k_j's.

Proposition. (Seshadri) *With the same notation as above,* $\deg(V_\rho) = -\sum_{i,j=1}^{N,r_i} k_{ij}\alpha'_{ij}$.

A.2.1.2. Semi-Stable Parabolic Bundles

Following Seshadri, by definition, a parabolic structure on a vector bundle E over a compact Riemann surface is given by the following data:
(1) a finite collection of points $P_1, \dots, P_N \in M$; and for each $P = P_i$,
(2) a flag $E_P = F_1 E_P \supset F_2 E_P \cdots \supset F_r E_P$; and
(3) a collection of parabolic weights $\alpha_1, \dots, \alpha_r$ attached to the filtration $F_1 E_P, \dots, F_r E_P$ such that $0 \leq \alpha_1 < \alpha_2 < \cdots < \alpha_r < 1$.
Often $k_1 = \dim F_1 E_P - \dim F_2 E_P, \dots, k_r = \dim F_r E_P$ are called the multiplicities of $\alpha_1, \dots, \alpha_r$; and a bundle E together with a parabolic structure

$$\left(P = P_i; E_P = F_1 E_P \supset F_2 E_P \cdots \supset F_r E_P; \alpha_1 = \alpha_{i1}, \dots, \alpha_r = \alpha_{ir_i}\right)_{i=1}^N$$

is called a parabolic bundle and is written as

$$\Sigma(E) := \Sigma := \Big(E; \big(P = P_i; E_P = F_1 E_P \supset F_2 E_P \cdots \supset F_r E_P; \\ \alpha_1 = \alpha_{i1}, \dots, \alpha_r = \alpha_{ir_i}\big)_{i=1}^N\Big).$$

Trivially, if W is a subbundle of E, then Σ induces a natural parabolic structure $\Sigma(W)$ on W.

For parabolic bundles, its associated parabolic degree is defined to be

$$\mathrm{para}\deg(\Sigma) := \deg E + \sum_{i=1}^N \left(\sum_{j=1}^{r_i} k_{ij}\alpha_{ij} \right).$$

So, in particular, if Σ is induced from a unitary representation of fundamental group of a Riemann surface, its associated para degree is zero. By definition, a parabolic bundle Σ is called (Mumford-Seshadri) semi-stable (resp. stable) if for any subbundle W of E,

$$\frac{\mathrm{para}\deg(\Sigma(W))}{\mathrm{rank}(W)} \quad \leq \quad (\text{resp.} \ <) \quad \frac{\mathrm{para}\deg(\Sigma(E))}{\mathrm{rank}(E)}.$$

Proposition. (Seshadri) *With the same notation as above, if Σ is induced from a unitary representation of the corresponding fundamental group, then Σ is semi-stable of degree zero. Moreover, if the representation is irreducible, then Σ is stable.*

A.2.1.3. Weil-Narasimhan-Seshadri Correspondence: A Micro Reciprocity Law

The real surprising result is the inverse of Proposition 2.1.2. The starting point for all this is the following classical result on line bundles: Over a compact Riemann surface, a line bundle is of degree zero if and only if it is flat, i.e., it is induced from a representation of the fundamental group. It is Weil who first generalized this to vector bundles in his fundamental paper on 'Generalisation des fonctions abeliennes' dated in 1938: Over a compact Riemann surface, a vector bundle is of degree zero if and only if it is induced from a representation of the fundamental group. (The reader may find a modern proof in Gunning's Princeton lecture notes in which all fundamental results on Riemann surfaces are used.)

It is said that Weil's primitive motivation is to develop a non-abelian CFT for Riemann surfaces. Granting this, clearly, the next step is to study what happens for general Riemann surfaces, which need not to be compact. This then leads to Weil and Toyama's theory on matrix divisors.

While all this seems to be essentially in the right direction, still many crucial points are missing in these pioneer studies.

Recall that the reciprocity law in CFT is essentially the one for finite field extensions and that for a finite group, in any equivalence class of (finite dimensional complex) representations there always exists a unitary one. Thus naturally we should strengthen Weil's theorem from any representation to that of unitary representation. For doing so, we meet a huge difficulty on algebraic side. That is, how to give a corresponding algebraic condition?! This is solved by Mumford with his famous intersection stability. In fact, not only Mumford introduced the intersection stability, he also studied the associated deformation theory via his fundamental work on GIT stability.

On the other hand, for Riemann surfaces, fundamental groups can be described very precisely, and hence deformations of the associated unitary representations can be quantitatively studied. All this, together with certain completeness for both unitary representations and Mumford's intersection stability, the so-called Langton's Principle, then leads Narasimhan and Seshadri to prove that over a compact Riemann surface, Mumford's semi-stable vector bundles of degree zero are naturally associated with *unitary* representations of the fundamental group of the surface. Later on, Seshadri first generalizes this result to π-bundles, and then to parabolic bundles.

Theorem. (Narasimhan-Seshadri, Seshadri) *There is a natural one-to-one correspondence between isomorphism classes of unitary representations of fundamental groups of punctured Riemann surfaces and Seshadri classes of semi-stable parabolic bundles of parabolic degree zero.*

We would like call this result the Weil-Narasimhan-Seshadri correspondence. Due to the fact that it reveals an intrinsic relation between fundamental groups and vector bundles, often we call it a *micro reciprocity law* as well. (Over higher dimensional compact Kähler manifold, a related correspondence is named the Kobayashi-Hitchin correspondence, which is established by Ulenberk-Yau. See also Donaldson for projective manifolds.)

Note also that, as stated above, the existing proof is based on

(a) Geometric invariant theory;

(b) Deformation of representations; and

(c) Completeness of semi-stable parabolic bundles – the Langton principle.

Consequently, all this is supposed to play a crucial role in our program for non-abelian CFT, the details of which will be discussed later.

While the Narasimhan-Seshadri correspondence is a kind of Reciprocity Law, it is only a micro one. Thus to find the genuine one, we need to do it globally. Thus, the following result for finite groups naturally enters into the picture: Any finite group is determined by its characters and vice versa. Hence, if we have a similar result for more general groups, we can establish our non-abelian CFT for Riemann surfaces. As usual, we are led to the theory of Tannakian categories. But before that we still need to make sure that coverings and parabolic bundles are under control. It is for this purpose, we introduce the Rationality discussion in our program.

A.2.2. Rationality: Geo-Ari Representations and Geo-Ari Bundles

A.2.2.1. Branched Coverings of Riemann Surfaces

The advantage in developing a non-abelian CFT for Riemann surfaces is that all the time we have concrete geometric models ready to use. As there is no additional cost and also for a possible development on non-abelian CFT for higher dimensional function fields over complex numbers, we next recall some basics for branched coverings of complex manifolds.

Let M be an n-dimensional connected complex manifold. A branched covering of M is by definition an n-dimensional irreducible normal complex space X together with a surjective holomorphic mapping $\pi : X \to M$ such that
(1) every fiber of π is discrete in X;
(2) $R_\pi := \{q \in X : \pi^* : \mathcal{O}_{\pi(q),M} \to \mathcal{O}_{q,X}$ is not isomorphic$\}$ and $B_\pi := \pi(R_\pi)$ are hypersurfaces of X and M respectively. As usual, we call R_π and B_π ramification locus and branch locus respectively;
(3) $\pi : X\backslash\pi^{-1}(B_\pi) \to M\backslash B_\pi$ is an unramified covering; and
(4) for any $p \in M$, there is a connected open neighborhood W of p in M such that for every connected component U of $\pi^{-1}(W)$, (i) $\pi^{-1}(p) \cap U = \{q\}$ is one point and (ii) $\pi_U := \pi|_U : U \to W$ is surjective and proper. Thus in particular, the induced map $\pi : X\backslash\pi^{-1}(B_\pi) \to M\backslash B_\pi$ is a topological covering and π_U is finite.

For example, if $\pi : X \to M$ is a surjective proper finite holomorphic map, π is a finite branched covering of M.

Two branch coverings $\pi : X \to M$ and $\pi' : X' \to M$ are said to be equivalent if there is a biholomorphic map $\phi : X \to X'$ such that $\pi = \pi' \circ \phi$. In this case we write $\pi \geq \pi'$ or $\pi' \leq \pi$. The set of all automorphisms of π forms a group G_π naturally. One checks easily that if we denote by π_1 the restriction of π to $X \backslash \pi^{-1}(B_\pi)$, then G_π is canonically isomorphic to G_{π_1}. By definition, π is called a Galois covering if G_π acts transitively on every fiber of π; and π is called abelian if π is Galois and G_π is abelian.

Theorem. *If $\pi : X \to M$ is a Galois covering, then*
(1) for every subgroup H of G_π, there is a branched covering $\pi_H : X/H \to M$ such that $\pi_H \leq \pi$;
(2) the correspondence $H \to \pi' = \pi_H$ gives a bijection between subgroups H and equivalence classes of branched coverings π' of M such that $\pi' \leq \pi$; and
(3) H is normal if and only if π_H is a Galois covering, for which G_{π_H} is isomorphic to G_π/H.

Note that for any branched covering $\pi : X \to M$, if p, q are points of B_π and $\pi^{-1}(B_\pi)$ respectively such that
(i) B_π is normally crossing at p, $q \in \pi^{-1}(p)$;
(ii) X is smooth at q; and
(iii) $\pi^{-1}(B_\pi)$ is normally crossing at q,
then, for a sufficiently small connected open neighborhood of p with a coordinate system $(w_1, \dots, w_n)$ such that $p = 0$ and $B_\pi \cap W = \{(w_1, \dots, w_n) : w_k \cdots w_n = 0\}$ for some k, there is a coordinate system $(z_1, \dots, z_n)$ in the connected component U of $\pi^{-1}(W)$ with $q \in U$ such that $q = 0$, $\pi^{-1}(B_\pi) \cap U = \{(z_1, \dots, z_n) \in U : z_k \cdots z_n = 0\}$ and $\pi_U(z_1, \dots, z_n) = (z_1, \dots, z_{k-1}, z_k^{e_k}, \dots, z_n^{e_n})$. Often we call e_j, $j = k, \dots, n$ the ramification index of the irreducible C_j of $\pi^{-1}(B_\pi)$ such that $C_j \cap U = \{(z_1, \dots, z_n) : z_j = 0\}$.

Suppose now that $B_\pi = D_1 \cup \cdots \cup D_N$ is the irreducible decomposition of B_π. Let $D = \sum_{i=1}^N e_i D_i$ be an effective divisor on M. By definition, the branched covering $\pi : X \to M$ is called branched at D (resp. at most at D) if for every irreducible component C of $\pi^{-1}(B)$

with $\pi(C) = D_j$, the ramification index of π at C is e_j (resp. divides e_j). In particular, a branched at D Galois covering $\pi : X \to M$ is called maximal if for any branched covering $\pi' : X' \to M$ which branches at most at D, $\pi \geq \pi'$.

While, in general, it is complicated to describe maximal branched covering for higher dimensional complex manifolds (see however [Na]), for Riemann surfaces, it may be simply stated as follows.

A result of Bundgaard-Nielsen-Fox says that there is no finite Galois covering $\pi : X \to M$ branched at $D = \sum_{j=1}^N e_j P_j$ if and only if either (i) $g = 0$ and $N = 1$ or (ii) $g = 0$ $N = 2$ and $e_1 \neq e_2$. Here we write D_j as P_j, $j = 1, \ldots, N$. Thus from now on, we always assume that we are not in these exceptions. Also we assume that $e_j \geq 2$. (Otherwise, we may omit it from the beginning as there is no ramification for the corresponding points P_j then.)

Let $J(D)$ be the smallest normal subgroup of $\pi_1(M^0)$ containing $S_1^{e_1}, \ldots, S_N^{e_N}$. Then (as used in the proof of the above result of Bundgaard-Nielsen-Fox,) the normal subgroup $J(D)$ satisfies the following:

Condition $(*)$: If $S_j^d \in J(D)$, then $d | e_j$ for $j = 1, \ldots, N$.

As a direct consequence, we have the following

Theorem. (Bundgaard-Nielsen-Fox) (1) *there is a maximal covering* $\tilde{\pi} : \tilde{M}(D) \to M$ *which branches at* D. *In particular,* $\tilde{M}(D)$ *is simply connected;*
(2) *there is a canonical one-to-one correspondence between subgroups (resp. normal subgroups)* H *of the quotient group* $\pi_1(M^0)/J(D)$ *and equivalence classes of branched coverings (resp. Galois coverings)* $\pi : X \to M$ *which branch at most at* D*;*
(3) $\pi : X \to M$ *branches at* D *if and only if the condition* $(*)$ *holds for* K *(here H=K/J(D)). That is, if* $S_j^d \in K$, *then* $d | e_j, i = 1, \ldots, N$.

A.2.2.2. Geo-Ari Representations and Geo-Ari Bundles

Motivated by the above discussion, we now introduce what we call geo-ari representations of fundamental groups.

As before, let M be a compact Riemann surface of genus g with marked points $P_1, \ldots, P_N$. Set $M^0 = M \backslash \{P_1, \ldots, P_N\}$. Then,

the fundamental group $\pi_1(M^0)$ is generated by $2g$ hyperbolic elements $A_1, B_1, \ldots, A_g, B_g$ and n parabolic elements $S_1, \ldots, S_N$ such that $[A_1, B_1] \cdots [A_g, B_g] S_1 \cdots S_N = 1$. Fix an effective divisor $D = \sum_{i=1}^N e_j P_j$ on M. By definition, a geometric-arithmetical representation, a geo-ari representation for short, of $\pi_1(M^0)$ along with D is a unitary representation $\rho : \pi_1(M^0) \to U(l)$ such that
(i) $\rho(S_i) = \mathrm{diag}\Big(\exp(2\pi i\, \beta_{i1}), \ldots, \exp(2\pi i\, \beta_{il})\Big)$ for all $i = 1, \ldots, N$; and
(ii) there exist integers $\gamma_{ij} > 0$ and $\delta_{ij} \geq 0$ such that
(a) $\gamma_{ij} | e_j$;
(b) $(\gamma_{ij}, \delta_{ij}) = 1$; and
(c) $\beta_{ij} = \frac{\delta_{ij}}{\gamma_{ij}}$ for all $i = 1, \ldots, N$ and $j = 1, \ldots, l$.

In parallel, we define a geo-ari bundle on M along D to be a parabolic degree zero semi-stable parabolic bundles

$$\Sigma = \Big(E; \big(P = P_i; E_P = F_1 E_P \supset F_2 E_P \cdots \supset F_r E_P; \\ \alpha_1 = \alpha_{i1}, \ldots, \alpha_r = \alpha_{ir_i}\big)_{i=1}^N\Big)$$

such that the following conditions are satisfied: there exist integers $\gamma_{ij} > 0$ and $\delta_{ij} \geq 0$ such that
(a) $\gamma_{ij} | e_j$;
(b) $(\gamma_{ij}, \delta_{ij}) = 1$; and
(c) $\alpha_{ij} = \frac{\delta_{ij}}{\gamma_{ij}}$ for all $i = 1, \ldots, N$ and $j = 1, \ldots, r_i$.

Obviously, by Theorem 2.1.3, we obtain the following

Theorem$'$. *There exists a natural one-to-one correspondence between isomorphic classes of geo-ari representations of $\pi_1(M^0)$ along with D and (Seshadri) equivalence classes of geo-ari bundles along D over M.*

Remark. We call this result the Harder-Narashimhan Correspondence, despite the fact that in the situation now, i.e., over complex numbers, the Harder-Narasimhan correspondence is simply a direct consequence of Narasimhan-Seshadri correspondence. Later we will see that when the constant field is finite, such a correspondence, first established by Harder-Narasimhan, is based on the vanishing of related Brauer groups.

Clearly, if $D' = \sum_{j=1} e'_j P_j$ is an effective divisor such that $e_j | e'_j$ for all $j = 1, \dots, N$, then geo-ari representations of $\pi_1(M^0)$ (resp. geo-ari bundles) along with D are also geo-ari representations of $\pi_1(M^0)$ (resp. geo-ari bundles) along with D'. (Usually, we write $D|D'$.) Thus if we denote $U(M;D)$ the category of equivalences classes of geo-ari representations of $\pi_1(M^0)$ along with D, (see, e.g., 2.3.1 below for a brief discussion on categories,) and $\mathcal{M}(M;D)$ the category of geo-ari bundles along D over M, then by using a result of Mehta-Seshadri, see e.g., Prop. 1.15 of [MS], we have the following

Proposition. *$U(M;D)$ and $\mathcal{M}(M;D)$ are equivalent abelian categories. Moreover if $D|D'$, then $U(M;D)$ and $\mathcal{M}(M;D)$ are abelian subcategories of $U(M;D')$ and $\mathcal{M}(M;D')$, respectively.*

A.2.3. KR^2-Trick and Completed Tannakian Categories

A.2.3.1. Completed Tannakian Category and van Kampen Completeness

Recall that a category **A** consists of two sets Obj and Arr of objects and morphisms with two associates dom and cod from morphisms to objects. Often an arrow f with $\mathrm{dom}(f) = x$ and $\mathrm{cod}(f) = y$ is written as $f : x \to y$. There is also a map called composition $\mathrm{Arr} \times \mathrm{Arr} \to \mathrm{Arr}$ defined for (f, g) when $\mathrm{dom}(g) = \mathrm{cod}(f)$ such that $\mathrm{dom}(g \circ f) = \mathrm{dom}(f)$ and $\mathrm{cod}(g \circ f) = \mathrm{cod}(g)$, $f \circ (g \circ h) = (f \circ g) \circ h$, and for every object x there is a morphism $1_x : x \to x$ such that $f \circ 1_x = 1_x \circ f = f$. Thus we may form the set $\mathrm{Hom}(x, y)$ by taking the collection of all $f : x \to y$. By definition, if moreover the hom sets are all abelian groups such that compositions are bilinear, we call it a preadditive category.

Among two categories, a functor $T : A \to B$ is defined to be a pair of maps $\mathrm{Obj}(A) \to \mathrm{Obj}(B)$ and $\mathrm{Arr}(A) \to \mathrm{Arr}(B)$ such that if $f : c \to c'$ is an morphism in A, then $T(f)$ is a morphism $T(f) : T(c) \to T(c')$ in B, and that $T(1_c) = 1_{T(c)}$, $T(g \circ f) = T(g) \circ T(f)$; and a natural transformation τ between two functors $S, T : A \to B$ is defined to be a collection of morphisms $\tau_c : T(c) \to S(c)$ in B such that for any $f : c \to c'$ in A, $\tau_c T(f) = S(f)\tau_{c'}$; if moreover all τ_c have inverses, then the natural transformation is called a functorial isomorphism.

Among categories, abelian categories are of special importance. By definition, an abelian category is a preadditive category such that
(1) there is a unique object called zero object such that it is the initial as well as the final object of the category;
(2) direct products exist;
(3) associated to any morphism are kernel and cokernel which are objects of the category as well; moreover, every monomorphism is the kernel of its cokernel, every epimorphism is the cokernel of its kernel; and
(4) every morphism can be factored into an epimorphism followed by a monomorphism.

To facilitate ensuing discussion, we introduce some new notion in category theory. By definition, an object x in an abelian category is called decomposable if there exist objects, y, z different from zero, such that $x = y \oplus z$; and x is called irreducible if it is not decomposable. Moreover, an abelian subcategory of an abelian category if called completed if
(i) for any object x, there is a unique finite decomposition $x = \oplus x_i$ with x_i irreducible, the irreducible components of x;
(ii) the subcategory contains all of its irreducible components of its objects.

Proposition 1. *The categories $U(M; D)$ and $\mathcal{M}(M; D)$ are completed abelian categories.*

The reader may prove this proposition from the following facts:
(a) Among two stable parabolic bundles of the same parabolic degree, homomorphisms are either zero or an isomorphism;
(b) There exist Jordan-Hölder filtrations for parabolic semi-stable bundles; and
(c) The associated Jordan-Hölder graded parabolic bundles in (b) is unique.

Remark. Motivated by this proposition, the reader may give a more abstract criterion to check when an abelian subcategory is completed.

Next, let us recall what a tensor category should be. By definition, a category is called a tensor category if there is an operation, called tensor product $A \otimes B$ for any two objects A, B of the category, such that the following conditions are satisfied:
(1) There are natural isomorphisms $S : A \otimes B \to B \otimes A$ and $T : (A \otimes B) \otimes C \to A \otimes (B \otimes C)$;
(2) S and T satisfy the so-called pentagon and hexagon axioms; and
(3) There is a unique identity object 1 such that $A \simeq A \otimes 1 \simeq 1 \otimes A$ for all object A.

Clearly, for tensor categories, we can introduce tensor operation for finitely many objects. Usually, there are many ways to do so, but the above conditions for tensor category implies that resulting objects from all these different ways are the same. Moreover, if for any objects x, y of A, $\mathrm{Hom}(x, y)$ is again an object in A satisfies the following conditions, we call A a rigid tensor category:
(4) there exists a morphism, the evaluation map, $\mathrm{ev}_{x,y} : \mathrm{Hom}(x, y) \otimes x \to y$ such that for any object t and any morphism $g : t \otimes x \to y$, there exists a unique morphism $f : t \to \mathrm{Hom}(x, y)$ such that the following commutative diagram commutes

$$\begin{array}{ccc} t \otimes x & \stackrel{f\otimes 1_x}{\to} & \mathrm{Hom}(x,y) \otimes x; \\ g \downarrow & \swarrow & \mathrm{ev}_{x,y} \\ y & & \end{array}$$

(5) There exists natural isomorphism

$$\mathrm{Hom}(x_1, y_1) \otimes \mathrm{Hom}(x_2, y_2) \simeq \mathrm{Hom}(x_1 \otimes x_2, y_1 \otimes y_2)$$

which is compactible with (4);
(6) For any object x, by (5), if we set $x^\vee := \mathrm{Hom}(x, 1)$, then there exists a natural isomorphism $x \simeq (x^\vee)^\vee$.

Now we are ready to introduce Tannakian categories. By definition, a Tannakian category is a category which is both an abelian category and a rigid tensor category such that the tensor operation is bilinear. Clearly then in such categories, $\mathrm{Hom}(x, y) \simeq x \otimes y^\vee$.

For example, the category Vec_k of finite dimensional vector spaces over a field k is a Tannakian category. By definition, a functor $\omega : A \to \mathrm{Vec}_k$ from a Tannakian category A to the category Vec_k is called a fiber functor, if ω is an exact faithful tensor functor. (Recall that faithful means there is a natural injection $\omega\big(\mathrm{Hom}_A(x,y)\big) \hookrightarrow \mathrm{Hom}_{\mathrm{Vec}_k}\big(\omega(x), \omega(y)\big)$ which is induced from ω.)

Associated to a fiber functor ω is naturally its automorphic group $\mathrm{Aut}^{\otimes}\omega$. Essentials of Tannakian categories may be summarized as the following:

Theorem 2. (Tannaka, Grothendieck, Saavedra Rivano) *Assume that $(A, \omega : A \to \mathrm{Ver}_k)$ consists of a Tannakian category and a fiber functor such that the field k is canonically isomorphic to $1_A^\vee$, then A is equivalent to the category of representations of the group* $\mathrm{Aut}^{\otimes}\omega$.

A proof of this theorem may be deduced from the following facts:
(1) Category of all representations forms naturally a Tannakian category together with a fiber functor;
(2) Knowledge of the representations of a group is equivalent to the knowledge of the group; and
(3) Any Tannakian category is in fact a clone of (1), via the so-called Tannaka Duality principle.

Now we are ready to introduce our own Tannakian categories which are completed and equipped with natural fiber functors. There are two of them, i.e., the one for geo-ari representations and the one for geo-ari bundles.

By Proposition 2.2.2, it is enough to introduce the fiber functors and show that geo-ari representations and geo-ari bundles are closed under the tensor operation. But all this is quite straightforward: By definition, tensors of unitary representations are again unitary, while for geo-ari bundles, the fiber functor may be defined by taking special fibers for the bundles at any point which is not a marked one. Note that morphisms between geo-ari bundles are either zero or isomorphisms, so the latest defined functor is faithful. Therefore, we obtain the following

Theorem 3. *$U(M;D)$ and hence $\mathcal{M}(M;D)$ are completed Tan-*

nakian categories equipped with natural fiber functors to $\mathrm{Ver}_{\mathbf{C}}$.

We end this brief discussion on Tannakian category by the so-called van Kampen completeness theorem, which plays a key role in the proof of the fundamental theorem of CFT for Riemann surfaces later.

For any group G, denote its associated Tannakian category of equivalence classes $[\rho]$ of unitary representations $\rho : G \to \mathrm{Aut}(V_\rho)$ by $U(G)$. Fix for all classes $[\rho]$ a representative ρ once and for all. A subset Z of $U(G)$ is said to contain sufficiently many representations if for any two distinct elements g_1, g_2 of G, there exists $[\rho]$ in Z such that $\rho(g_1) \neq \rho(g_2)$.

van Kampen Completeness Theorem. *If G is compact, then, as a completed Tannakian category, $U(G)$ may be generated by any collection of objects which contains sufficiently many representations.*

A.2.3.2. KR^2-Trick

While it is quite nice to have a proof of Theorem 3 in 2.3.1, we are by no means satisfied with it. Recall that Theorem 3 consists of two aspects: the algebraic one for bundles and the analytic one for unitary representations. So, what we need to do should be the follows;

(1) From analytic point of view, to prove that
  (a) the tensor operation is closed; and
  (b) the so-called forgetful functor is faithful; and
(2) From algebraic point of view, to show that
  (a) the tensor operation is closed; and
  (b) the functor introduced in 2.3.1 is faithful.

However in our approach outlined in 2.3.1, only (1.a) and (2.b) are shown. That is to say, with the help of the so-called Narasimhan-Seshadri correspondence, a micro reciprocity law, we make no distinction between algebraic and analytic aspects. Thus, a purely analytic proof for (1.a) and a purely algebraic proof for (2.a) should be pursued.

The proof of (1.a) is simple since unitary representations for fundamental groups, or better for quotients of fundamental groups by (normal) subgroups generated by weighted parabolic generators have semi-simplifications. We leave the details to the reader. Thus the real challenge here is an algebraic proof of (2.a), i.e., a proof for that tensor products of geo-ari bundles are again geo-ari bundles. It is here we should use another central concept in GIT, the so-called instability flag of Mumford-Kempf (with the refined version given by Ramanan and Ramanathan). This goes as follows.

In his study of GIT stabilities, Mumford conjectures that if a point is not semi-stable, then there should exist a parabolic subgroup which takes responsibility, in the view of the so-called Hilbert-Mumford criterion. This is confirmed by Kempf. (In Kempf's study, as suggested by Mumford, the rationality problem is also treated successfully, at least when constant fields are perfect.) Kempf's result then motivates Ramanan and Ramanathan to show that even though, for the original action, the corresponding point may possibly be not semi-stable, but if a certain type well-controlled modification is allowed (with the aim to cancel the instability contribution in the original action), a new (yet well-associated) point could be constructed such that with respect to the natural induced action this new point becomes semi-stable. As a direct consequence, in the case for semi-stable vector bundles without parabolic structures, since the new well-associated action may be associated to the intersection stability condition for bundles naturally, we can then obtain an algebraic proof of (2.a) for bundles.

Remark. Over complex numbers, the rationality is not a serious problem. However, over finite fields, this turns to be a difficult one. We then need to consider the Frobenius, or better p^n reduction, to tackle the rationality problem.

Therefore, to make our non-abelian CFT for function fields over complex numbers pure, we need to find a more general version, what we call the KR^2-trick, of the above result of Kempf, and Ramanan and Ramanathan, which works for parabolic bundles. For this purpose, we may follow a more down-to-earth approach of Faltings and Totaro ([FW] and [To]).

A.2.4. Non-Abelian CFT for Function Fields over Complex Numbers

A.2.4.1. Micro Reciprocity Law, Tannakian Duality and the Reciprocity Map

Let M be a compact Riemann surface of genus g with marks $P_1, \ldots, P_N$. Set $M^0 := M\backslash\{P_1, \ldots, P_N\}$. Then, it is well known that the fundamental group $\pi_1(M^0)$ is generated by $2g$ hyperboilic generators $A_1, B_1, \ldots, A_g, B_g$ and N parabolic generators $S_1, \ldots, S_N$ which satisfy one single relation $[A_1, B_1] \cdots [A_g, B_g] \cdot S_1 \cdots S_N = 1$. Moreover, with respect to an effective divisor $D = \sum_{i=1}^N e_i P_i$, we have a Tannakian category $U(M; D)$ consisting of equivalent classes of unitary representations $[\rho : \pi_1(M^0) \to \mathrm{Aut}(V_\rho)]$ of $\pi_1(M^0)$ such that

(i) $\rho(S_i) = \mathrm{diag}\Big(\exp(2\pi i\, \beta_{i1}), \ldots, \exp(2\pi i\, \beta_{il})\Big)$ for all $i = 1, \ldots, N$; and

(ii) there exist integers $\gamma_{ij} > 0$ and $\delta_{ij} \geq 0$ such that

(a) $\gamma_{ij} | e_j$;

(b) $(\gamma_{ij}, \delta_{ij}) = 1$; and

(c) $\beta_{ij} = \frac{\delta_{ij}}{\gamma_{ij}}$ for all $i = 1, \ldots, N$ and $j = 1, \ldots, l$.

Now in each equivalence class $[\rho]$, fix a unitary representation, denoted also by ρ, once and hence for all. Clearly, ρ induces a unitary representation of the group $\pi_1(M^0)/J(D)$, where $J(D)$ denotes the normal subgroup of $\pi_1(M^0)$ generated by $S_1^{e_1}, \ldots, S_N^{e_N}$. Call this latest representation ρ as well. Then, for any element $g \in \pi_1(M^0)/J(D)$, $\rho(g)$ induces for each object $[\rho : \pi_1(M^0) \to \mathrm{Aut}(V_\rho)]$ in $U(M; D)$ an automorphism of V_ρ. As a direct consequence, we obtain a natural group morphism from $\pi_1(M^0)/J(D)$ to the automorphism group of the corresponding fiber functor $U(M; D) \to \mathrm{Vec}_{\mathbf{C}}$.

On the other hand, for the Tannakian category $\mathcal{M}(M; D)$ of geo-ari bundles on M along D together with the fiber functor $\omega(M; D) : \mathcal{M}(M; D) \to \mathrm{Ver}_{\mathbf{C}}$, by Tannakian Duality, we conclude that $\omega(M; D)$ is equivalent to the Tannakian category of the representations of $\mathrm{Aut}^{\otimes}\omega$. Therefore, by the Narasimhan-Seshadri correspondence (and the Harder-Narasimhan correspondence), the so-called micro reciprocity law, we obtain a canonical group morphism

$$\Omega(D) : \pi_1(M^0)/J(D) \to \mathrm{Aut}^{\otimes}\big(\omega(M;D)\big).$$

We will call $\Omega(D)$ the reciprocity map associated with (M, D).

A.2.4.2. Non-Abelian CFT

By definition, a subcategory S of a Tannakian category with respect to a fiber functor ω is called a *finitely completed Tannakian subcategory*, if

(1) it is a completed Tannakian subcategory;

(2) there exist finitely many objects which generated S as an abelian tensor subcategory; and

(3) $\mathrm{Aut}^{\otimes}(\omega|_S)$ is a finite group.

With this, by using the van Kampen completeness theorem and the above reciprocity map, we then can manage to obtain the following fundamental theorem on non-abelian class field theory for function fields over complex numbers.

Non-Abelian CFT for Riemann Surfaces. (Weng [We5]) (1) (Existence and Conductor Theorem) *There is a natural one-to-one correspondence* $\omega_{M,D}$ *between*

$$\{\mathbf{S} : \text{finitely completed Tannakian subcategory of } \mathcal{M}(M;D)\}$$

and

$$\{\pi : M' \to M : \text{finite Galois covering branched at most at } D\};$$

(2) (Reciprocity Law) *There is a natural group isomorphism*

$$\mathrm{Aut}^{\otimes}\big(\omega(M;D)\big|_{\mathbf{S}}\big) \simeq \mathrm{Gal}\,\big(\omega_{M,D}(\mathbf{S})\big).$$

We end this discussion by pointing out that, as an application, one may use this fundamental theorem to solve the geometric inverse Galois problem.

A.2.5. Classical (Abelian) CFT: An Example of Kwada-Tata and Kawada

A.2.5.1. Class Formation

Classical abelian class field theory was first formulated axiomatically in Artin-Tate seminar in terms of cohomology of groups. As an example of this formulation, later Kawada-Tate [KT] and Kawada [Ka] studied function fields over complex numbers. To make a comparison between this classical approach of CFT and what we outlined above, next we recall the works of Kawada-Tate and Kawada.

Let k_0 be a given ground field and Ω a fixed infinite separable normal algebraic extension of k_0. Let $\mathcal{R}$ be the set of all finite extensions of k_0 in Ω. By definition we call a collection $\{E(K) : K \in \mathcal{R}\}$ of abelian groups $E(K)$ a formulation if the following conditions are satisfied:
F1. If $k \subset K$ then there is an injective morphism $\phi_{k/K} : E(k) \hookrightarrow E(K)$;
F2. If $k \subset l \subset K$, then $\phi_{l/K} \circ \phi_{k/l} = \phi_{k/K}$;
F3. If K/k is normal and $G = \mathrm{Gal}(K/k)$ is its Galois group, then G acts on $E(K)$ and $\phi_{k/K}(E(k)) = E(K)^G$; and
F4. If $k \subset L \subset K$ and $L/k, K/k$ are both normal, then the Galois group $F = \mathrm{Gal}(L/k)$ is a quotient group of $G = \mathrm{Gal}(K/k)$. Furthermore, denote by $\lambda_{G/F} : G \to F$ the canonical quotient map. Then, $\sigma \circ \phi_{L/K}(f) = \phi_{L/K} \circ (\lambda_{G/F}\sigma)(f)$ for any $\sigma \in G$ and $f \in E(L)$.

Moreover a formation is called a class formation if it satisfies the following additional conditions on group cohomology:
C1: $H^1(G, E(K)) = 0$; and
C2: $H^2(G, E(K)) \simeq [K : k_0]\mathbf{Z}$.

By a theorem of Tate, C1 and C2 imply and hence are equivalent to the following stronger condition
C. In a class formation, for all $r \in \mathbf{Z}$, $H^r(G, E(K)) \simeq H^{r-2}(G, \mathbf{Z})$ for all r. In particular, if a 2-cocycle $\alpha(K)$ generates the cyclic group $H^2(G, E(K))$, then the above isomorphisms are all induced by the cup-product $g \mapsto \alpha(K) \cup g$.

Note that by definition, $H^{-2}(G, \mathbf{Z}) = G^{\mathrm{ab}} := G/[G, G]$, and the

0-th cohomology group $H^0\big(G, E(K)\big)$ is nothing but $E(K)^G/T_G E(K)$, (where $T_G a = \sum_{\sigma\in G}\sigma(a)$). So for a class formation, we obtain then the reciprocity law

$$E(k)/\phi^{-1}\big(T_G E(K)\big) \simeq G^{\mathrm{ab}}.$$

Furthermore, for a class formation, with respect to the so-called res (restriction), infl (inflation) and ver (Verlagerung) operations of group cohomology, we have the follows;
Case 1. For $k \subset l \subset K$ with K/k normal, $G := \mathrm{Gal}(K/k)$ and $H := \mathrm{Gal}(K/l)$, in $H^2\big(G, E(K)\big), \mathrm{ver}_{H/G} \circ \mathrm{res}_{G/H} = [G : H] \cdot \mathbf{1}$. Consequently, $\mathrm{res}_{G/H}$ is surjective and $\mathrm{ver}_{H/G}$ is injective;
Case 2. For $k \subset L \subset K$ with L/k, K/k normal, and $G := \mathrm{Gal}(K/k)$, $H := \mathrm{Gal}(K/L)$, we have the exact sequences

$$0 \to H^2\big(F, E(L)\big) \overset{\mathrm{infl}}{\to} H^2\big(G, E(K)\big) \overset{\mathrm{res}}{\to} H^2\big(H, E(K)\big).$$

Therefore, there exist 2-cocycles $\big\{\alpha(K)\big\}$ of $G(K/k)$ over $E(K)$ such that
D1. $\alpha(K/k)$ are generators of the cyclic groups $H^2\big(G, E(K)\big)$;
D2. In Case 1,

$$\mathrm{res}_{G/H}(\alpha(K/k)) \sim f(K/l); \qquad \mathrm{ver}_{H/G}(\alpha(K/k)) \sim [G : H] \cdot \alpha(K/k);$$

D3. In Case 2, $\mathrm{infl}_{F/G}\alpha(L/k) \sim [K : L] \cdot \alpha(K/k)$. Here $\sim$ means cohomologous.

We call such a system $\big\{\alpha(K/k)\big\}$ the canonical 2-cocycle of $G(K/k)$ over $E(K)$, with which we can write down the reciprocity law as follows: Introduce $(\frac{K/k}{\sigma}) \in E(k)$, where $\sigma \in G := \mathrm{Gal}(K/k)$ for normal extension K/k by

$$\left(\frac{K/k}{\sigma}\right) := \phi_{k/K}^{-1}\Big(\prod_{\rho\in G}\alpha_{K/k}(\rho,\sigma)\Big) \mod \phi_{k/K}^{-1}\big(T_G E(K/k)\big).$$

Then by a result of Nakayama, the symbol $\big(a, K/k\big) \in G^{\mathrm{ab}}$, $a \in E(k)$ defined by

$$\big(a, K/k\big) := \sigma \bmod [G, G] \text{ when } \left(\frac{K/k}{\sigma}\right) = a \bmod \phi_{k/K}^{-1}\big(T_G E(K)\big),$$

satisfies all properties of the norm residue symbol in number theory.

Let $k \in \mathcal{R}$ be fixed and $\Omega^a(k)$ be the maximal abelian extension of k in Ω. Set $\mathcal{R}^a(k)$ be the collection of finite abelian extensions of k (in Ω^a). Then for $K \in \mathcal{R}^a(k)$ define a subgroup $A(K/k)$ of $E(k)$ to be $\phi_{k/K}^{-1}\big(T_{\rm Gal}(K/k)E(K)\big)$. By definition, a subgroup F of $E(k)$ is called admissible if $F = A(K/k)$ for some $K \in \mathcal{R}^a(k)$. Denote the set of all admissible subgroups of $E(k)$ by $\mathcal{U}\big(E(k)\big)$. Then one checks, by the properties of norm residue symbol, that
(1) (Combination Axiom) $A\big((K_1 \cdot K_2)/k\big) = A\big(K_1/k\big) \cap A\big(K_2/k\big)$ and $A\big((K_1 \cap K_2)/k\big) = A\big(K_1/k\big) \cdot A\big(K_2/k\big)$;
(2) (Ordering Axiom) $A\big(K_1/k\big) \supset A\big(K_2/k\big)$ if and only if $K_1 \subset K_2$;
(3) (Uniqueness Axiom) $A\big(K_1/k\big) = A\big(K_2/k\big)$ if and only if $K_1 = K_2 \in \mathcal{R}^a(k)$.

Also, let $\Gamma(k)$ be the compact Galois group $\Omega^a(k)/k$, then $\Gamma(k)$ is the inverse limit group of $\big\{G(K/k) : K \in \mathcal{R}^a(k)\big\}$. Moreover, for $a \in E(k)$, the limit, called the generalized norm residue symbol, $\big(a,k\big) := \lim_{K\in\mathcal{R}^a(k)} \big(a,K/k\big) \in \Gamma(k)$ exists. Set $\mathcal{T}(k) := \big(E(k),k\big) \subset \Gamma(k)$ and $\mathcal{R}(k) := \big\{(a \in E(k) : (a,k) = 1\big\} \subset E(k)$, then the mapping $a \mapsto \big(a,k\big)$ induces an isomorphism $E(k)/\mathcal{R}(k) \simeq \mathcal{T}(k) \subset \Gamma(k)$ with $\mathcal{T}(k)$ dense in $\Gamma(k)$.

For examples, classical (abelian) CFTs for global fields are all class formations:
(A) For a number field k, we may take $E(k)$ to be the idele class group C_k and $\phi_{k/K}$ the natural inclusion. In particular, $\mathcal{R}(k)$ is then the connected component of the unity of C_k, $\Gamma(k) = \mathcal{T}(k)$, and $\mathcal{U}\big(E(k)\big)$ is the set of all open subgroup of $E(k)$ of finite index;
(B) For a function field k of one variable over a finite field, we may take $E(k)$ to be the idele class group C_k and $\phi_{k/K}$ the natural inclusion. In particular, $\mathcal{R}(k) = 1$, $\Gamma(k)/\mathcal{T}(k)$ is a uniquely divisible group isomorphic to $\hat{\mathbf{Z}}/\mathbf{Z}$, and $\mathcal{U}\big(E(k)\big)$ is the set of all open subgroup of $E(k)$ of finite index.

A.2.5.2. The Work of Kawada and Tate

Choose k_0 to be an algebraic function field of one variable over complex numbers with Ω the algebraic closure of k_0. Let $D(k)$ be the group of all fractional divisors $\prod_v P_v^{r_v}$, where $r_v \in \mathbf{Q}$, and $r_v = 0$ for almost all v, and $P(k)$ the group of all principle divisors. Let $\mathbf{D}(k) = D(k)/P(k)$.

Define $E(k) := \mathbf{D}(k)^\vee = \mathrm{Hom}\big(\mathbf{D}(k), \mathbf{R}/\mathbf{Z}\big)$, the group of characters of $\mathbf{D}(k)$, then $\big\{E(k)\big\}$ with the conorm map $\phi_{k/K}$ is a class formation such that the norm residue map $\Phi_k : E(k) \to A(k)$ is surjective but $F(k) := \mathrm{Ker}\,\Phi_k \neq 0$. Consequently, if $T\mathbf{D}(k)$ denotes the torsion subgroup of $\mathbf{D}(k)$, $\big\{E(k)^* = E(k)/F(k) = \big(T\mathbf{D}(k)\big)^\vee\big\}$ gives also a class formation.

On the other hand, clearly,
(i) the character of k_0 is zero;
(ii) k_0 contains all the roots of unity; and
(iii) for any finite normal extension K/k with $k \supset k_0$ and K/k finite, $N_{K/k}K = k$.
So, by applying the so-called Kummer theory, $E(k) := \big(k^* \otimes \mathbf{Q}/\mathbf{Z}\big)^\vee$, and the conorm $\phi_{k/K} : E(k) \to E(K)$, i.e., $\phi_{k/K}(\chi)(A) = \chi\big(N_{K/k}A\big)$ for $\chi \in E(k), A \in K^* \otimes \mathbf{Q}/\mathbf{Z}$, give a class formation with $E(k) = A(k)$.

Thus, in particular, by comparing these class formations, we obtain a canonical isomorphism $k^* \otimes \mathbf{Q}/\mathbf{Z} \simeq T\mathbf{D}(k)$. Moreover, if Ω_ϕ is a maximal unramified extension of k_0, $t := t\big(\Omega_\phi/k\big)$ and $E(k)_\phi := \big(\mathbf{D}_\phi(k)\big)^\vee$ with $\mathbf{D}_\phi(k)$ the divisor class group of the usual sense, i.e., with integral coefficients, then $\big\{E_\phi(k)\big\}$ gives a class formation for t_ϕ. On the other hand, for a fixed finite set $S \neq \emptyset$ of prime divisors of k_0, let Ω_S be the maximal S-ramified extension of k_0. Put $t_S := t\big(\Omega_S/k_0\big)$ and $E(k)_S := \big(\mathbf{D}_S(k)\big)^\vee$ with $\mathbf{D}_S(k)$ the S-fractional divisor class group of k. Then $\big\{E_S(k)\big\}$ is a class formation for t_S, such that $\mathrm{Im}\,\Phi_k = A(k)$ and $\mathrm{Ker}\,\Phi_k$ is the connected component of $E(k)_S$ which is not zero. Similarly, if we take $E(k)_S^* := \big(T\mathbf{D}_S(k)\big)^\vee$, then $\big\{E(k)_S^*\big\}$ forms also a class formation for t_S such that $E(k)_S^* \simeq A(k)$.

A.2.5.3. Abelian CFT for Riemann Surfaces in Terms of Geo-Ari Bundles

However, it is via the third approach of a class formation for Riemann surfaces due to Kawada [Ka] that these classical results are related with our approach to the CFT. It goes as follows:

For $k_0 \subset k$, let $S(k)$ denote the set of prime divisors of k which are extensions of a prime divisor of k_0 contained in S. Let $R(k)$ be the Riemann surface of k and $R_S(k) = R(k)\backslash S(k)$. Let $E(k)$ be the one-dimensional integral homology group $H_1\big(R_S(k), \mathbf{Z}\big)$ of $R_S(k)$, and define $\phi_{k/K} : E(k) \to E(K)$ by $\gamma \mapsto V\gamma$ where $V\gamma$ is the covering path of $\gamma \in R_S(k)$ on the unramified covering surface $R_S(K)$ of $R_S(k)$. Then $\{E(k)\}$ forms again a class formation for t_S with $\mathrm{Im}\,\Phi_k$ dense in $A(k)$ and $\mathrm{Ker}\,\Phi_k = 0$. Indeed, one may obtain this by looking at the canonical pairing, the micro reciprocity law in this context, $H_1\big(R_S(k), \mathbf{Z}\big) \times T\mathbf{D}_S(k) \to \mathbf{S}^1 \subset \mathbf{C}$ defined by $(\eta, A)_S \mapsto \exp\big(\int_\eta d\log A\big)$ for 1-cycle η on $R_S(k)$ and $A \in T\mathbf{D}_S(k)$, where $d\log A$ denotes the abelian differential of the third kind on $R_S(k)$ corresponding to a divisor A.

On the other hand, over a compact Riemann surface M with punctures $P_1, \ldots, P_N$ with respect to an effective divisor $D = \sum e_j P_j$, we may introduce the group $\mathrm{Div}_0^{\mathbf{Q}}\big(M; D\big)$ of degree zero $\mathbf{Q}$-divisors along D on M by collecting all degree zero $\mathbf{Q}$-divisors of the form $\sum_j \frac{a_j}{e_j} P_j + E$ with $a_j \in \mathbf{Z}$ and E an ordinary integral divisor on M. Denote the induced (rational equivalence) divisor class group $\mathrm{Cl}_0^{\mathbf{Q}}\big(M; D\big)$. Then, $\mathrm{Cl}_0^{\mathbf{Q}}\big(M; D\big)$ is simply the collection of all geo-ari bundles of rank 1 introduced in 2.2.2. Hence Theorem 2.4.2 then implies the following

Theorem. *There is a one-to-one and onto correspondence between the set of all isomorphism classes of finite abelian coverings* $\pi : X \to M$ *branched at most at* D *and the set of all finite subgroups* S *of* $\mathrm{Cl}_0^{\mathbf{Q}}\big(M; D\big)$. *Moreover, the correspondence* $\pi \mapsto S(\pi)$ *satisfies that*

(i) (Reciprocity Law) $S(\pi) \simeq G_\pi$; and

(ii) (Ordering Theorem) $\pi \leq \pi'$ *if and only if* $S(\pi) \subset S(\pi')$.

Therefore, it seems to be very crucial to understand the precise relation between Kawada-Tate's results and this latest theorem, in order to develop a (non-abelian) CFT for global fields.

A.3. Towards Non-Abelian CFT for Global Fields

A.3.1. Weil-Narasimhan-Seshadri Type Correspondence

A.3.1.1. Geometric Representations

From what discussed above, in order to develop a non-abelian CFT for local and global fields, the first step should be the one to establish a micro reciprocity law, i.e., a Weil-Narasimhan-Seshadri type correspondence. Therefore, we are supposed to

(1) introduce suitable classes of representations of Galois groups;

(2) find corresponding classes for bundles in terms of intersection;

(3) establish natural correspondences between classes in (1) and (2).

Hence, it is then more practical to divide the problem into two. Namely, a general one in the sense of Weil Correspondence, and a refined one in the sense of Narasimhan-Seshadri correspondence.

We start with the Weil Correspondence. Here, we are then supposed to first introduce a general notion for representation of Galois groups such that, naturally, associated to such representations are special vector bundles together with additional structure.

As an example, let us explain what we have in mind in the case for number fields. So, let F be a number field with a finite subset S of places of F, including all Archimedean places. Denote the corresponding Galois group by $G_{F,S}$. Naturally, by a representation, it should be first a continuous group homolorphism $\rho : G_{F,S} \to \mathrm{GL}_n(\mathbf{A}_{F,S})$, where $\mathbf{A}_{F,S}$ denotes the ring of S-adeles. Moreover, among others, we should assume that

(a) for all places v of F, the induced representations $\rho_v : G_{F,S} \to \mathrm{GL}_n(F_v)$ are unramified almost everywhere;

(b) for a fixed places p of $\mathbf{Q}$, there should be a compatibility condition for all places v above p.

In addition, it also seems to be very natural to assume that

(c) for all induced $\rho_v : G_v \to \mathrm{GL}_n(F_v)$, there are invariant $\mathcal{O}_v$-lattices M_v of F_v which are induced from a global $\mathcal{O}_F$-lattice M over F; and

(d) at places $v \in S$, there should be naturally a certain weighted filtration induced by the action of Frobenius.

Thus in particular, associated to such a representation, is naturally a well-defined vector bundle equipped with a parabolic structure. We are expecting that geometric bundles are of Arakelov degree zero.

With Weil type correspondence, we are able to enter the level of Narasimhan-Seshadri type correspondence. Then we are expect to overcome certain essential difficulties. Chiefly, what should be a natural analog of being unitary? A suitable candidate seems to be that of Fountaine's semi-stability at finite places and (unitary) at infinite places. However, we are not very sure about this, as somehow we believe that the condition of semi-stable representation is too restricted. For this reason, we propose the follows;

(e) for all places v, the images of the induced representation $\rho_v : G_v \to \mathrm{GL}_n(F_v)$ are contained in maximal compact subgroups; moreover, certain compatibility conditions are satisfied by ρ_v for all v over a fixed place of $\mathbf{Q}$;

(f) there should be a natural deformation theory for these representations such that

(i) the size of all equivalence classes of these representations can be controlled;

(ii) a certain completeness holds.

Here the reader may find the work of Rapoport and Zink [RZ] and the lecture notes of Tilouine [Ti] very useful. The cases for local fields and function fields may be similarly discussed. If successful, we call such a representation a geometric representation.

A.3.1.2. Semi-Stability in Terms of Intersection

This is the algebraic side of the micro reciprocity law. We here only study what happens for function fields over finite fields, while leave a detailed definition for number fields in Part (B).

With this restriction of fields, the situation becomes much simpler: We may use the existing stability condition of Seshadri for parabolic bundles.

That is to say, what we care of here are the so-called parabolic semi-stable bundles of parabolic degree zero defined on algebraic curves over finite fields, introduced by Seshadri. However, in doing so, we are afraid that our program leads to only a non-abelian

CFT with tame ramifications. So it seems that, to deal with wild ramifications, additional works are needed. Therefore, in general, for this algebraic aspect, we propose the following:
(0) Once with intersection semi-stability, we should be able to establish (i) an analog of the existence and uniqueness of Harder-Narasimhan type filtration; (ii) the existence of Jordan-Hölder type filtration and the uniqueness of the associated graded Jordal-Hölder objects; and (iii) morphisms of stable objects are either zero or isomorphisms;
(1) The so-called Langton's Completeness principle should hold for such intersection semi-stabilities;
(2) The intersection semi-stability should be naturally related with a GIT stability. As a direct consequence, then moduli spaces can be formed naturally, and by (1), are indeed compact;
(3) With the help of the Frobenius, or better, modulo p^n-reduction for all n, we should be able to define a special subset (of points) of moduli spaces, tensor products of whose corresponding bundles are represented again by points in this special subset.

If successful, we will call these objects geometric bundles.

A.3.1.3. Weil-Narasimhan-Seshadri Type Correspondence

For a Weil type correspondence, from the classical proof, say, the one given in Gunning's Princeton lecture notes, we should develop an analog of the de Rham, the Dolbeault cohomologies as well as a Hodge type theory. While, in general, it is out of reach, note that our base is of dimension one, it is still hopeful. For example, for curves over finite fields, we understand that all this is known to experts.

Next, let us consider a Narasimhan-Seshadri type correspondence. Here, we should establish a natural correspondence between equivalence classes of geometric representations and (Seshadri) equivalence classes of geometric bundles. Hence, the key points are the follows:
(1) By definition, a geometric representation should naturally give a geometric bundle;
(2) A Weil type correspondence holds. In particular, this implies that, by definition, geometric bundles come naturally from representations of Galois groups;

(3) Representations resulting from geometric bundles in (2) should be geometric. To establish this, we should use the deformation theories of geometric representations and geometric bundles as proposed in 3.1.1 and 3.1.2 above. For example, the final justification should be based on the compactness of both geometric representations and geometric bundles, via a direct counting. (In geometry, the count is possible since we know the structure of the fundamental group. But in arithmetic, this is quite difficult. Here the work of Fried and Völklein [FV] seems to be quite helpful.)

A.3.2. Harder-Narasimhan Correspondence

This is designed to select special subclasses of geometric representations and geometric bundles so as to get what we call geo-ari representations and geo-ari bundles. There are two main reasons for doing this. The first is based on the facts that for function fields, there are examples of stable bundles whose tensor products are no longer semi-stable; and more importantly that if the bundles and their associated Frobenius twists are all semi-stable, then, so are their tensor products. The second comes from our construction of non-abelian zeta functions in Part (B). To define such non-abelian zeta functions, we use only rational points (over constant fields) of the moduli spaces, based on a result of Harder and Narasimhan, which guarantees the coincidence of the rationality of moduli points and the rationality of the corresponding bundles.

Thus, for example, for function fields over finite fields, a geo-ari bundle is defined to a geometric bundle which satisfies not only all the conditions for geo-ari bundles over Riemann surfaces, but an additional one, which says that all its Frobenius twists are semi-stable as well.

So for our more general purpose, the key points are as follows;
(1) to give a proper definition of geo-ari representations so that they form a natural abelian category;
(2) to give a suitable definition of geo-ari bundles so that they are closed under tensor product; and
(3) to establish a natural correspondence between (1) and (2). If successful, we call such a result a Harder-Narasimhan type correspondence, even we here substantially board the meaning of [HN].

A.3.3. KR^2**-Trick**

This is specially designed to offer an algebraic proof of the following key statement appeared as 3.2(2): tensor products of geo-ari bundles are again geo-ari bundles.

We start with geo-ari bundles over function fields. When no parabolic structures is involved, this is solved by Kempf and Ramanan-Ramanthan. In their proof, the key points are as follows.
(1) Rationality and uniqueness of instability flags, where the Frobenius twists are used;
(2) Existence of a GIT stable modification associated to any non semi-stable one; and
(3) Existence of GIT points for bundles such that GIT stability implies intersection stability by definition;
(4) Relation between the GIT modification in (2) and the intersection stability, where the intersection stability for each component of the tensor product plays a key rule.

More precisely, if the tensor were not stable in terms of intersection, by (3), the associated GIT point would not be GIT stable. Thus from (2) there exists a GIT stable modification which is well understood by (1). Therefore, finally, by (4), i.e., the intersection stability of each components, we may finally conclude the intersection stability for the tensor product with the help of (2).

Thus, the problem left here is to see how the work of Ramanan-Rananthan can be developed to deal with parabolic structure. In theory, this may be done by working on product of varieties instead of just a single one. As this process is closed related to the construction of moduli space of parabolic semi-stable bundles, an expert should be able to carry the details out.

However, there is a more elementary approach, essentially given in the supplementary works of Faltings and Totaro. It goes as follows. First, as what Faltings does, write parabolic subbundles for the tensor in terms of weighted filtrations on the fibers (supported over points which are disjoint from punctures); then introduce a GIT stability with respect to weighted filtrations. So there are two possibilities: if the weighted filtration resulted from the parabolic subbundle is GIT stable, then by (3) above, we get the intersection stability of the tensor product. Otherwise, by (1) and (2) above, we may obtain a well-associated modification which is then GIT stable. This then by applying (4) completes the proof.

We would like to draw reader's attention to a relation between the stabilities of generic and special points due to Mumford: GIT stability for objects over the generic point implies that over almost all but finitely many special points, the associated points are GIT semi-stable.

All in all the basic principle here is that if an object is not stable, then there should be a parabolic subgroup which takes responsibility. In this sense, we have already had a very satisfied result for lattices. For details, please see the global bridge between algebraic and analytic truncations in B.4.

A.3.4. Tannakian Category Theory over Arbitrary Bases

To establish a non-abelian CFT for function fields over complex numbers, we apply a standard theory of Tannakian category. However, for global fields, a much more general theory is needed. Say for number fields, a theory of Tannakian categories over the ring of integers should be developed. Key points for this new type of Tannakian categories are supposed to satisfy the follows.
(1) Automorphisms of fiber functors should give us a group. Particularly, whether fiber functors are to categories of vector spaces is not really important. Moreover, fiber functors should be faithful exact and tensor, among others;
(2) An analog of Tannakian duality holds; and
(3) Tannakian category should contain the so-called finitely completed Tannakian subcategories with respect to which an analog of van Kampen Completeness Theorem holds.

If successful, by (1) and (2), using Narasimhan-Seshadri and Harder-Narasimhan correspondences, we get naturally a reciprocity map. This then by (3), leads to a completed non-abelian CFT.

A.3.5. Non-Abelian CFT for Local and Global Fields

All in all, from the above discussion, what we then expect is the following conjecture, or better,

Working Hypothesis. *For local and global fields,*
(1) there are well-defined geometric representations and geometric bundles such that a Weil-Narasimhan-Seshadri type correspondence holds;
(2) there are refined well-defined geo-ari representations and geo-ari bundles such that a Harder-Narasimhan type correspondence holds;
(3) there is well-defined GIT type stability such that the intersection stability as appeared in the definition of geo-ari bundles could be understood in terms of this new GIT stability. Moreover, an analog of KR^2*-trick works;*
(4) there is a well-established Tannakian type category theory, for which a Tannaka type duality and van Kampen type completeness theorem hold; moreover the category of geo-ari objects is naturally such a Tannakian type category.
Consequently, the fundamental results in non-abelian class field theory, such as existence theorem, conductor theorem and reciprocity law, hold.

A.3.6. Non-Commutative Euler Product

In the classical CFT for number fields, reciprocity law can be used to define L-functions in terms of Euler product formalism. Thus, we expect that non-abelian reciprocity law will help us to understand non-abelian L-functions. However, there is a difficulty in doing so: in abelian theory, we do not really care about the order of Frobenius appearing in the Euler product since Galois groups involved are commutative. This then leads naturally to the consideration of non-abelian Euler products.

It is for this purpose, motivated by the non-abelian CFT for function fields over complex numbers, we propose to study algebraic relations among images of the Frobenius classes of absolute Galois groups in finite (subquotient) groups in Σ_r. Here Σ_r is defined to be the collection of all finite groups G satisfying
(1) all irreducible unitary representations of G are of rank $\leq r$;
(2) G admits at least one rank r irreducible unitary representation.

Examples. (1) When $r = 1$, Σ_r is simply the collection of all finite

abelian groups and hence the algebraic relations we are seeking for are simply the commutative law;
(2) Groups G in Σ_r may be characterized by the following property: For any $2r$ elements $x_1, x_2, \ldots, x_{2r}$, in the group ring $\mathbb{Z}[G]$,

$$\sum_{\sigma \in S_{2r}} \mathrm{sgn}(\sigma) \cdot x_{\sigma(1)} x_{\sigma(2)} \cdots x_{\sigma(2r)} = 0,$$

where S_{2r} denotes the symmetric groups on $2r$ symbols, and $\mathrm{sgn}(\sigma)$ denotes the sign of σ. (In particular, if $r = 1$ we get the commutative law $x_1 x_2 = x_2 x_1$.)

As such, then non-abelian Euler product is supposed to be a formalism which exposes intrinsic relations for algebraic structures among Galois groups in Σ_r whose specializations give non-abelian L-functions.

B. Non-Abelian L-Functions

B.1. Non-Abelian Zeta Functions for Curves

B.1.1. Local Non-Abelian Zeta Functions

B.1.1.1. Artin Zeta Functions for Curves

We start this part by recalling the construction and basic properties of classical Artin zeta functions for curves defined over finite fields.

Let C be a projective irreducible reduced regular curve of genus g defined over a finite field $k := \mathbf{F}_q$ with q elements. Then the arithmetic degree $d(P)$ of a closed point P on C is defined to be $d(P) := \big[k(P) : k\big]$, where $k(P)$ denotes the residue class field of C at P. So $q^{d(P)}$ is nothing but the number $N(P)$ of elements in $k(P)$. Extending this to all divisors, we define the Artin zeta function $\zeta_C(s)$ for curve C over k by setting

$$\zeta_C(s) := \prod_P \frac{1}{1-N(P)^{-s}} = \sum_{D\geq 0} N(D)^{-s} = \sum_{D\geq 0} \left(q^{-s}\right)^{d(D)}, \ \mathrm{Re}(s) > 1,$$

where the sum is taken over all effective divisors D on C. As usual, set

$$t := q^{-s} \qquad \text{and} \qquad Z_C(t) := \sum_{D\geq 0} t^{d(D)}.$$

Note that the number of positive divisors which are rational equivalent to D is $\frac{q^{h^0(C,D)}-1}{q-1}$,

$$\begin{aligned}\zeta_C(t) = \sum_{D\geq 0} \left(q^{-s}\right)^{d(D)} &= \sum_{d\geq 0}\sum_{\mathcal{D}}\sum_{D\in\mathcal{D}} \left(q^{-s}\right)^d \\ &= \sum_{d\geq 0}\sum_{\mathcal{D}} \frac{q^{h^0(C,\mathcal{D})}-1}{q-1} \cdot \left(q^{-s}\right)^d,\end{aligned}$$

where the sum $\sum_{\mathcal{D}}$ is taken over all rational divisor classes of degree d. Hence by the duality, the Riemann-Roch and a vanishing theorem, we have the following

Theorem. (see e.g. [Me]) *The Artin zeta function $\zeta_C(s)$ is well-defined for* $\mathrm{Re}\, s > 1$*, and admits a meromorphic continuation to the whole complex s-plane. Moreover,*

$$\zeta_C(s) = N(K_C)^{1/2-s}\zeta_C(1-s),$$

and there exists a polynomial $P_C(t) \in \mathbb{Z}[T]$ of degree $2g$ such that

$$Z_C(t) = \frac{P_C(t)}{(1-t)(1-qt)}.$$

B.1.1.2. Two Different Generalizations: Weil Zeta Functions and A New Approach

Based on the so-called reciprocity law, Artin zeta function may be written as

$$Z_C(t) = \exp\Big(\sum_{m\geq 1} \frac{N_m}{m} \cdot t^m\Big)$$

where $N_m := \#C(\mathbf{F}_{q^m})$ denotes the number of $\mathbf{F}_{q^m}$-rational points of C. This then leads to a far reaching generalization to the so-called Weil zeta functions of higher dimensional varieties defined over finite fields, the study of which dominates what we call Arithmetic Geometry in the second half of 20th century.

On the other hand, Artin zeta functions may also be naturally interpreted in a different way which leads to a non-abelian theory for zeta functions, based on the follows;
(1) $\sum_{\mathcal{D}}$ may be viewed as a sum over degree d Picard group of the curve;
(2) Picard groups are moduli spaces of (semi-stable) line bundles;
(3) All the terms appeared in Artin zeta functions, such as h^0 and degree, make sense for vector bundles as well; and
(4) $q-1$ in the denominator is the order of the automorphic group of a line bundle.

In other words, the Artin zeta function $\zeta_C(s)$ may be rewritten as

$$\zeta_C(s) := \sum_{d\geq 0} \sum_{L\in \mathrm{Pic}^d(C)} \frac{q^{h^0(C,L)}-1}{\#\mathrm{Aut}(L)} \cdot \left(q^{-s}\right)^{d(L)}.$$

As such, these abelian zetas are ready to be naturally extended to their non-abelian counterparts.

B.1.1.3. Moduli Spaces of Semi-Stable Bundles

Assume now that C is defined over an algebraically closed field $\bar{k}$. Then according to Mumford [M], a vector bundle V on C is called semi-stable (resp. stable) if for any proper subbundle V' of V,

$$\mu(V') := \frac{d(V')}{r(V')} \leq (\text{resp.} <) \frac{d(V)}{r(V)} =: \mu(V).$$

Here d denotes the degree and r denotes the rank.

Proposition 1. *Let V be a vector bundle over C. Then*
(a) ([HN]) *there exists a unique filtration of subbundles of V, the so-called Harder-Narasimhan filtration of V,*

$$\{0\} = V_0 \subset V_1 \subset V_2 \subset \cdots \subset V_{s-1} \subset V_s = V$$

such that V_i/V_{i-1} is semi-stable and $\mu(V_i/V_{i-1}) > \mu(V_{i+1}/V_i)$;
(b) (see e.g. [Se2]) *if moreover V is semi-stable, there exists a filtration of subbundles of V, a Jordan-Hölder filtration of V,*

$$\{0\} = V^{t+1} \subset V^t \subset \cdots \subset V^1 \subset V^0 = V$$

s.t. V^i/V^{i+1} is stable and $\mu(V^i/V^{i+1}) = \mu(V)$. Moreover, $\mathrm{Gr}(V) := \oplus_{i=0}^t V^i/V^{i+1}$, the associated (Jordan-Hölder) graded bundle of V, is determined uniquely by V.

Following Seshadri, two semi-stable vector bundles V and W are called S-equivalent, if their associated Jordan-Hölder graded bundles are isomorphic, i.e., $\mathrm{Gr}(V) \simeq \mathrm{Gr}(W)$. Applying Mumford's general result on geometric invariant theory ([M]), Narasimhan and Seshadri establish the following

Theorem 2. ([NS], [Se1]) *Let C be a regular, reduced, irreducible projective curve defined over an algebraically closed field. Then over the set $\mathcal{M}_{C,r}(d)$ of S-equivalence classes of rank r and degree d semi-stable vector bundles over C, there is a natural normal, projective algebraic variety structure.*

Now assume that C is defined over a finite field k. Then there is a notion of k-rational bundles over C, i.e., bundles which are defined over k. Moreover, from geometric invariant theory, projective varieties $\mathcal{M}_{C,r}(d)$ are defined over a certain finite extension of k. Thus it also makes sense to talk about k-rational points of these moduli spaces too. The relation between these two types of rationality is given by Harder and Narasimhan based on a discussion about Brauer groups:

Proposition 3. ([HN]) *There exists a finite field $\mathbf{F}_q$ such that for any d, the subset of $\mathbf{F}_q$-rational points of $\mathcal{M}_{C,r}(d)$ consists exactly of all S-equivalence classes of $\mathbf{F}_q$-rational bundles in $\mathcal{M}_{C,r}(d)$.*

From now on, without loss of generality, we always assume that finite fields $\mathbf{F}_q$ (with q elements) satisfy the property in this proposition. Also for simplicity, we write $\mathcal{M}_{C,r}(d)$ for $\mathcal{M}_{C,r}(d)(\mathbf{F}_q)$, the subset of $\mathbf{F}_q$-rational points, and call them moduli spaces by an abuse of notation.

B.1.1.4. Local Non-Abelian Zeta Functions

For C a regular, reduced, irreducible projective curve over $\mathbf{F}_q$, define its *rank r non-abelian zeta function* $\zeta_{C,r,\mathbf{F}_q}(s)$ by setting

$$\zeta_{C,r,\mathbf{F}_q}(s) := \sum_{V\in[V]\in\mathcal{M}_{C,r}(d),d\geq 0} \frac{q^{h^0(C,V)}-1}{\#\mathrm{Aut}(V)}\cdot(q^{-s})^{d(V)}, \qquad \mathrm{Re}(s)>1.$$

Clearly, we have the following

Lemma. *$\zeta_{C,1,\mathbf{F}_q}(s)$ is nothing but the classical Artin zeta function for curve C.*

Remark. In the definition of ζ above, the sum is taken over all elements in Seshadri classes. This may be modified by taking only one representative for each class. Another modification, as suggested by Deninger, is to consider only stable objects. We leave the details to the reader.

B.1.1.5. Basic Properties for Non-Abelian Zeta Functions

Many basic properties for classical Artin zeta functions are satisfied by our non-abelian zeta functions as well. More precisely, by using a vanishing theorem, Serre duality, and the Riemann-Roch, we obtain the following

Theorem. ([We4]) (1) *The non-abelian zeta function* $\zeta_{C,r,\mathbf{F}_q}(s)$ *is well-defined for* $\mathrm{Re}(s) > 1$, *and admits a meromorphic continuation to the whole complex* s*-plane;*
(2) (**Rationality**) *Set* $t := q^{-s}$ *and introduce the non-abelian* Z*-function of* C *by*

$$\zeta_{C,r,\mathbf{F}_q}(s) =: Z_{C,r,\mathbf{F}_q}(t) := \sum_{V\in[V]\in\mathcal{M}_{C,r}(d),d\geq 0} \frac{q^{h^0(C,V)}-1}{\#\mathrm{Aut}(V)}\cdot t^{d(V)},\ |t|<1.$$

Then there exists a polynomial $P_{C,r,\mathbf{F}_q}(s) \in \mathbf{Q}[t]$ *such that*

$$Z_{C,r,\mathbf{F}_q}(t) = \frac{P_{C,r,\mathbf{F}_q}(t)}{(1-t^r)(1-q^r t^r)};$$

(3) (**Functional Equation**) *Set the rank* r *non-abelian* ξ*-function* $\xi_{C,r,\mathbf{F}_q}(s)$ *by*

$$\xi_{C,r,\mathbf{F}_q}(s) := \zeta_{C,r,\mathbf{F}_q}(s)\cdot (q^s)^{r(g-1)}.$$

Then

$$\xi_{C,r,\mathbf{F}_q}(s) = \xi_{C,r,\mathbf{F}_q}(1-s).$$

B.1.2. Global Non-Abelian Zeta Functions for Curves

B.1.2.1. Definition

Let $\mathcal{C}$ be a regular, reduced, irreducible projective curve of genus g defined over a number field F. Let S_{bad} be the collection of all infinite places and these finite places of F at which $\mathcal{C}$ does not have good reduction. As usual, a place v of F is called good if $v \notin S_{\mathrm{bad}}$.

For any good place v of F, the v-reduction of $\mathcal{C}$, denoted as $\mathcal{C}_v$, gives a regular, reduced, irreducible projective curve defined over the residue field $F(v)$ of F at v. Denote the cardinal number of $F(v)$ by q_v. Then, by 1.1, we obtain the associated rank r non-abelian zeta function $\zeta_{C_v,r,\mathbf{F}_{q_v}}(s)$. Moreover, from the rationality of $\zeta_{C_v,r,\mathbf{F}_{q_v}}(s)$, there exists a degree $2rg$ polynomial $P_{C_v,r,\mathbf{F}_{q_v}}(t) \in \mathbf{Q}[t]$ such that

$$Z_{C_v,r,\mathbf{F}_{q_v}}(t) = \frac{P_{C_v,r,\mathbf{F}_{q_v}}(t)}{(1-t^r)(1-q^r t^r)}.$$

Clearly, $P_{C_v,r,\mathbf{F}_{q_v}}(0) \neq 0$. Set

$$\tilde{P}_{C_v,r,F(v)}(t) := \frac{P_{C_v,r,F(v)}(t)}{P_{C_v,r,F(v)}(0)}.$$

Now by definition, *the rank r non-abelian zeta function $\zeta_{C,r,F}(s)$ of C* over F is the following Euler product

$$\zeta_{C,r,F}(s) := \prod_{v:\text{good}} \frac{1}{\tilde{P}_{C_v,r,\mathbf{F}_{q_v}}(q_v^{-s})}, \qquad \mathrm{Re}(s) \gg 0.$$

Clearly, when $r = 1$, $\zeta_{C,r,F}(s)$ coincides with the classical Hasse-Weil zeta function for C over F.

Conjecture. *For a regular, reduced, irreducible projective curve C of genus g defined over a number field F, its associated rank r global non-abelian zeta function $\zeta_{C,r,F}(s)$ admits a meromorphic continuation to the whole complex s-plane.*

Recall that even when $r = 1$, i.e., for the classical Hasse-Weil zeta functions, this statement, as a part of a series of high profile conjectures is still open. On the other hand, using a result of (Harder-Narasimhan) Siegel on Tamagawa numbers of SL_r, the ugly yet very precise formula for local zeta function in 1.2.1, Clifford Lemma for semi-stable bundles, and Weil's theorem on the Riemann hypothesis for Artin zeta functions, we have the following

Proposition. ([We4]) *When* $\mathrm{Re}(s) > 1+g+(r^2-r)(g-1)$, $\zeta_{C,r,F}(s)$ *converges.*

B.1.2.2. Working Hypothesis

Like in the theory for abelian zeta functions, we want to use our non-abelian zeta functions to study non-abelian aspect of arithmetic of curves.

For doing so, we then also need to introduce local factors for 'bad' places. This may be done as follows. For Γ-factors, we take these coming from the functional equation for $\zeta_F(rs)\cdot\zeta_F\big(r(s-1)\big)$, where $\zeta_F(s)$ denotes the standard Dedekind zeta function for F; while for finite bad factors, first, use the semi-stable reduction for curves to find a semi-stable model for $\mathcal{C}$, then use Seshadri's moduli spaces of parabolic bundles to construct polynomials for singular fibers, which usually have degree lower than $2rg$. With all this being done, we then can introduce a completed rank r non-abelian zeta function for $\mathcal{C}$ over F, or better, the completed rank r non-abelian zeta function $\xi_{X,r,\mathcal{O}_F}(s)$ for a semi-stable model $X\to\mathrm{Spec}(\mathcal{O}_F)$ of $\mathcal{C}$. Here $\mathcal{O}_F$ denotes the ring of integers of F. (If necessary, we take a finite extension of F.)

Working Hypothesis. $\xi_{X,r,\mathcal{O}_F}(s)$ *admits a unique meromorphic continuation to the whole complex s-plane and satisfies the functional equation*

$$\xi_{X,r,\mathcal{O}_F}(s)=\varepsilon\cdot\xi_{X,r,\mathcal{O}_F}\left(1+\frac{1}{r}-s\right)$$

with $|\varepsilon|=1$.

Moreover, we expect that for certain classes of curves, the inverse Mellin transform of our non-abelian zeta functions are naturally associated to certain modular forms of weight $1+\frac{1}{r}$.

To end this general discussion, we remark that as they stand, these global zeta functions are expected to reflect arithmetic aspect of moduli spaces of semi-stable hermitian vector sheaves over arithmetic surfaces. (The definition of semi-stability is clear, as we have the notation of arithmetic ample line bundle and hermitian vector sheaves – in general we do have problem to develop an Arakelov theory for coherent sheaves since there is no Bott-Chern theory for coherent sheaves over complex manifolds, but for our limited purpose here we can restrict ourselves to hermitian vector bundles at infinity. That is to say, over integral models, coherent sheaves are allowed, but at infinite places, these sheaves are supposed to be locally free.)

B.1.3. Refined Brill-Noether Locus: Towards a Reciprocity Law

B.1.3.1 Results of Atiyah

Let E be an elliptic curve defined over $\overline{\mathbf{F}_q}$. Recall that a vector bundle V on E is called indecomposable if V is not a direct sum of two proper subbundles, and that every vector bundle on E may be written as a direct sum of indecomposable bundles, whose summands and multiplicities are uniquely determined. Thus to understand vector bundles, it suffices to study indecomposable ones. To this end, we have the following result of Atiyah. In the sequel, for simplicity, we always assume that the characteristic of $\mathbf{F}_q$ is strictly bigger than the rank of V.

Theorem 1. ([At]) (a) *For any $r \geq 1$, there is a unique indecomposable vector bundle I_r of rank r over E, all of whose Jordan-Hölder constituents are isomorphic to $\mathcal{O}_E$. Moreover, the bundle I_r has a canonical filtration*

$$\{0\} \subset F^1 \subset \cdots \subset F^r = I_r$$

with $F^i = I_i$ and $F^{i+1}/F^i = \mathcal{O}_E$;
(b) *For any $r \geq 1$ and any integer a, relative prime to r and each line bundle λ over E of degree a, there exists up to isomorphism a unique indecomposable bundle $W_r(a;\lambda)$ over E of rank r with λ the determinant;*
(c) *The bundle $I_r\big(W_{r'}(a;\lambda)\big) = I_r \otimes W_{r'}\big(a;\lambda\big)$ is indecomposable and every indecomposable bundle is isomorphic to $I_r\big(W_{r'}(a;\lambda)\big)$ for a suitable choice of r, r', λ. Every bundle V over E is a direct sum of vector bundles of the form $I_{r_i}\big(W_{r_i'}(a_i;\lambda_i)\big)$, for suitable choices of r_i, r_i' and λ_i. Moreover, the triples $\big(r_i, r_i', \lambda_i\big)$ are uniquely specified up to permutation by the isomorphism type of V.*

Here note in particular that $W_r\big(0,\lambda\big) \simeq \lambda$, and that indeed $I_r\big(W_{r'}(a;\lambda)\big)$ is the unique indecomposable bundle of rank rr' such that all of whose successive quotients in the Jordan-Hölder filtration are isomorphic to $W_{r'}(a;\lambda)$. Clearly, $I_r\big(W_{r'}(a;\lambda)\big)$ is semi-stable with $\mu\big(I_r(W_{r'}(a;\lambda))\big)=a/r'$.

Theorem 2. (a) (Atiyah) *Every bundle V over E is isomorphic to a direct sum $\oplus_i V_i$ of semi-stable bundles, where $\mu(V_i)>\mu(V_{i+1})$;*
(b) (Atiyah) *Let V be a semi-stable bundle over E with slop $\mu(V)=a/r'$ where r' is a positive integer and a is an integer relatively prime to r'. Then V is a direct sum of bundles of the form $I_r\big(W_{r'}(a;\lambda)\big)$, where λ is a line bundle of degree a;*
(c) (Atiyah, Mumford-Seshadri) *There exists a natural projective algebraic variety structure on*

$$\mathcal{M}_{E,r}(\lambda):=\{V:\mathrm{semi-stable},\mathrm{det}V=\lambda,\mathrm{rank}(V)=r\}/\sim_S.$$

Moreover, if $\lambda\in\mathrm{Pic}^0(E)$, then $\mathcal{M}_{E,r}(\lambda)$ is simply the projective space $\mathbf{P}^{r-1}_{\overline{\mathbf{F}}_q}$.

B.1.3.2. Refined Brill-Noether Locus

Note that if V is semi-stable with strictly positive degree d, then $h^0(E,V)=d$. Hence standard Brill-Noether locus is either the whole space or empty. In this way, we are led to the case $d=0$.

For this, recall that for $\lambda\in\mathrm{Pic}^0(E)$,

$$\mathcal{M}_{E,r}(\lambda)=\{V:\mathrm{semi-stable},\mathrm{rank}(V)=r,\mathrm{det}(V)=\lambda\}/\sim_S$$

is identified with

$$\{V=\oplus_{i=1}^r L_i:\otimes_i L_i=\lambda, L_i\in\mathrm{Pic}^0(E), i=1,\ldots,r\}/\sim_{\mathrm{iso}}\simeq\mathbf{P}^{r-1}$$

where $/\sim_{\mathrm{iso}}$ means modulo isomorphisms.

Now introduce the standard Brill-Noether locus

$$W^a_{E,r}(\lambda):=\big\{[V]\in\mathcal{M}_{E,r}(\lambda):h^0(E,\mathrm{gr}(V))\geq a\big\}$$

and its 'stratification' $W^a_{E,r}(\lambda)^0$

$$:=\big\{[V]\in W_{E,r}(\lambda):h^0(E,\mathrm{gr}(V))=a\big\}=W^a_{E,r}(\lambda)\backslash\cup_{b\geq a+1}W^b_{E,r}(\lambda).$$

One checks easily that $W^a_{E,r}(\lambda) \simeq \mathbf{P}^{(r-a)-1}$,

$$W^{a+1}_{E,r+1}(\lambda) \simeq W^a_{E,r}(\lambda), \quad \text{and} \quad W^{a+1}_{E,r+1}(\lambda)^0 \simeq W^a_{E,r}(\lambda)^0.$$

The Brill-Noether theory is based on the consideration of h^0. But in the case for elliptic curves, for arithmetic consideration, such a theory is not fine enough: *not only h^0 plays a crucial role, the automorphism groups are important as well.* Based on this, we introduce, for a fixed $(k+1)$-tuple non-negative integers $(a_0; a_1, \ldots, a_k)$, the subvariety of $W^{a_0}_{E,r}$ by setting

$$W^{a_0;a_1,\ldots,a_k}_{E,r}(\lambda) := \Big\{ [V] \in W^{a_0}_{E,r}(\lambda) \,:\, \mathrm{gr}(V) = \mathcal{O}_E^{(a_0)} \oplus \oplus_{i=1}^k L_i^{(a_i)}, \\ \otimes_i L_i^{\otimes a_i} = \lambda,\ L_i \in \mathrm{Pic}^0(E),\ 1 \leq i \leq k \Big\}.$$

Moreover, we define the associated 'stratification' by setting

$$W^{a_0;a_1,\ldots,a_k}_{E,r}(\lambda)^0 := \big\{ [V] \in W^{a_0;a_1,\ldots,a_k}_{E,r}(\lambda),\ \#\{\mathcal{O}_E, L_1, \ldots, L_k\} = k+1 \big\}.$$

Easily we have

$$W^{a_0+1;a_1,\ldots,a_k}_{E,r+1}(\lambda) \simeq W^{a_0;a_1,\ldots,a_k}_{E,r}(\lambda),$$
$$W^{a_0+1;a_1,\ldots,a_k}_{E,r+1}(\lambda)^0 \simeq W^{a_0;a_1,\ldots,a_k}_{E,r}(\lambda)^0,$$

and $\mathcal{M}_{E,r}(\lambda) = \cup_{a_0;a_1,\ldots,a_k} W^{a_0;a_1,\ldots,a_k}_{E,r}(\lambda)^0$.

Now recall that for elliptic curves E,
(1) The quotient space $E^{(n)}/S_n$ is isomorphic to the $\mathbf{P}^{n-1}$-bundle over E; and
(2) The quotient of $E^{(n-1)}/S_n$ is isomorphic to $\mathbf{P}^{(n-1)}$. Here we embed $E^{(n-1)}$ as a subspace of $E^{(n)}$ under the map $(x_1, \ldots, x_{n-1}) \mapsto (x_1, \ldots, x_{n-1}, x_n)$ with $x_n = \lambda - (x_1 + x_2 + \cdots + x_{n-1})$.

Proposition. ([We3]) *Regroup $(a_0; a_1, \ldots, a_k)$ as $(a_0; b_1^{(s_1)}, \ldots, b_l^{(s_l)})$ where we assume that $b_1 > b_2 > \cdots > b_l$ and $s_1, s_2, \ldots, s_l \in \mathbf{Z}_{>0}$, then*
(1) if $b_l = 1$, $W^{a_0;a_1,\ldots,a_k}_{E,r}(\lambda) \simeq \prod_{i=1}^{l-1} \mathbf{P}_E^{s_i-1} \times \mathbf{P}^{s_l}$;
(2) if $b_l > 1$, $W^{a_0;a_1,\ldots,a_k}_{E,r}(\lambda) \simeq \prod_{i=1}^{l} \mathbf{P}_E^{s_i}$.

Remark. When $\lambda = \mathcal{O}_E$, refined Brill-Noether loci $W_{E,r}^{a_0;a_1,\ldots,a_k}(\mathcal{O}_E)$ are isomorphic to products of (copies of) projective bundles over E and (copies of) projective spaces, which are special subvarieties in $\mathcal{M}_{E,r}(\mathcal{O}_E) = \mathbf{P}^{r-1}$. It appears that intersections among these refined Brill-Noether loci are quite interesting. So define the Brill-Noether tautological ring $\mathbf{BN}_{E,r}(\mathcal{O}_E)$ to be the subring generated by all the associated refined Brill-Noether loci (in the corresponding Chow ring). What can we say about it?

Examples. (1) If $r = 2$, then this ring consists of only two elements: the 0-dimensional one $W_{E,2}^{2;0}(\mathcal{O}_E) = \{[\mathcal{O}_E \oplus \mathcal{O}_E]\}$ and the whole space $\mathbf{P}^1$. So everything is simple;
(2) If $r = 3$, then (generators of) this ring contains five elements: 2 of 0-dimensional objects: $W_{E,3}^{3;0}(\mathcal{O}_E) = \{[\mathcal{O}_E^{(3)}]\}$ and $W_{E,3}^{1;2}(\mathcal{O}_E) = \{[\mathcal{O}_E \oplus T_2^{(2)}] : T_2 \in E_2\}$ containing 4 elements; 2 of 1-dimensional objects: $W_{E,3}^{1;1,1} = \{[\mathcal{O}_E \oplus L \oplus L^{-1}] : L \in \mathrm{Pic}^0(E)\} \simeq \mathbf{P}^1$, a degree 2 projective line contained in $\mathbf{P}^2 = \mathcal{M}_{E,3}(\mathcal{O}_E)$; and $W_{E,3}^{0;2,1} = \{[L^{(2)} \oplus L^{-2}] : L \in \mathrm{Pic}^0(E)\}$ a degree 3 curve which is isomorphic to E; and finally the whole space. Moreover, the intersection of $W_{E,3}^{1;1,1} = \mathbf{P}^1$ and $W_{E,3}^{0;2,1} = E$ are supported on 0-dimensional locus $W_{E,3}^{1;1,1}$, with the multiplicity 3 on the single point locus $W_{E,3}^{3;0}(\mathcal{O}_E)$ and 1 on the complement of the points in $W_{E,3}^{1;1,1}$.

B.1.3.3. Towards a Reciprocity Law

To measure the Brill-Noether locus, we introduce

$$\alpha_{E,r}(\lambda) := \sum_{V \in [V] \in \mathcal{M}_{E,r}(\lambda)} \frac{q^{h^0(E,V)}}{\#\mathrm{Aut}(V)},$$

$$\alpha_{E,r}^{a_0+1;a_1,\ldots,a_k}(\lambda) := \sum_{V \in [V] \in W_{E,r}^{a_0+1;a_1,\ldots,a_k}(\lambda)^0} \frac{q^{h^0(E,V)}}{\#\mathrm{Aut}(V)},$$

$$\beta_{E,r}^{a_0;a_1,\ldots,a_k}(\lambda) := \sum_{V \in [V] \in W_{E,r}^{a_0;a_1,\ldots,a_k}(\lambda)^0} \frac{1}{\#\mathrm{Aut}(V)}.$$

Before going further, we remark that above, we write $V \in [V]$ in the summation. This is because in each S-equivalence class $[V]$, there are usually more than one vector bundles V. For example, $[\mathcal{O}_E^{(4)}]$ consists of $\mathcal{O}_E^{(4)}$, $\mathcal{O}_E^{(2)} \oplus I_2$, $I_2 \oplus I_2$, $\mathcal{O}_E \oplus I_3$, and I_4.
Theorem. ([HN] & [DR]) *For all* $\lambda, \lambda' \in \mathrm{Pic}^d(E)$, $\beta_{E,r}(\lambda) = \beta_{E,r}(\lambda')$. *Moreover,*

$$N_1 \cdot \beta_{E,r}(\lambda) = \frac{N_1}{q-1} \cdot \prod_{i=2}^{r} \zeta_E(i) - \sum_{\Sigma_1^k r_i = r, \Sigma_i d_i = d, \frac{d_1}{r_1} > \cdots > \frac{d_k}{r_k}, k \geq 2} \prod_i \beta_{E,r_i}(d_i) \frac{1}{q^{\Sigma_{i<j}(r_j d_i - r_i d_j)}}.$$

Here N_1 *denotes* $\#E (:= \#E(\mathbf{F}_q))$.

Thus, we are lead to introduce the γ-series invariants $\gamma_{E,r}(\lambda)$ and $\gamma_{E,r}^{a_0+1;a_1,\ldots,a_k}(\lambda)$ by setting $\gamma := \alpha - \beta$. That is to say,

$$\gamma_{E,r}(\lambda) := \sum_{V \in [V] \in \mathcal{M}_{E,r}(\lambda)} \frac{q^{h^0(E,V)} - 1}{\#\mathrm{Aut}(V)},$$

$$\gamma_{E,r}^{a_0;a_1,\ldots,a_k}(\lambda) := \sum_{V \in [V] \in W_{E,r}^{a_0;a_1,\ldots,a_k}(\lambda)^0} \frac{q^{h^0(E,V)} - 1}{\#\mathrm{Aut}(V)}.$$

Based on the above theorem and lower rank examples using the stratification just introduced, we make the following

Conjecture. *For all* $\lambda \in \mathrm{Pic}^0(E)$, $\alpha_{E,r}(\lambda) = \alpha_{E,r}(\mathcal{O}_E)$. *Hence also* $\gamma_{E,r}(\lambda) = \gamma_{E,r}(\mathcal{O}_E)$.

Appendix to B.1: Weierstrass Groups

Motivated by a certain construction of Euler systems for elliptic curves in terms of elements in K_2 using torsion points, we here introduce what we call Weierstrass groups using Weierstrass divisors for curves.

1. Weierstrass Divisors

(1.1) Let M be a compact Riemann surface of genus $g \geq 2$. Denote its degree d Picard variety by $\mathrm{Pic}^d(M)$. Fix a Poincaré line bundle $\mathcal{P}_d$ on $M \times \mathrm{Pic}^d(M)$. (One checks easily that our constructions do not depend on a particular choice of the Poincaré line bundle.) Let Θ be the theta divisor of $\mathrm{Pic}^{g-1}(M)$, i.e., the image of the natural map $M^{g-1} \to \mathrm{Pic}^{g-1}(M)$ defined by $(P_1, \ldots, P_{g-1}) \mapsto [\mathcal{O}_M(P_1 + \cdots + P_{g-1})]$. Here $[\cdot]$ denotes the class defined by $\cdot$. We will view the theta divisor as a pair $(\mathcal{O}_{\mathrm{Pic}^{g-1}(M)}(\Theta), \mathbf{1}_\Theta)$ with $\mathbf{1}_\Theta$ the defining section of Θ via the structure exact sequence $0 \to \mathcal{O}_{\mathrm{Pic}^{g-1}(M)} \to \mathcal{O}_{\mathrm{Pic}^{g-1}(M)}(\Theta)$.

Denote by p_i the i-th projection of $M \times M$ to M, $i = 1, 2$. Then for any degree $d = g - 1 + n$ line bundle on M, we get a line bundle $p_1^* L(-n\Delta)$ on $M \times M$ which has relative p_2-degree $g - 1$. Here, Δ denotes the diagonal divisor on $M \times M$. Hence, we get a classifying map $\phi_L : M \to \mathrm{Pic}^{g-1}(M)$ which makes the following diagram commute:

$$\begin{array}{ccc} M \times M & \rightarrow & M \times \mathrm{Pic}^{g-1}(M) \\ p_2 \downarrow & & \downarrow \pi \\ M & \stackrel{\phi_L}{\rightarrow} & \mathrm{Pic}^{g-1}(M). \end{array}$$

One checks that there are canonical isomorphisms

$$\lambda_\pi(\mathcal{P}^{g-1}) \simeq \mathcal{O}_{\mathrm{Pic}^{g-1}(M)}(-\Theta)$$

and

$$\lambda_{p_2}(p_1^* L(-n\Delta)) \simeq \phi_L^* \mathcal{O}_{\mathrm{Pic}^{g-1}(M)}(-\Theta).$$

Here, λ_π (resp. λ_{p_2}) denotes the Grothendieck-Mumford cohomology determinant with respect to π (resp. p_2). (See e.g., [L].)

Thus, $\phi_L^* \mathbf{1}_\Theta$ gives a canonical holomorphic section of the dual of the line bundle $\lambda_{p_2}(p_1^* L(-n\Delta))$, which in turn gives an effective divisor $W_L(M)$ on M, the so-called *Weierstrass divisor associated to* L.

Example. With the same notation as above, take $L = K_M^{\otimes m}$ with K_M the canonical line bundle of M and $m \in \mathbf{Z}$. Then we get an effective divisor $W_{K_M^{\otimes m}}(M)$ on M, which will be called the *m-th Weier-*

strass divisor associated to M. For simplicity, denote $W_{K_M^{\otimes m}}(M)$ (resp. $\phi_{K_M^{\otimes m}}$) by $W_m(M)$ (resp. ϕ_m).

One checks easily that the degree of $W_m(M)$ is $g(g-1)^2(2m-1)^2$ and we have an isomorphism $\mathcal{O}_M(W_m(M)) \simeq K_M^{\otimes g(g-1)(2m-1)^2/2}$. Thus, in particular,

$$f_{m,n} := \frac{(\phi_m^* \mathbf{1}_\Theta)^{\otimes(2n-1)^2}}{(\phi_n^* \mathbf{1}_\Theta)^{\otimes(2m-1)^2}}$$

gives a canonical meromorphic function on M for all $m, n \in \mathbf{Z}$.

Remark. We may also assume that $m \in \frac{1}{2}\mathbf{Z}$. Furthermore, this construction has a relative version as well, for which we assume that $f : \mathcal{X} \to B$ is a semi-stable family of curves of genus $g \geq 2$. In that case, we get an effective divisor $(\mathcal{O}_\mathcal{X}(W_m(f)), \mathbf{1}_{W_m(f)})$ and canonical isomorphism $(\mathcal{O}_\mathcal{X}(W_m(f)), \mathbf{1}_{W_m(f)})$

$$\simeq (\mathcal{O}_\mathcal{X}(W_1(f)), \mathbf{1}_{W_1(f)})^{\otimes(2m-1)^2} \otimes (\mathcal{O}_\mathcal{X}(W_{\frac{1}{2}}(f)), \mathbf{1}_{W_{\frac{1}{2}}(f)})^{\otimes 4m(1-m)}.$$

The proof may be given by using Deligne-Riemann-Roch theorem, which in general, implies that we have the following canonical isomorphism:

$$(\mathcal{O}_\mathcal{X}(W_L(f)), \mathbf{1}_{W_L(f)}) \otimes f^*\lambda_f(L) \simeq L^{\otimes n} \otimes K_f^{\otimes n(n-1)/2}.$$

(See e.g. [Bu].) To allow m be a half integer, we then should assume that f has a spin structure. Certainly, without using spin structure, a modified canonical isomorphism, valid for integers, can be given.

2. K-Groups

(2.1) Let M be a compact Riemann surface of genus $g \geq 2$. Then by the localization theorem, we get the following exact sequence for K-groups

$$K_2(M) \xrightarrow{\lambda} K_2(\mathbf{C}(M)) \overset{\coprod_{p\in M}\partial_p}{\to} \coprod_{p\in M} \mathbf{C}_p^*.$$

Note that the middle term may also be written as $K_2(\mathbf{C}(M\backslash S))$ for any finite subset S of M, we see that naturally by a theorem of

Matsumoto, the Steinberg symbol $\{f_{m,n}, f_{m',n'}\}$ gives a well-defined element in $K_2(\mathbf{C}(M))$. Denote the subgroup generated by all these $\{f_{m,n}, f_{m',n'}\}$ with $m, n, m', n' \in \mathbf{Z}_{>0}$ in $K_2(\mathbf{C}(M))$ as $\Sigma(M)$.

Definition. With the same notation as above, the *first Weierstrass group* $W_I(M)$ of M is defined to be the λ-pull-back of $\Sigma(M)$, i.e., the subgroup $\lambda^{-1}(\Sigma(M))$ of $K_2(M)$.

(2.2) For simplicity, now let C be a regular projective irreducible curve of genus $g \geq 2$ defined over $\mathbf{Q}$. Assume that C has a semi-stable regular module X over $\mathbf{Z}$ as well. Then we have a natural morphism $K_2(X) \xrightarrow{\phi} K_2(M)$. Here $M := C(\mathbf{C})$.

Conjecture I. *With the same notation as above,* $\phi\big(K_2(X)\big)_{\mathbf{Q}} = W_I(M)_{\mathbf{Q}}$.

3. Generalized Jacobians

(3.1) Let C be a projective, regular, irreducible curve. Then for any effective divisor D, one may canonically construct the so-called generalized Jacobian $J_D(C)$ together with a rational map $f_D : C \to J_D(C)$.

More precisely, let C_D be the group of classes of divisors prime to D modulo these which can be written as $\mathrm{div}(f)$. Let C_D^0 be the subgroup of C_D which consists of all elements of degree zero. For each p_i in the support of D, the invertible elements modulo those congruent to 1 (mod D) form an algebraic group R_{D,p_i} of dimension n_i, where n_i is the multiplicity of p_i in D. Let R_D be the product of these R_{D,p_i}. One checks easily that $\mathbf{G}_m$, the multiplicative group of constants naturally embeds into R_D. It is a classical result that we then have the short exact sequence

$$0 \to R_D/\mathbf{G}_m \to C_D^0 \to J \to 0$$

where J denotes the standard Jacobian of C. (See e.g., [S].) Denote $R_D/\mathbf{G}_m$ simply by $\mathbf{R}_D$.

Now the map f_D extends naturally to a bijection from C_D^0 to J_D. In this way the commutative algebraic group J_D becomes an extension as algebraic groups of the standard Jacobian by the group $\mathbf{R}_D$.

Example. Take the field of constants as $\mathbf{C}$ and $D = W_m(M)$, the m-th Weierstrass divisor of a compact Riemann surface M of genus $g \geq 2$. By (1.1), $W_m(M)$ is effective. So we get the associated generalized Jacobian $J_{W_m(M)}$. Denote it by $WJ_m(M)$ and call it the m-th *Weierstras-Jacobian* of M. For example, if $m = 0$, then $WJ_0(M) = J(M)$ is the standard Jacobian of M. Moreover, one knows that the dimension of $R_{W_m(M),p}$ is at most $g(g+1)/2$. For later use denote $\mathbf{R}_{W_m(M)}$ simply by $\mathbf{R}_m$.

(3.2) The above construction works on any base field as well. We leave the detail to the reader while point out that if the curve is defined over a field F, then its associated m-th Weierstrass divisor is rational over the same field as well. (Obviously, this is not true for the so-called Weierstrass points, which behavior in a rather random way.) As a consequence, by the construction of the generalized Jacobian, we see that the m-th Weierstrass-Jacobians are also defined over F. (See e.g., [S].)

4. Galois Cohomology Groups

(4.1) Let K be a perfect field, $\overline{K}$ be an algebraic closure of K and $G_{\overline{K}/K}$ be the Galois group of $\overline{K}$ over K. Then for any $G_{\overline{K}/K}$-module M, we have the Galois cohomology groups $H^0(G_{\overline{K}/K}, M)$ and $H^1(G_{\overline{K}/K}, M)$ such that if

$$0 \to M_1 \to M_2 \to M_3 \to 0$$

is an exact sequence of $G_{\overline{K}/K}$-modules, then we get a natural long exact sequence

$$\begin{aligned} 0 \to H^0(G_{\overline{K}/K}, M_1) &\to H^0(G_{\overline{K}/K}, M_2) \to H^0(G_{\overline{K}/K}, M_3) \\ &\to H^1(G_{\overline{K}/K}, M_1) \to H^1(G_{\overline{K}/K}, M_2) \to H^1(G_{\overline{K}/K}, M_3). \end{aligned}$$

Moreover, if G is a subgroup of $G_{\overline{K}/K}$ of finite index or a finite subgroup, then M is naturally a G-module. This leads a restriction map on cohomology res : $H^1(G_{\overline{K}/K}, M) \to H^1(G, M)$.

(4.2) Now let C be a projective, regular irreducible curve defined over a number field K. Then for each place p of K, fix an extension of p to $\overline{K}$, which then gives an embedding $\overline{K} \subset \overline{K}_p$ for the p-adic completion K_p of K and a decomposition group $G_p \subset G_{\overline{K}/K}$.

Apply the construction in (3.1) to the short exact sequence

$$0 \to \mathbf{R}_m \to WJ_m(C) \to J(C) \to 0$$

over K. Then we have the following long exact sequence

$$\begin{aligned} 0 \to \mathbf{R}_m(K) \to WJ_m(K) \to J(K) &\to H^1(G_{\overline{K}/K}, \mathbf{R}_m(K)) \\ \to H^1(G_{\overline{K}/K}, WJ_m(K)) &\stackrel{\psi}{\to} H^1(G_{\overline{K}/K}, J(K)). \end{aligned}$$

Similarly, for each place p of K, we have the following exact sequence

$$\begin{aligned} &0 \to \mathbf{R}_m(K_p) \to WJ_m(K_p) \to J(K_p) \\ &\quad \to H^1(G_p, \mathbf{R}_m(K_p)) \to H^1(G_p, WJ_m(K_p)) \stackrel{\psi_p}{\to} H^1(G_p, J(K_p)). \end{aligned}$$

Natural inclusions $G_p \subset G_{\overline{K}/K}$ and $\overline{K} \subset \overline{K}_p$ give restriction maps on cohomology, so we arrive at a natural morphism

$$\Phi_m : \psi\Big(H^1(G_{\overline{K}/K}, WJ_m(K))\Big) \to \prod_{p \in M_K} \psi_p\Big(H^1(G_p, WJ_m(K_p))\Big).$$

Here M_K denotes the set of all places over K.

Definition. *The second Weierstrass group* $W_{II}(C)$ of C is defined to be the subgroup of $H^1(G_{\overline{K}/K}, J(C)(K))$ generated by all $\mathrm{Ker}\,\Phi_m$, the kernel of Φ_m, i.e., $W_{II}(C) := \langle \mathrm{Ker}\,\Phi_m : m \in \mathbf{Z}_{>0}\rangle_{\mathbf{Z}}$.

Conjecture II. *The second Weierstrass group $W_{II}(C)$ is finite.*

5. Deligne-Beilinson Cohomology

(5.1) Let C be a projective regular curve of genus g. Let P be a finite set of C. For simplicity, assume that all of them are defined over $\mathbf{R}$.

Then we have the associated Deligne-Belinsion cohomology group $H^1_{\mathcal{D}}(C\backslash P, \mathbf{R}(1))$ which leads to the following short exact sequence:

$$0 \to \mathbf{R} \to H^1_{\mathcal{D}}(C\backslash P, \mathbf{R}(1)) \stackrel{\text{div}}{\to} \mathbf{R}[P]^0 \to 0$$

where $\mathbf{R}[P]^0$ denotes (the group of degree zero divisors with support on $P)_{\mathbf{R}}$.

The standard cup product on Deligne-Beilinson cohomology leads to a well-defined map:

$$\cup : H^1_{\mathcal{D}}(C\backslash P, \mathbf{R}(1)) \times H^1_{\mathcal{D}}(C\backslash P, \mathbf{R}(1)) \to H^2_{\mathcal{D}}(C\backslash P, \mathbf{R}(2)).$$

Furthermore, by Hodge theory, there is a canonical short exact sequence

$$H^1(C\backslash P, \mathbf{R}(1)) \cap F^1(C\backslash P) \hookrightarrow H^2_{\mathcal{D}}(C\backslash P, \mathbf{R}(2)) \stackrel{p_{\mathcal{D}}}{\twoheadrightarrow} H^2_{\mathcal{D}}(C, \mathbf{R}(2))$$

where F^1 denotes the F^1-term of the Hodge filtration on $H^1(C\backslash P, \mathbf{C})$.

All this then leads to a well-defined morphism

$$[\cdot,\cdot]_{\mathcal{D}} : \wedge^2 \mathbf{R}[P]^0 \to H^2_{\mathcal{D}}(C, \mathbf{R}(2)) = H^1(C, \mathbf{R}(1))$$

which make the associated diagram coming from the above two short exact sequences commute. (See e.g. [Be1].)

(5.2) Now applying the above construction with P being the union of the supports of W_1, W_m and W_n for $m, n > 0$. Thus for fixed m, n, in $\mathbf{R}[P]^0$, we get two elements $\text{div}(f_{1,m})$ and $\text{div}(f_{1,n})$. This then gives $[\text{div}(f_{1,m}), \text{div}(f_{1,n})]_{\mathcal{D}} \in H^1(X, \mathbf{R}(1))$. Thus, by a simple argument using the Stokes formula, we obtain the following

Lemma. *For any holomorphic differential 1-form ω on C, we have*

$$\begin{aligned}&\langle [\text{div}(f_{1,m}), \text{div}(f_{1,n})]_{\mathcal{D}}, \omega\rangle \\ :=& -\frac{1}{2\pi\sqrt{-1}} \int [\text{div}(f_{1,m}), \text{div}(f_{1,n})]_{\mathcal{D}} \wedge \bar{\omega} \\ =& -\frac{1}{2\pi\sqrt{-1}} \int g(\text{div}(f_{1,m}), z) dg(\text{div}(f_{1,n}), z) \wedge \bar{\omega},\end{aligned}$$

Here $g(D, z)$ denotes the Green's function of D with respect to any fixed normalized (possibly singular) volume form of quasi-hyperbolic type. (See e.g., [We1].)

(5.3) With exactly the same notation as in (4.2), then in $H^1(X, \mathbf{R}(1))$ we get a collection of elements $[\mathrm{div}(f_{1,m}), \mathrm{div}(f_{1,n})]_{\mathcal{D}}$ for $m, n \in \mathbf{Z}_{>0}$.
Definition. For a curve C defined over $\mathbf{Z}$, define the - *first quasi-Weierstrass group* $W'_{-I}(C)$ of C as the subgroup of $H^1(X, \mathbf{R}(1))$ generated by $[\mathrm{div}(f_{1,m}), \mathrm{div}(f_{1,n})]_{\mathcal{D}}$ for all $m, n \in \mathbf{Z}_{>0}$ and call $W'_{-I}(C)_{\mathbf{Q}}$ the - *first Weierstrass group* $W_{-I}(C)$ of C. That is to say, $W_{-I}(C) := \langle [\mathrm{div}(f_{1,m}), \mathrm{div}(f_{1,n})]_{\mathcal{D}} : m, n \in \mathbf{Z}_{>0} \rangle_{\mathbf{Q}}$.

Conjecture III. $W_{-1}(C)_{\mathbf{R}} = H^1(X, \mathbf{R}(1))$.

That is, Weierstrass divisors should give a new rational structure for $H^1(X, \mathbf{R}(1))$, and hence the corresponding regulator should give the leading coefficient of the L-function of C at $s = 0$, up to rationals.

B.2. New Non-Abelian Zeta Functions for Number Fields

B.2.1. Iwasawa's ICM Talk on Dedekind Zeta Functions

As for function fields, here we start with a discussion on abelian zeta functions for number fields, i.e., Dedekind zeta functions. However, we will not adopt the classical approach, rather we would like to recall Iwasawa's interpretation in our language. (Iwasawa's original choice of certain auxiliary functions do not naturally lead to any meaningful cohomology, we make some subtle changes here.)

Let F be a number field. Denote by S the collection of all (unequivalent) normalized places of F. Set S_∞ to be the collection of all Archimedean places of F and $S_{\mathrm{fin}} := S \backslash S_\infty$.

Denote by $\mathbf{I}$ the idele group of F, $N : \mathbf{I} \to \mathbf{R}_{\geq 0}$ and $\deg : \mathbf{I} \to \mathbf{R}$ the norm map and the degree map on ideles respectively. Also introduce the following subgroups of $\mathbf{I}$:

$$\begin{aligned}
\mathbf{I}^0 &:= \{a = (a_v) \in \mathbf{I} : \deg(a) = 0\}, \\
F^* &:= \{a = (a_v) \in \mathbf{I}^0 : a_v = \alpha \in F \backslash \{0\}, \forall v \in S\}, \\
U &:= \{a = (a_v) \in \mathbf{I}^0 : |a_v|_v = 1 \forall v \in S\}, \\
\mathbf{I}_{\mathrm{fin}} &:= \{a = (a_v) \in \mathbf{I} : a_v = 1 \forall v \in S_\infty\}, \\
\mathbf{I}_\infty &:= \{a = (a_v) \in \mathbf{I} : a_v = 1 \forall v \in S_{\mathrm{fin}}\}.
\end{aligned}$$

Set $U_{\text{fin}} := U \cap \mathbf{I}_{\text{fin}}$. Then, with respect to the natural topology on $\mathbf{I}$, we have
(1) $F \hookrightarrow \mathbf{I}$ is discrete and $\mathbf{I}^0/F^*$ is compact. Write $\text{Pic}(F) = \mathbf{I}/F^*$;
(2) $U \hookrightarrow \mathbf{I}$ is compact;
(3) $U_{\text{fin}} \hookrightarrow \mathbf{I}_{\text{fin}}$ is both open and compact. Moreover, the morphism $I : [a = (a_v)] \mapsto I(a) := \prod_{v \in S_{\text{fin}}} P_v^{\text{ord}_v(a_v)}$ induces an isomorphism between $\mathbf{I}_{\text{fin}}/U_{\text{fin}}$ and the ideal group of F, where P_v denotes the maximal ideal of the ring of integers $\mathcal{O}_F$ corresponding to the place v; and, $N(a) = \prod_{v \in S} |a_v|_v^{N_v := [F_v : \mathbf{Q}_p]} = N(I(a))^{-1}$ with $N(I(a))$ the norm of the ideal $I(a)$;
(4) $\mathbf{I} = \mathbf{I}_{\text{fin}} \times \mathbf{I}_\infty$. Hence we may write an idele a as $a = a_{\text{fin}} \cdot a_\infty$ with $a_{\text{fin}} \in \mathbf{I}_{\text{fin}}$ and $a_\infty \in \mathbf{I}_\infty$ respectively. In particular, if $d\mu(a)$ denotes the normalized Haar measure on $\mathbf{I}$ as say in Weil's Basic Number Theory, we have

$$d\mu(a) = d\mu(a_{\text{fin}}) \cdot d\mu(a_\infty)$$

corresponding to the decomposition $\mathbf{I} = \mathbf{I}_{\text{fin}} \times \mathbf{I}_\infty$.

Set

$$e(a_{\text{fin}}) := \begin{cases} 1, & \text{if } I(a_{\text{fin}}) \subset \mathcal{O}_F; \\ 0, & \text{otherwise,} \end{cases}$$

$$e(a_\infty) = \exp\Big(-\pi \sum_{v:\mathbf{R}} a_v^2 - 2\pi \sum_{v:\mathbf{C}} |a_v|^2 \Big)$$

and

$$e(a_{\text{fin}} \cdot a_\infty) = e(a_{\text{fin}}) \cdot e(a_\infty),$$

for $a_{\text{fin}} \in \mathbf{I}_{\text{fin}}$ and $a_\infty \in \mathbf{I}_\infty$, Denote by Δ_F (the absolute values of) the discriminant of F, r_1 and r_2 the number of real and complex places in S_∞ as usual.

Now we are ready to write down Iwasawa's interpretation of the Dedekind zeta function for F in the form suitable for our later study. This goes as follows.

For $s \in \mathbf{C}, \text{Re}(s) > 1$,

$$\xi_F(s) := \Delta_F^{\frac{s}{2}} \Big(2\pi^{-\frac{s}{2}} \Gamma\Big(\frac{s}{2}\Big)\Big)^{r_1} \Big((2\pi)^{-s} \Gamma(s) \Big)^{r_2} \sum_{0 \neq I \subset \mathcal{O}_F} N(I)^{-s}$$

$$
\begin{aligned}
&= \Delta_F^{\frac{s}{2}} \cdot \sum_{0 \neq I \subset \mathcal{O}_F} N(I)^{-s} \\
&\qquad \int_{t_v \in F_v, v \in S_\infty} \prod_v |t_v|_v^s \exp\Big(-\pi \sum_{v:\mathbf{R}} t_v^2 - 2\pi \sum_{v:\mathbf{C}} |t_v|^2 \Big) \prod_v d^* t_v \\
&= \Delta_F^{\frac{s}{2}} \cdot \sum_{0 \neq I \subset \mathcal{O}_F} N(I)^{-s} \int_{\mathbf{I}_\infty} N(a_\infty)^s e(a_\infty) d\mu(a_\infty) \\
&= \Delta_F^{\frac{s}{2}} \cdot \frac{1}{\mathrm{vol}(U_{\mathrm{fin}})} \\
&\qquad \cdot \Big(\int_{\mathbf{I}_{\mathrm{fin}}} N(a_{\mathrm{fin}})^s e(a_{\mathrm{fin}}) d\mu(a_{\mathrm{fin}}) \cdot \int_{\mathbf{I}_\infty} N(a_\infty)^s e(a_\infty) d\mu(a_\infty) \Big) \\
&= \frac{1}{\mathrm{vol}(U_{\mathrm{fin}})} \cdot \Delta_F^{\frac{s}{2}} \\
&\qquad \cdot \int_{\mathbf{I}_{\mathrm{fin}} \times \mathbf{I}_\infty} \Big(N(a_{\mathrm{fin}}) N(a_\infty) \Big)^s \Big(e(a_{\mathrm{fin}}) e(a_\infty) \Big) \Big(d\mu(a_{\mathrm{fin}}) d\mu(a_\infty) \Big) \\
&= \frac{1}{\mathrm{vol}(U_{\mathrm{fin}})} \cdot \Delta_F^{\frac{s}{2}} \cdot \int_{\mathbf{I}} N(a)^s e(a) d\mu(a).
\end{aligned}
$$

Now denote by $d\mu([a])$ the induced Haar measure on the Picard group $\mathrm{Pic}(F) := \mathbf{I}/F^*$. Note that

$$
e(a_{\mathrm{fin}} a_\infty) := \begin{cases} e(a_\infty), & \text{if } I(a_{\mathrm{fin}}) \subset \mathcal{O}_F, \\ 0, & \text{otherwise}, \end{cases}
$$

and that by (1) above, $F^* \hookrightarrow \mathbf{I}$ is discrete, (hence taking integration over F^* means taking summation), we get

$$
\begin{aligned}
&\Big(\mathrm{vol}(U_{\mathrm{fin}}) \cdot \Delta_F^{\frac{-s}{2}} \Big) \cdot \xi_F(s) \\
&= \int_{\mathbf{I}/F^*} \Big(\int_{F^*} N(\alpha a)^s e(\alpha a) d\alpha \Big) d\mu([a]) \\
&= \int_{\mathbf{I}/F^*} \Big(\sum_{\alpha \in F^*} e(\alpha a) \Big) \cdot N([a])^s d\mu([a]) \\
&\qquad \text{(by the product formula)}
\end{aligned}
$$

$$
\begin{aligned}
&= \int_{\mathbf{I}/F^*} N([a])^s d\mu([a]) \cdot \Big(\sum_{\alpha \in F^*} e(\alpha a_{\mathrm{fin}}) e(\alpha a_\infty) \Big) \\
&= \int_{\mathbf{I}/F^*} N([a])^s d\mu([a]) \cdot \Big(\sum_{\alpha \in F^*, I(\alpha a_{\mathrm{fin}}) \subset \mathcal{O}_F} e(\alpha a_\infty) \Big) \\
&= \int_{\mathbf{I}/F^*} N([a])^s d\mu([a]) \cdot \Big(\sum_{\alpha \in F^*, \alpha a_v \subset \mathcal{O}_v, \forall v \in S_{\mathrm{fin}}} e(\alpha a_\infty) \Big).
\end{aligned}
$$

For an idele $L = (a_v)$, define the 0-th algebraic cohomology group of the idele L^{-1} by

$$
H^0(F, L^{-1}) := \{\alpha \in F^*, \alpha a_v \subset \mathcal{O}_v, \forall v \in S_{\mathrm{fin}}\} = I(L^{-1});
$$

moreover for the associated idele class, introduce its associated geometric 0-th geo-ari cohomology via

$$
\begin{aligned}
& h^0(F, L^{-1}) \\
:= & \log \Big(\sum_{\alpha \in H^0(F, L^{-1}) \setminus \{0\}} \exp \Big(-\pi \sum_{v:\mathbf{R}} |g_v \alpha|^2 - 2\pi \sum_{v:\mathbf{C}} |g_v \alpha|^2 \Big) \Big).
\end{aligned}
$$

With this, then easily, we have

$$
\begin{aligned}
& \Big(\mathrm{vol}(U_{\mathrm{fin}}) \cdot \Delta_F^{\frac{-s}{2}} \Big) \cdot \xi_F(s) \\
= & \int_{\mathrm{Pic}(F)} \frac{1}{w_F} \sum_{\alpha \in H^0(F, L^{-1}) \setminus \{0\}} \exp \Big(-\pi \sum_{v:\mathbf{R}} |g_v \alpha|^2 - 2\pi \sum_{v:\mathbf{C}} |g_v \alpha|^2 \Big) \\
& \qquad \cdot N(L)^s d\mu(L) \\
= & \int_{\mathrm{Pic}(F)} \frac{1}{w_F} \Big(e^{h^0(F, L^{-1})} - 1 \Big) N(L)^s d\mu(L) \\
= & \frac{1}{w_F} \cdot \int_{\mathrm{Pic}(F)} \Big(e^{h^0(F, L)} - 1 \Big) N(L)^{-s} d\mu(L) \\
= & \int_{\mathrm{Pic}(F)} \frac{e^{h^0(F, L)} - 1}{\mathrm{Aut}(L)} \cdot \Big(e^{\deg(L)} \Big)^{-s} d\mu(L).
\end{aligned}
$$

Here w_F denotes the number of units of F, and $\mathrm{Aut}(L)(= w_F)$ denotes the number of automorphisms of L.

It is very clear that, formally, this version of (the completed) Dedekind zeta functions for number fields stands exactly the same as our interpretation of Artin zeta functions for function fields. So to introduce a non-abelian zeta function for number fields, the key points are as follows:

(1) A suitable stability in terms of intersection for bundles over number fields should be introduced;

(2) Stable bundles over number fields should form moduli spaces, over which there exist natural measures; and

(3) There should be a geo-ari cohomology such that duality and Riemann-Roch type results hold.

B.2.2. A Geo-Arithmetical Cohomology

B.2.2.1. Cohomology Groups

For the number field F, denote by $\mathbb{A} = \mathbb{A}_F$ the ring of adeles of F, by $\mathrm{GL}_r(\mathbb{A})$ the rank r general linear group over $\mathbb{A}$, and write $\mathbb{A} := \mathbb{A}_{\mathrm{fin}} \oplus \mathbb{A}_\infty$ and $\mathrm{GL}_r(\mathbb{A}) := \mathrm{GL}_r(\mathbb{A})_{\mathrm{fin}} \times \mathrm{GL}_r(\mathbb{A})_\infty$ according to finite and infinite partition. Also for simplicity, we often use v (resp. σ) to denote elements in S_{fin} (resp. S_∞).

For any $g = (g_{\mathrm{fin}} : g_\infty) = (g_v; g_\sigma) \in \mathrm{GL}_r(\mathbb{A})$, define an injective morphism $i(g) := i(g_\infty) : F^r \to \mathbb{A}^r$ by $f \mapsto (f; g_\sigma \cdot f)$. Let $F^r(g) := \mathrm{Im}\big(i(g)\big)$ and set

$$\mathbb{A}^r(g) := \left\{ (a_v; a_\sigma) \in \mathbb{A}^r : \begin{array}{c} \text{(i) } g_v(a_v) \in \mathcal{O}_v^r,\ \forall v; \\ \text{(ii) } \exists f \in F^r \text{ such that} \\ \text{(ii.a) } g_v(f) \in \mathcal{O}_v^r,\ \forall v; \quad \text{(ii.b) } (f; a_\sigma) = i(g_\infty)(f) \end{array} \right\}$$

Then we have the following 9-diagram with exact columns and rows:

$$\begin{array}{ccccc} 0 & & 0 & & 0 \\ \downarrow & & \downarrow & & \downarrow \\ \mathbb{A}^r(g)\cap F^r(g) & \rightarrowtail & \mathbb{A}^r(g) & \twoheadrightarrow & \mathbb{A}^r(g)/\mathbb{A}^r(g)\cap F^r(g) \\ \downarrow & & \downarrow & & \downarrow \\ F^r(g) & \rightarrowtail & \mathbb{A}^r & \twoheadrightarrow & \mathbb{A}^r/F^r(g) \\ \downarrow & & \downarrow & & \downarrow \\ F^r(g)/\mathbb{A}^r(g)\cap F^r(g) & \rightarrowtail & \mathbb{A}^r/\mathbb{A}^r(g) & \twoheadrightarrow & \mathbb{A}^r/\mathbb{A}^r(g)+F^r(g) \\ \downarrow & & \downarrow & & \downarrow \\ 0 & & 0 & & 0. \end{array}$$

Guided by C.4 of this paper, which itself is motivated by Weil's adelic cohomology theory for divisors over algebraic curves, (see e.g, [S]), we introduce the following

Definition. ([We6]) *For any $g \in \mathrm{GL}_r(\mathbb{A})$, define its 0-th and 1-st geo-arithmetical cohomology groups by*

$$H^0(\mathbb{A}_F, g) := \mathbb{A}^r(g) \cap F^r(g), \text{ and } H^1(\mathbb{A}_F, g) := \mathbb{A}^r/\mathbb{A}^r(g) + F^r(g).$$

Proposition 1. ([We6]) (Serre Duality = Pontrjagin Duality) *As locally compact groups,*

$$H^1(\mathbb{A}_F, g) \simeq H^0(\widehat{\mathbb{A}_F, \kappa_F \otimes} g^{-1}).$$

Here κ_F denotes an idelic dualizing element of F, and $\hat{\ }$ the Pontrjagin dual. In particular, H^0 is discrete and H^1 is compact.

Remark. For $v \in S_{\mathrm{fin}}$, denote by ∂_v the local different of F_v, the v-completion of F, and by $\mathcal{O}_v$ the valuation ring with π_v a local parameter. Then $\partial_v =: \pi_v^{\mathrm{ord}_v(\partial_v)} \cdot \mathcal{O}_v$. We call $\kappa_F := (\partial_v^{\mathrm{ord}_v(\partial_v)}; 1) \in \mathbb{I}_F := \mathrm{GL}_1(\mathbb{A})$ an idelic dualizing element of F.

B.2.2.2 Geo-Arithmetical Counts

Motivated by the Pontrjagin duality and that the dimensions of a vector space and its dual are the same, one basic principle we adopt in counting locally compact groups is the following:

Counting Axiom. *If $\#_{\rm ga}$ counts a certain class of locally compact groups G, then $\#_{\rm ga}(G) = \#_{\rm ga}(\hat{G})$.*

Practically, our counts of arithmetic cohomology groups are based on the Fourier inverse formula, or more accurately, the Plancherel formula in Fourier analysis over locally compact groups. (See e.g, [Fo].)

While any reasonable test function on $\mathbb{A}^r$ would do, as a continuation of a more traditional mathematics and also for simplicity, we set $f := \prod_v f_v \cdot \prod_\sigma f_\sigma$, where
(i) f_v is the characteristic function of $\mathcal{O}_v^r$;
(ii.a) $f_\sigma(x_\sigma) := e^{-\pi|x_\sigma|^2/2}$ if σ is real; and
(ii.b) $f_\sigma(x_\sigma) := e^{-\pi|x_\sigma|^2}$ if σ is complex.
Moreover, we take the following normalization for the Haar measure dx, which we call *standard*, on $\mathbb{A}$: locally

for v, dx is the measure for which $\mathcal{O}_v$ gets the measure $N(\partial_v)^{-1/2}$; while for σ real (resp. complex), dx is the ordinary Lebesgue measure(resp. twice the ordinary Lebesgue measure).

Definition. ([We6]) (1) *The geo-arithmetical counts of the 0-th and the 1-st cohomology groups for $g \in \mathrm{GL}_r(\mathbb{A})$ are*

$$\#_{\rm ga}\big(H^0(\mathbb{A}_F, g)\big) :=\#_{\rm ga}\Big(H^0(\mathbb{A}_F, g); f, dx\Big) := \int_{H^0(\mathbb{A}_F,g)} |f(x)|^2 dx;$$

$$\#_{\rm ga}\big(H^1(\mathbb{A}_F, g)\big) :=\#_{\rm ga}\Big(H^1(\mathbb{A}_F, g); \hat{f}, d\xi\Big) := \int_{H^1(\mathbb{A}_F,g)} |\hat{f}(\xi)|^2 d\xi.$$

Here dx denotes (the restriction of) the standard Haar measure on $\mathbb{A}$, $d\xi$ (the induced quotient measure from) the dual measure (with respect to χ), and $\hat{f}$ the corresponding Fourier transform of f;
(2) *The 0-th and the 1-st geo-arithmetical cohomologies of $g \in \mathrm{GL}_r(\mathbb{A})$ are*

$$h^0(\mathbb{A}_F, g) := \log\Big(\#_{\rm ga}\big(H^0(\mathbb{A}_F, g)\big)\Big);$$

$$h^1(\mathbb{A}_F, g) := \log\Big(\#_{\rm ga}\big(H^1(\mathbb{A}_F, g)\big)\Big).$$

Remark. Even groups $H^0(\mathbb{A}_F, g)$ and $H^1(\mathbb{A}_F, g)$ do depend on $g \in \mathrm{GL}_r(\mathbb{A})$, their counts $h^0(\mathbb{A}_F, g)$ and $h^1(\mathbb{A}_F, g)$ depend only on the class of $g \in \mathrm{GL}_r(F)\backslash\mathrm{GL}_r(\mathbb{A})$ by definition.

B.2.2.3. Serre Duality and Riemann-Roch

For the arithmetic cohomologies just introduced, we have the following

Theorem. ([We6]) For any $g \in \mathrm{GL}_r(F)\backslash\mathrm{GL}_r(\mathbb{A})$,
(1) (Serre Duality) $h^1(\mathbb{A}_F, g) = h^0(\mathbb{A}_F, \kappa_F \otimes g^{-1})$;
(2) (Riemann-Roch Theorem)

$$h^0(\mathbb{A}_F, g) - h^1(\mathbb{A}_F, g) = \deg(g) - \frac{r}{2} \cdot \log|\Delta_F|.$$

Often, we also write $e^{h^i(\mathbb{A}_F,g)} := H^i_{\mathrm{ga}}(F, g),\ i = 1, 2.$

Remarks. (1) Our work here is motivated by the works of Weil [W4], Tate [T], Iwasawa [Iw1,2], van der Geer-Schoof [GS], and Li [Li], as well as the works of Lang [L1,2], Arakelov [L3], Szpiro [L3], Parshin [Pa], Moreno [Mo], Neukirch [Ne], Deninger [De 1,2], Connes [Co], and Borisov [Bor]. Also, it would be extremely interesting if one could relate the work here with that of Connes [Co] and Deninger [De1,2,3].
(2) One may apply the discussion here to wider classes of (multiplicative) characters and test functions so as to obtain a more general class of L-functions as done in Tate's Thesis. But we later would take a substantially different approach.
(3) Instead of working on S_F, the set of equivalence classes of valuations of F, or better, normalized valuations over F, one may also work on the set of *all* valuations of F, which in particular is parametrized by $S_F \times \mathbb{R}$ to get a much more refined geo-arithmetical cohomology theory.

To understand the above theory, we suggest the reader use the language of lattices (whose details will be recalled in the following section): Denote by $\mathcal{O}_F$ the ring of integers of F. Then for an $\mathcal{O}_F$-lattice Λ of rank r, its geo-arithmetical cohomology groups are defined to be the locally compact groups

$$H^0(F, \Lambda) := \Lambda \qquad \text{and} \qquad H^1(F, \Lambda) := \left(\mathbb{R}^{r_1} \times \mathbb{C}^{r_2}\right)^r / \Lambda.$$

Consequently, then

(i) $H^0(F,\Lambda)$ is discrete while $H^1(F,\Lambda)$ is compact;
(ii) The Serre duality coincides with the Pontrajin duality, say in the case $F=\mathbb{Q}$,

$$\Big(\big(\mathbb{R}^{r_1}\times\mathbb{C}^{r_2}\big)^r/\Lambda\Big)^\vee\simeq\Lambda;$$

(iii) the 0-th cohomology is then simply given by

$$h^0(F,\Lambda):=\log\Big(\sum_{\mathbf{x}\in\mathbf{\Lambda}}\exp\Big(-\pi\sqrt{-1}\sum_{\sigma:\mathbb{R}}\|\mathbf{x}\|_\sigma-\mathbf{2}\pi\sqrt{-\mathbf{1}}\sum_{\tau:\mathbb{C}}\|\mathbf{x}\|_\tau\Big)\Big),$$

a definition due to van der Geer and Schoof [GS].

B.2.3. Non-Abelian Zeta Functions

B.2.3.1. Space of $\mathcal{O}_K$-Lattices

B.2.3.1.1. Projective $\mathcal{O}_K$-modules

Let K be an algebraic number field. Denote by $\mathcal{O}_K$ the ring of integers of K. Then an $\mathcal{O}_K$-module M is called *projective* if there exists an $\mathcal{O}_K$-module N such that $M\oplus N$ is a free $\mathcal{O}_K$-module. Easily, for a fractional ideal $\mathfrak{a}$,

$$P_\mathfrak{a}:=P_{r;\mathfrak{a}}:=\mathcal{O}_K^{r-1}\oplus\mathfrak{a}$$

is a rank r projective $\mathcal{O}_K$-module. The nice thing is that such types of projective $\mathcal{O}_K$-modules, up to isomorphism, give all rank r projective $\mathcal{O}_K$-modules.

Proposition. (1) *For fractional ideals $\mathfrak{a}$ and $\mathfrak{b}$, $P_{r;\mathfrak{a}}\simeq P_{r;\mathfrak{b}}$ if and only if $\mathfrak{a}\simeq\mathfrak{b}$;*
(2) *For a rank r projective $\mathcal{O}_K$-module P, there exists a fractional ideal $\mathfrak{a}$ such that $P\simeq P_\mathfrak{a}$.*

B.2.3.1.2. $\mathcal{O}_K$-Lattices

Let σ be an Archimedean place of K, and K_σ be the σ-completion of K. It is well known that K_σ is equal to either $\mathbb{R}$, or $\mathbb{C}$. Accordingly, we call σ (to be) real or complex, write sometimes in terms of $\sigma:\mathbb{R}$ or $\sigma:\mathbb{C}$ accordingly.

Recall also that a finite dimensional K_σ-vector space V_σ is called a *metrized space* if it is equipped with an inner product.

By definition, an $\mathcal{O}_K$-*lattice* Λ consists of
(1) a projective $\mathcal{O}_K$-module $P = P(\Lambda)$ of finite rank; and
(2) an inner product on the vector space $V_\sigma := P \otimes_{\mathcal{O}_K} K_\sigma$ for each of the Archmidean place σ of K.
Set $V = P \otimes_{\mathbb{Z}} \mathbb{R}$, then $V = \prod_{\sigma \in S_\infty} V_\sigma$, where S_∞ denotes the collection of all (inequivalent) Archimedean places of K. Indeed, this is a direct consequence of the fact that as a $\mathbb{Z}$-module, an $\mathcal{O}_K$-ideal is of rank $n = r_1 + 2r_2$ where $n = [K : \mathbb{Q}]$, r_1 denotes the number of real places and r_2 denotes the number of complex places (in S_∞).

B.2.3.1.3. Space of $\mathcal{O}_K$-Lattices

Let P be a rank r projective $\mathcal{O}_K$-module. Denote by $GL(P) := \mathrm{Aut}_{\mathcal{O}_K}(P)$. Let $\widetilde{\mathbf{\Lambda}} := \widetilde{\mathbf{\Lambda}}(P)$ be the space of $\mathcal{O}_K$-lattices Λ whose underlying $\mathcal{O}_K$-module is P. For $\sigma \in S_\infty$, let $\widetilde{\mathbf{\Lambda}}_\sigma$ be the space of inner products on V_σ; if a basis is chosen for V_σ as a real or a complex vector space according to whether σ is real or complex, $\widetilde{\mathbf{\Lambda}}_\sigma$ may be realized as an open set of a real or complex vector space. We have $\widetilde{\mathbf{\Lambda}} = \prod_{\sigma \in S_\infty} \widetilde{\mathbf{\Lambda}}_\sigma$ and this provides us with a natural topology on $\widetilde{\mathbf{\Lambda}}$.

Consider $GL(P)$ to act on P from the left. Given $\Lambda \in \widetilde{\mathbf{\Lambda}}$ and $u, w \in V_\sigma$, let $\langle u, w\rangle_{\Lambda,\sigma}$ or $\langle u, w\rangle_{\rho_\Lambda(\sigma)}$ denote the value of the inner product on the vectors u and w associated to the lattice Λ. Then, if $A \in GL(P)$, we may define a new lattice $A \cdot \Lambda$ in $\widetilde{\mathbf{\Lambda}}$ by the following formula

$$\langle u, w\rangle_{A\cdot\Lambda,\sigma} := \langle A^{-1} \cdot u, A^{-1} \cdot w\rangle_{\Lambda,\sigma}.$$

This defines an action of $GL(P)$ on $\widetilde{\mathbf{\Lambda}}$ from the left. Clearly, then the map $v \mapsto Av$ gives an isometry $\Lambda \cong A \cdot \Lambda$ of the lattices. (By an isometry here, we mean an isomorphism of $\mathcal{O}_K$-modules for the underlying $\mathcal{O}_K$-modules subjecting the condition that the isomorphism also keeps the inner product unchanged.) Conversely, suppose that $A : \Lambda_1 \cong \Lambda_2$ is an isometry of $\mathcal{O}_K$-lattices, each of which is in $\widetilde{\mathbf{\Lambda}}$. Then, A defines an element, also denoted by A, of $GL(P)$. Clearly $\Lambda_2 \cong A \cdot \Lambda_1$.

Therefore, the orbit set $GL(P)\backslash\widetilde{\mathbf{\Lambda}}(P)$ can be regarded as the set of isometry classes of $\mathcal{O}_K$-lattices whose underlying $\mathcal{O}_K$-modules are isomorphic to P.

B.2.3.2. Geometric Truncation

B.2.3.2.1. Semi-Stable Lattices

Let Λ be an $\mathcal{O}_K$-lattice with underlying $\mathcal{O}_K$-module P. Then any submodule $P_1 \subset P$ can be made into an $\mathcal{O}_K$-lattice by restricting the inner product on each V_σ to the subspace $V_{1,\sigma} := P_1 \otimes_K K_\sigma$. Call the resulting $\mathcal{O}_K$-lattice $\Lambda_1 := \Lambda \cap P_1$ and write $\Lambda_1 \subset \Lambda$. If moreover, P/P_1 is projective, we say that Λ_1 is a *sublattice* of Λ.

The orthogonal projections $\pi_\sigma : V_\sigma \to V_{1,\sigma}^\perp$ to the orthogonal complement $V_{1,\sigma}^\perp$ of $V_{1,\sigma}$ in V_σ provide isomorphisms $(P/P_1) \otimes_{\mathcal{O}_K} K_\sigma \simeq V_{1,\sigma}^\perp$, which can be used to make P/P_1 into an $\mathcal{O}_K$-lattice. We call this resulting lattice the *quotient lattice* of Λ by Λ_1, and denote it by Λ/Λ_1.

There is a procedure called *restriction of scalars* which makes an $\mathcal{O}_K$-lattice into a standard $\mathbb{Z}$-lattice. Recall that $V = \Lambda \otimes_{\mathbb{Z}} \mathbb{R} = \prod_{\sigma \in S_\infty} V_\sigma$. Define an inner product on the real vector space V by

$$\langle u, w \rangle_\infty := \sum_{\sigma:\mathbb{R}} \langle u_\sigma, w_\sigma \rangle_\sigma + \sum_{\sigma:\mathbb{C}} \mathrm{Re}\langle u_\sigma, w_\sigma \rangle_\sigma.$$

Let $\mathrm{Res}_{K/\mathbb{Q}}\Lambda$ denote the $\mathbb{Z}$-lattice obtained by equipped P, regarding as a $\mathbb{Z}$-module, with this inner product (at the unique infinite place ∞ of $\mathbb{Q}$).

We let $\mathrm{rk}(\Lambda)$ denote the $\mathcal{O}_K$-module rank of P (or of Λ), and define the *Lebesgue volume* of Λ, denoted by $\mathrm{Vol}_{\mathrm{Leb}}(\Lambda)$, to be the (co)volume of the lattice $\mathrm{Res}_{K/\mathbb{Q}}\Lambda$ inside its inner product space V.

Clearly, if P' is a submodule of finite index in P, then

$$\mathrm{Vol}_{\mathrm{Leb}}(\Lambda') = [P : P']\mathrm{Vol}_{\mathrm{Leb}}(\Lambda),$$

where $\Lambda' = \Lambda \cap P$ is the lattice induced from P'.

Examples. Take $P = \mathcal{O}_K$ and for each place σ, let $\{1\}$ be an orthonormal basis of $V_\sigma = K_\sigma$, i.e., equipped $V_\sigma = \mathbb{R}$ or $\mathbb{C}$ with the standard Lebesgue measure. This makes $\mathcal{O}_K$ into an $\mathcal{O}_K$-lattice $\overline{\mathcal{O}_K} = (\mathcal{O}_K, \mathbf{1})$ in a natural way. It is a well-known fact, see e.g., [L1], that

$$\mathrm{Vol}_{\mathrm{Leb}}\left(\overline{\mathcal{O}_K}\right) = 2^{-r_2} \cdot \sqrt{\Delta_F},$$

where Δ_F denotes the absolute value of the discriminant of K.

More generally, take $P = \mathfrak{a}$ as a fractional idea of K and equip the same inner product as above on V_σ. Then $\mathfrak{a}$ becomes an $\mathcal{O}_K$-lattice $\overline{\mathfrak{a}} = (\mathfrak{a}, \mathbf{1})$ in a natural way with $\mathrm{rk}(\mathfrak{a}) = 1$. It is a well-known fact, see e.g., [Ne], that

$$\mathrm{Vol}_{\mathrm{Leb}}\Big(\overline{\mathfrak{a}}\Big) = 2^{-r_2} \cdot \Big(N(\mathfrak{a}) \cdot \sqrt{\Delta_K}\Big),$$

where $N(\mathfrak{a})$ denote the norm of $\mathfrak{a}$.

Due to the appearence of the factor 2^{-r_2}, we also define the *canonical volume* of Λ, denoted by $\mathrm{Vol}_{\mathrm{can}}(\Lambda)$ or simply by $\mathrm{Vol}(\Lambda)$, to be $2^{r_2 \mathrm{rk}(\Lambda)} \mathrm{Vol}_{\mathrm{Leb}}(\Lambda)$. So in particular,

$$\mathrm{Vol}\Big(\overline{\mathfrak{a}}\Big) = N(\mathfrak{a}) \cdot \sqrt{\Delta_K} \qquad \text{and} \qquad \mathrm{Vol}\Big(\overline{\mathcal{O}_K}\Big) = \sqrt{\Delta_K}.$$

Now we are ready to introduce our first key definition.

Definition. An $\mathcal{O}_K$ lattice Λ is called *semi-stable* (resp. *stable*) if for any proper sublattice Λ_1 of Λ,

$$\mathrm{Vol}(\Lambda_1)^{\mathrm{rk}(\Lambda)} \geq \text{(resp. >)}\, \mathrm{Vol}(\Lambda)^{\mathrm{rk}(\Lambda_1)}.$$

Clearly the last inequality is equivalent to

$$\mathrm{Vol}_{\mathrm{Leb}}(\Lambda_1)^{\mathrm{rk}(\Lambda)} \geq \mathrm{Vol}_{\mathrm{Leb}}(\Lambda)^{\mathrm{rk}(\Lambda_1)}.$$

So it does not matter which volume, the canonical one or the Lebesgue one, we use.

Remark. The canonical measures has an advantage theoretically. For example, we have the following

Arakelov-Riemann-Roch Formula: For an $\mathcal{O}_K$-lattice Λ of rank r,

$$-\log\Big(\mathrm{Vol}(\Lambda)\Big) = \deg(\Lambda) - \frac{r}{2}\log \Delta_K.$$

(For the reader who does not know the definition of the Arakelov degree, he or she may simply take this relation as the definition.)

B.2.3.2.2. Canonical Filtration

Based on stability, we may introduce a more general geometric truncation for the space of lattices. For this, we start with the following well-known

Lemma. *For a fixed $\mathcal{O}$-lattice Λ, $\Big\{\mathrm{Vol}(\Lambda_1)\,:\,\Lambda_1\subset\Lambda\Big\}\subset\mathbb{R}_{\geq 0}$ is discrete and bounded from below.*

Sketch of proof. It is based on the follows:
1) If a lattice Λ_1 induced from a submodule of the lattice Λ has the minimal volume among all lattices induced from submodules of Λ of the same rank, then Λ_1 is a sublattice;
2) Taking wedge product, we may further assume that the rank is 1. It is then clear.

As a direct consequence, we have the following
Proposition. ([We2]) *Let Λ be an $\mathcal{O}$-lattice. Then*
(1) (**Canonical Filtration**) *There exists a unique filtration of proper sublattices*

$$0=\Lambda_0\subset\Lambda_1\subset\cdots\subset\Lambda_s=\Lambda$$

such that

(i) *for all $i=1,\ldots,s$, Λ_i/Λ_{i-1} is semi-stable; and*

(ii) *for all $j=1,\ldots,s-1$,*

$$\Big(\mathrm{Vol}(\Lambda_{j+1}/\Lambda_j)\Big)^{\mathrm{rk}(\Lambda_j/\Lambda_{j-1})}>\Big(\mathrm{Vol}(\Lambda_j/\Lambda_{j-1})\Big)^{\mathrm{rk}(\Lambda_{j+1}/\Lambda_j)};$$

(2) (**Jordan-Hölder Filtration**) *If moreover, Λ is semi-stable, then there exists a filtration of proper sublattices,*

$$0=\Lambda^{t+1}\subset\Lambda^t\subset\cdots\subset\Lambda^0=\Lambda$$

such that

(i) *for all $k=0,\ldots,t$, Λ^k/Λ^{k+1} is stable; and*

(ii) *for all $l=1,\ldots,t$,*

$$\Big(\mathrm{Vol}(\Lambda^l/\Lambda^{l+1})\Big)^{\mathrm{rk}(\Lambda^{l-1}/\Lambda^l)}=\Big(\mathrm{Vol}(\Lambda^{l-1}/\Lambda^l)\Big)^{\mathrm{rk}(\Lambda^l/\Lambda^{l+1})}.$$

Furthermore, up to isometry, the lattice $\mathrm{Gr}(\Lambda) := \oplus_{k=0}^{t} \Lambda^k/\Lambda^{k+1}$ *is uniquely determined by* Λ.

Sketch of Proof. Existence is clear and the uniqueness of Jordan-Hölder graded lattice is fairly standard. Let us look at the uniqueness of canonical filtration, an analogue of the well-known Harder-Narasimhan filtration for vector bundles, a bit carefully: Existence of two such filtrations will lead to a contradiction by applying fact that for any two sublattices Λ_1, Λ_2 of Λ,

$$\mathrm{Vol}\Big(\Lambda_1/(\Lambda_1 \cap \Lambda_2)\Big) \geq \mathrm{Vol}\Big((\Lambda_1 + \Lambda_2)/\Lambda_2\Big).$$

B.2.3.2.3. Compactness

For a lattice $\Lambda \in \widetilde{\mathbf{\Lambda}}$ with the associated canonical filtration

$$0 = \overline{\Lambda}_0 \subset \overline{\Lambda}_1 \subset \cdots \subset \overline{\Lambda}_s = \Lambda$$

define the associated *canonical polygone* $\overline{p}_\Lambda : [0, r] \to \mathbb{R}$ by the following conditions:

(1) $\overline{p}_\Lambda(0) = \overline{p}_\Lambda(r) = 0$;

(2) $\overline{p}_\Lambda$ is affine over the closed interval $[\mathrm{rk}\overline{\Lambda}_i, \mathrm{rk}\overline{\Lambda}_{i+1}]$; and

(3) $\overline{p}_\Lambda(\mathrm{rk}\overline{\Lambda}_i) = \deg(\overline{\Lambda}_i) - \mathrm{rk}(\overline{\Lambda}_i) \cdot \frac{\deg(\overline{\Lambda})}{r}$.

Clearly, the canonical polygon is well-defined on the space $\widetilde{\mathbf{\Lambda}}$.

To go further, let us introduce an operation among the lattices in $\widetilde{\mathbf{\Lambda}}$. If T is a positive real number, then from Λ, we can produce a new $\mathcal{O}_K$-lattice called $\Lambda[T]$ by multiplying each of the inner products on Λ, or better, on Λ_σ for $\sigma \in S_\infty$, by T^2. Obviously, the $[T]$-construction changes volumes of lattices in the following way

$$\mathrm{Vol}(\Lambda[T]) = T^{\mathrm{rk}(\Lambda)\cdot[K:\mathbb{Q}]} \cdot \mathrm{Vol}(\Lambda).$$

That is to say, the $[T]$-construction naturally fixes a specific volume for a certain family of lattices, while does not really change the 'essential' structures of lattices involved.

Accordingly, let $\mathbf{\Lambda} = \mathbf{\Lambda}(P)$ be the quotient topological space of $\widetilde{\mathbf{\Lambda}}$ modulo the equivalence relation $\Lambda \sim \Lambda[T]$. Since canonical polygon is invariant under the scaling operation and hence descends to this quotient space $\mathbf{\Lambda}$.

With all this, we are ready to state the following fundamental

Theorem. *For any fixed convex polygon* $p : [0, r] \to \mathbb{R}$, *the subset* $\Big\{[\Lambda] \in \mathbf{\Lambda} : \overline{p}_\Lambda \leq p\Big\}$ *is compact in* $\mathbf{\Lambda}$.

Sketch of Proof. This is based on the follows:
(1) Dirichlet's Unit Theorem; and
(2) Minkowski's Reduction Theory: Due to the fact that volumes of lattices involved are fixed, semi-stability condition implies that the first Minkowski successive minimums of these lattices admit a natural lower bound away from 0 (depending only on r). Hence by the standard reduction theory, see e.g., Borel [Bo1,2], the subset $\Big\{[\Lambda] \in \mathbf{\Lambda} : \overline{p}_\Lambda \leq p\Big\}$ is compact.

B.2.3.2.4. Canonical Polygons and Geo-Arithmetical Truncation: Adelic Version

Let $X = \operatorname{Spec} \mathcal{O}_F$. If E is a vector sheaf of rank r over X, i.e, a locally free $\mathcal{O}_F$-sheaf of rank r, denote by E_F the fiber of E at the generic point $\operatorname{Spec}(F) \hookrightarrow \operatorname{Spec} \mathcal{O}_F$ of X (E_F is an F-vector space of dimension r), and for each $v \in S_f := S_{\text{fin}}$, set $E_{\mathcal{O}_v} := H^0(\operatorname{Spec} \mathcal{O}_{F_v}, E)$ a free $\mathcal{O}_v$-module of rank r. In particular, we have a canonical isomorphism:

$$\operatorname{can}_v : F_v \otimes_{\mathcal{O}_v} E_{\mathcal{O}_v} \simeq F_v \otimes_F E_F.$$

Now if E is a vector sheaf of rank r over X equipped with a basis $\alpha_F : F^r \simeq E_F$ of its generic fiber and a basis $\alpha_{\mathcal{O}_v} : \mathcal{O}_v^r \simeq E_{\mathcal{O}_v}$ for any $v \in S_f$, the elements $g_v := (F_v \otimes_F \alpha_F)^{-1} \circ \operatorname{can}_v \circ (F_v \otimes_{\mathcal{O}_v} \alpha_{\mathcal{O}_v}) \in GL_r(F_v)$ for all $v \in S_f$ define an element $g_{\mathbb{A}} := (g_v)_{v \in S_f}$ of $GL_r(\mathbb{A}_f)$, i.e, for almost every v we have $g_v \in GL_r(\mathcal{O}_v)$. By this construction, we obtain a bijection from the set of isomorphism classes of triples $(E; \alpha_F; (\alpha_{\mathcal{O}_v})_{v \in S_f})$ as above onto $GL_r(\mathbb{A}_f)$. Moreover, if $r \in GL_r(F), k \in GL_r(\mathcal{O}_F)$ and if this bijection maps the triple $(E; \alpha_F; (\alpha_{\mathcal{O}_v})_{v \in S_f})$ onto $g_{\mathbb{A}}$, the same map maps the triple

$(E;\alpha_F\circ r^{-1};(\alpha_{\mathcal{O}_v}\circ k_v)_{v\in S_f})$ onto $rg_{\mathbb{A}}k$. Therefore the above bijection induces a bijection between the set of isomorphism classes of vector sheaves of rank r on $\operatorname{Spec}\mathcal{O}_F$ and the double coset space $GL_r(F)\backslash GL_r(\mathbb{A}_f)/GL_r(\mathcal{O}_F)$.

More generally, let $r=r_1+\cdots+r_s$ be a partition $I=(r_1,\ldots,r_s)$ of r and let P_I be the corresponding standard parabolic subgroup of GL_r. Then we have a natural bijection from the set of isomorphism classes of triple $(E_*;\alpha_{*,F}:(\alpha_{*,\mathcal{O}_v})_{v\in S_f})$ onto $P_I(\mathbb{A}_f)$, where $E_*:=\big((0)=E_0\subset E_1\subset\cdots\subset E_s\big)$ is a filtration of vector sheaves of rank $(r_1,r_1+r_2,\cdots,r_1+r_2+\cdots+r_s=r)$ over X, (i.e, each E_j is a vector sheaf of rank $r_1+r_2+\cdots+r_j$ over X and each quotient E_j/E_{j-1} is torsion free,) which is equipped with an isomorphism of filtrations of F-vector spaces

$$\alpha_{*,F}:\big((0)=F_0\subset F^{r_1}\subset\cdots\subset F^{r_1+r_2+\cdots+r_s=r}\big)\simeq(E_*)_F,$$

and with an isomorphism of filtrations of free $\mathcal{O}_v$-modules

$$\alpha_{*,\mathcal{O}_v}:\big((0)\subset\mathcal{O}_v^{r_1}\subset\cdots\subset\mathcal{O}_v^{r_1+r_2+\cdots+r_s=r}\big)\simeq(E_*)_{\mathcal{O}_v},$$

for every $v\in S_f$. Moreover this bijection induces a bijection between the set of isomorphism classes of the filtrations of vector sheaves of rank $(r_1,r_1+r_2,\cdots,r_1+r_2+\cdots+r_s=r)$ over X and the double coset space $P_I(F)\backslash P_I(\mathbb{A}_f)/P_I(\mathcal{O}_F)$. The natural embedding $P_I(\mathbb{A}_f)\hookrightarrow P_I(\mathbb{A})$ (resp. canonical projections $P_I(\mathbb{A}_f)\to M_I(\mathbb{A}_f)\to GL_{r_j}(\mathbb{A}_f)$ for $j=1,\ldots,s$, where M_I denotes the standard Levi of P_I) admits the modular interpretation

$$(E_*;\alpha_{*,F}:(\alpha_{*,\mathcal{O}_v})_{v\in S_f})\mapsto(E_s;\alpha_{s,F}:(\alpha_{s,\mathcal{O}_v})_{v\in S_f})$$

(resp.

$$(E_*;\alpha_{*,F}:(\alpha_{*,\mathcal{O}_v})_{v\in S_f})\mapsto(\mathrm{gr}_j(E_*);\mathrm{gr}_j(\alpha_{*,F}),\mathrm{gr}_j(\alpha_{*,\mathcal{O}_v})_{v\in S_f}),$$

where $\mathrm{gr}_j(E_*):=E_j/E_{j-1}$, $\mathrm{gr}_j(\alpha_{*,F}):F^{r_j}\simeq\mathrm{gr}_j(E_*)_F$ and $\mathrm{gr}_j(\alpha_{*,\mathcal{O}_v}):{\mathcal{O}_v}^{r_j}\simeq\mathrm{gr}_j(E_*)_{\mathcal{O}_v}$, $v\in S_f$ are induced by $\alpha_{*,F}$ and $\alpha_{*,\mathcal{O}_v}$ respectively.)

Moreover, any $g=(g_f;g_\infty)\in GL_r(\mathbb{A}_f)\times GL_r(\mathbb{A}_\infty)=GL_r(\mathbb{A})$ gives first a rank r vector sheaf E_g on $\operatorname{Spec}\mathcal{O}_F$, which via the embedding $E_F\hookrightarrow(\mathbb{R}^{r_1}\times\mathbb{C}^{r_2})^r$ gives a discrete subgroup, a free rank r

$\mathcal{O}_F$-module. In particular, $g_\infty = (g_\sigma)$ then induces a natural metric on E_g by twisting the standard one on $(\mathbb{R}^{r_1} \times \mathbb{C}^{r_2})^r$ via the linear transformation induced from g_∞. As a direct consequence, see e.g, [L3],

$$\deg(E_g, \rho_g) = -\log\big(N(\det g)\big)$$

with $N : GL_1(\mathbb{A}_F) = \mathbb{I}_F \to \mathbb{R}_{>0}$ the standard norm map of the idelic group of F.

With this, for $g = (g_f; g_\infty) \in GL_r(\mathbb{A})$ and a parabolic subgroup Q of GL_r, denote by $E_*^{g;Q}$ the filtration of the vector sheaf E_{g_f} induced by the parabolic subgroup Q. Then we have a filtration of hermitian vector sheaves $(E_*^{g;Q}, \rho_*^{g;Q})$ with the hermitian metrics $\rho_j^{g;Q}$ on $E_j^{g;Q}$ obtained via the restrictions of ρ_{g_∞}.

Now introduce an associated polygon $p_Q^g : [0, r] \to \mathbb{R}$ by the following 3 conditions:
(i) $p_Q^g(0) = p_Q^g(r) = 0$;
(ii) p_Q^g is affine on the interval $[\mathrm{rank}E_{i-1}^{g;Q}, \mathrm{rank}E_i^{g;Q}]$; and
(iii) for all indices i,

$$p_Q^g(\mathrm{rank}E_i^{g;Q}) = \deg(E_i^{g;Q}, \rho_i^{g;Q}) - \frac{\mathrm{rank}E_i^{g;Q}}{r} \cdot \deg(E_g, \rho_g).$$

Then, there is a unique convex polygon $\bar{p}^g$, the canonical polygon associated with g, which bounds all p_Q^g from above for all parabolic subgroups Q for GL_r. Moreover there exists a parabolic subgroup $\bar{Q}^g$ such that $p_{\bar{Q}^g}^g = \bar{p}^g$.

Hence we have the following fundamental

Main Lemma. *For any fixed polygon $p : [0, r] \to \mathbb{R}$ and any $d \in \mathbb{R}$, the subset*

$$\{g \in GL_r(F)\backslash GL_r(\mathbb{A}) : \deg g = d, \bar{p}^g \le p\}$$

is compact.

Similarly yet more generally, for a fixed parabolic subgroup P of GL_r and $g \in GL_r(\mathbb{A})$, there is a unique maximal element $\bar{p}_P^g$ among all p_Q^g, where Q runs over all parabolic subgroups of GL_r which are contained in P. And we have

Main Lemma′. *For any fixed polygon $p : [0, r] \to \mathbb{R}$, $d \in \mathbb{R}$ and any standard parabolic subgroup P of GL_r, the subset*

$$\{g \in GL_r(F)\backslash GL_r(\mathbb{A}) : \deg g = d, \bar{p}_P^g \leq p, p_P^g \geq -p\}$$

is compact.

B.2.3.3. Non-Abelian Zeta Functions for Number Fields

For the number field F with (absolute value of) discriminant Δ_F, denote by $\mathcal{M}_{F,r}$ the moduli space of semi-stable $\mathcal{O}_F$-lattices of rank r. Then $\mathcal{M}_{F,r} := \dot{\cup}_{T \in \mathbb{R}_+} \mathcal{M}_{F,r}[T]$ where $\mathcal{M}_{F,r}[T] := \mathcal{M}_{F,r}(\log T)$ denotes the volume T part. For an adele element $g \in \mathrm{GL}_r(\mathbb{A})$, denote its associated $\mathcal{O}_F$-lattice of rank r by $\Pi(g)$. Clearly, Π factors through $\mathrm{GL}_r(F)\backslash\mathrm{GL}_r(\mathbb{A})$. Denote by $\mathcal{M}_{\mathbb{A}_F,r} \subset \mathrm{GL}_r(F)\backslash\mathrm{GL}_r(\mathbb{A})$ the inverse image of $\mathcal{M}_{F,r}$. It is well known that fibers of $\Pi : \mathcal{M}_{\mathbb{A}_F,r} \to \mathcal{M}_{F,r}$ are all compact. Hence so is $\mathcal{M}_{\mathbb{A}_F,r}[T] := \Pi^{-1}\Big(\mathcal{M}_{F,r}[T]\Big)$. Consequently, we obtain natural measures $d\mu$ on $\mathcal{M}_{F,r}$ and $\mathcal{M}_{\mathbb{A}_F,r}[T]$ induced from that on $\mathrm{GL}_r(\mathbb{A})$. In particular, being compact, $\mathcal{M}_{\mathbb{A}_F,r}[\Delta_F^{\frac{r}{2}}]$ is of finite volume.

For any $\Lambda \in \mathcal{M}_{\mathbb{A}_F,r}$, set $h^i(F, \Lambda) := h^i(\mathbb{A}_F, g)$ where g is an element of $\mathrm{GL}_r(\mathbb{A})$ whose associated $\mathcal{O}_F$-lattice coincides with Λ. This is well-defined since for any $a \in \mathrm{GL}_r(F)$, $h^i(\mathbb{A}_F, a \cdot g) = h^i(\mathbb{A}_F, g)$ by definition. Indeed, as said earlier,

$$h^0(F, \Lambda) = \log\Bigg(\sum_{\mathbf{x} \in \mathbf{\Lambda}} \exp\Bigg(-\pi\sqrt{-1}\sum_{\sigma:\mathbb{R}} \|\mathbf{x}\|_\sigma - \mathbf{2}\pi\sqrt{-\mathbf{1}}\sum_{\tau:\mathbb{C}} \|\mathbf{x}\|_\tau\Bigg)\Bigg).$$

Define the (completed) rank r non-abelian zeta function $\xi_{F,r}(s)$ of F by

$$\xi_{F,r}(s) := \Big(\Delta_F^{\frac{r}{2}}\Big)^s \int_{\Lambda \in \mathcal{M}_{F,r}} \Big(e^{h^0(F,\Lambda)} - 1\Big)\big(e^{-s}\big)^{\deg(\Lambda)} \cdot d\mu, \quad \mathrm{Re}(s) > 1.$$

Following 2.1, i.e., Iwasawa's interpretation of Dedekind zeta functions, we see that $\xi_{F,1}(s)$ is essentially the completed Dedekind zeta function $\xi_F(s)$ for F.

Remarks. (1) The terms appeared above, such as degree, 0-th cohomology, moduli spaces, and the Tamagawa type measure are all

canonically associated to number fields. So non-abelian zeta functions are naturally there. By exposing them, we hope that then non-abelian arithmetic properties of number fields can be obtained systematically.
(2) The map $\Pi : \mathcal{M}_{\mathbf{A}_F,r}(d) \to \mathcal{M}_{F,r}(d)$ may be understood as a kind of moment map. Thus non-abelian zetas may be understood as a kind of algebraic version of Feynman type integral. On the other hand, in Part A, we propose a Weil-Narasimhan-Seshadri type correspondence, a micro reciprocity law. So it is not unreasonable to expect that these zetas cab be written in terms of analytic Feynman type integrals, i.e., via global Galois representations.

B.2.3.4. Basic Properties

Just as Dedekind zeta functions, non-abelian zetas are canonical as well.

Theorem. ([We6]) (1) *Non-abelian zeta function*

$$\xi_{F,r}(s) := \left(\Delta_F^{\frac{r}{2}}\right)^s \int_{\Lambda\in\mathcal{M}_{F,r}} \left(e^{h^0(F,\Lambda)} - 1\right)\left(e^{-s}\right)^{\deg(\Lambda)} \cdot d\mu$$

converges absolutely and uniformly when $\mathrm{Re}(s) \geq 1+\delta$ *for any* $\delta > 0$. *Moreover,* $\xi_{F,r}(s)$ *admits a unique meromorphic continuation to the whole complex s-plane;*
(2) *The extended* $\xi_{F,r}(s)$ *satisfies the functional equation* $\xi_{F,r}(s) = \xi_{F,r}(1-s)$*;* and
(3) *The extended* $\xi_{F,r}(s)$ *has two singularities, simple poles at* $s = 0\,1$ *with*

$$\mathrm{Res}_{s=0}\,\xi_{F,r}(s) \;=\; \mathrm{Res}_{s=0}\,\xi_{F,r}(s) \;=\; \mathrm{Vol}\Big(\mathcal{M}_{F,r}[\Delta_F^{\frac{r}{2}}]\Big).$$

Remarks. (1) With respect to fixed real constants A, B, C, α and β, introduce a formal zeta integration

$$Z_{F,r;A,B,C;\alpha,\beta}(s) := \left(|\Delta_F|^{-\frac{rB}{2}}\right)^s \cdot$$
$$\int_{E\in\mathcal{M}_{\mathbb{A}_F,r}} \left(\left(e^{h^0(F,E)}\right)^A \cdot \left(e^{\deg(E)}\right)^{Bs+C} - \left(e^{\deg(E)}\right)^{\alpha s+\beta}\right) d\mu(E).$$

Formally, from the duality and Riemann-Roch, we obtain a functional equation

$$Z_{F,r;A,B,C;\alpha,\beta}(s) = Z_{F,r;A,B,C;\alpha,\beta}\Big(-s-\frac{A+2C}{B}\Big).$$

Moreover, with $A > 0$ and the compatibility conditions: $\alpha = B$ and $\beta = C$, these integrals are convergent. Thus, by a change of variables, we may further assume that $B = 1$ and $C = 0$ without loss of generality to get $Z_{F,r;A}(s) := Z_{F,r;A,-1,0;-1,0}(s)$. This is a version of two variables non-abelian zetas, and $\xi_{F,r}(s) = Z_{F,r;1}(s)$. In particular, the above theorem holds for $Z_{F,r;A}(s), A > 0$.
(2) The global non-abelian invariant $\mathrm{Vol}\Big(\mathcal{M}_{F,r}[\Delta_F^{\frac{r}{2}}]\Big)$ for F can be calculated in terms of special values $\xi_F(2), \xi_F(3), \ldots, \xi_F(r)$ of the Dedekind zeta. For details, see B.5.6.

B.2.4. Examples

B.2.4.1. Epstein Zeta Functions and Non-Abelian Zeta Functions

Recall that the rank r non-abelian zeta function $\xi_{\mathbb{Q},r}(s)$ of $\mathbb{Q}$ is given by

$$\xi_{\mathbb{Q},r}(s) = \int_{\mathcal{M}_{\mathbb{Q},r}} \Big(e^{h^0(\mathbb{Q},\Lambda)} - 1\Big)\cdot \big(e^{-s}\big)^{\deg(\Lambda)}\, d\mu(\Lambda), \qquad \mathrm{Re}(s) > 1.$$

Here as before,

$$h^0(\mathbb{Q},\Lambda) := \log\left(\sum_{x\in\Lambda} \exp\big(-\pi|x|^2\big)\right), \ \deg(\Lambda) = -\log\mathrm{Vol}\big(\mathbb{R}^r/\Lambda\big).$$

Theorem′. (i) $\xi_{\mathbb{Q},1}(s)$ coincides with the (completed) Riemann-zeta function;
(ii) $\xi_{\mathbb{Q},r}(s)$ can be meromorphically extended to the whole complex plane;
(iii) $\xi_{\mathbb{Q},r}(s)$ satisfies the functional equation $\xi_{\mathbb{Q},r}(s) = \xi_{\mathbb{Q},r}(1-s)$;
(iv) $\xi_{\mathbb{Q},r}(s)$ has two singularities at $s = 0, 1$ only, all simple poles with the same residues $\mathrm{Vol}\,(\mathcal{M}_{\mathbb{Q},r}[1])$.

Decomposite according to their volumes, $\mathcal{M}_{\mathbb{Q},r} = \cup_{T>0}\mathcal{M}_{\mathbb{Q},r}[T]$, and there is a natural morphism $\mathcal{M}_{\mathbb{Q},r}[T] \to \mathcal{M}_{\mathbb{Q},r}[1]$, $\Lambda \mapsto T^{\frac{1}{r}} \cdot \Lambda$. Consequently,

$$\begin{aligned}\xi_{\mathbb{Q},r}(s) &= \int_{\cup_{T>0}\mathcal{M}_{\mathbb{Q},r}[T]} \left(e^{h^0(\mathbb{Q},\Lambda)} - 1\right) \cdot \left(e^{-s}\right)^{\deg(\Lambda)} d\mu(\Lambda) \\ &= \int_0^\infty T^s \frac{dT}{T} \int_{\mathcal{M}_{\mathbb{Q},r}[1]} \left(e^{h^0(\mathbb{Q},T^{\frac{1}{r}}\cdot\Lambda)} - 1\right) \cdot d\mu(\Lambda).\end{aligned}$$

But $h^0(\mathbb{Q}, T^{\frac{1}{r}} \cdot \Lambda) = \log\left(\sum_{x\in\Lambda} \exp\left(-\pi|x|^2 \cdot T^{\frac{2}{r}}\right)\right)$ and

$$\int_0^\infty e^{-AT^B} T^s \frac{dT}{T} = \frac{1}{B} \cdot A^{-\frac{s}{B}} \cdot \Gamma\left(\frac{s}{B}\right) \qquad \text{for } B \neq 0,$$

we have $\xi_{\mathbb{Q},r}(s) = \frac{r}{2} \cdot \pi^{-\frac{r}{2}s} \Gamma(\frac{r}{2}s) \cdot \int_{\mathcal{M}_{\mathbb{Q},r}[1]} \left(\sum_{x\in\Lambda\backslash\{0\}} |x|^{-rs}\right) \cdot d\mu_1(\Lambda)$. Accordingly, introduce the completed Epstein zeta function for Λ by

$$\hat{E}(\Lambda; s) := \pi^{-s}\Gamma(s) \cdot \sum_{x\in\Lambda\backslash\{0\}} |x|^{-2s}.$$

Proposition. (Eisenstein series and Zeta Functions)

$$\xi_{\mathbb{Q},r}(s) = \frac{r}{2} \int_{\mathcal{M}_{\mathbb{Q},r}[1]} \hat{E}(\Lambda, \frac{r}{2}s)\, d\mu_1(\Lambda).$$

B.2.4.2. Rankin-Selberg Method: An Example with SL_2

Consider the action of $\mathrm{SL}(2, \mathbb{Z})$ on the upper half plane $\mathcal{H}$. Then a standard 'fundamental domain' is given by $D = \{z = x + iy \in \mathcal{H} : |x| \leq \frac{1}{2}, y > 0, x^2 + y^2 \geq 1\}$. Recall also the completed standard Eisenstein series

$$\hat{E}(z; s) := \pi^{-s}\Gamma(s) \cdot \sum_{(m,n)\in\mathbb{Z}^2\backslash\{(0,0)\}} \frac{y^s}{|mz + n|^{2s}}.$$

Naturally, we are led to consider the integral $\int_D \hat{E}(z, s) \frac{dx\, dy}{y^2}$. However, this integration diverges. Indeed, near the only cusp $y = \infty$, $\hat{E}(z, s)$ has the Fourier expansion

$$\hat{E}(z; s) = \sum_{n=-\infty}^{\infty} a_n(y, s) e^{2\pi i n x}$$

with

$$a_n(y,s) = \begin{cases} \xi(2s)y^s + \xi(2-2s)y^{1-s}, & \text{if } n = 0; \\ 2|n|^{s-\frac{1}{2}}\sigma_{1-2s}(|n|)\sqrt{y}K_{s-\frac{1}{2}}(2\pi|n|y), & \text{if } n \neq 0, \end{cases}$$

where $\xi(s)$ is the completed Riemann zeta function, $\sigma_s(n) := \sum_{d|n} d^s$, and $K_s(y) := \frac{1}{2}\int_0^\infty e^{-y(t+\frac{1}{t})/2}t^s\frac{dt}{t}$ is the K-Bessel function. Moreover,

$$|K_s(y)| \leq e^{-y/2}K_{\mathrm{Re}(s)}(2), \text{ if } y > 4, \qquad \text{and} \qquad K_s = K_{-s}.$$

So $a_{n\neq 0}(y,s)$ decay exponentially, and the problematic term comes from $a_0(y,s)$, which is of slow growth.

Therefore, to make the original integration meaningful, we need to cut off the slow growth part. There are two ways to do so: one is geometric and hence rather direct and simple; while the other is analytic, and hence rather technical and traditional, dated back to Rankin-Selberg.

(a) **Geometric Truncation**

Draw a horizontal line $y = T \geq 1$ and set

$$D_T = \{z = x+iy \in D : y \leq T\}, \qquad D^T = \{z = x+iy \in D : y \geq T\}.$$

Then $D = D_T \cup D^T$. Introduce a well-defined integration

$$I_T^{\mathrm{Geo}}(s) := \int_{D_T} \hat{E}(z,s)\,\frac{dx\,dy}{y^2}.$$

(b) **Analytic Truncation**

Define a truncated Eisenstein series $\hat{E}_T(z;s)$ by

$$\hat{E}_T(z;s) := \begin{cases} \hat{E}(z;s), & \text{if } y \leq T; \\ \hat{E}(z,s) - a_0(y;s), & \text{if } y > T. \end{cases}$$

Introduce a well-defined integration

$$I_T^{\mathrm{Ana}}(s) := \int_D \hat{E}_T(z;s)\,\frac{dx\,dy}{y^2}.$$

With this, from the Rankin-Selberg method, one checks that we have the following:

Proposition. (Analytic Truncation = Geometric Truncation in Rank 2)

$$I_T^{\mathrm{Geo}}(s) = \xi(2s)\frac{T^{s-1}}{s-1} - \xi(2s-1)\frac{T^{-s}}{s} = I_T^{\mathrm{Ana}}(s).$$

Each of the above two integrations has its own merit: for the geometric one, we keep the Eisenstein series unchanged, while for the analytic one, we keep the original fundamental domain of $\mathcal{H}$ under $\mathrm{SL}(2,\mathbb{Z})$ as it is.

Note that the nice point about the fundamental domain is that it admits a modular interpretation. Thus it would be very idealistic if we could at the same time keep the Eisenstein series unchanged, while offer some integration domains which appear naturally in certain moduli problems. Guided by this, in B.3, we will introduce non-abelian L-functions using integrations of Eisenstein series over generalized moduli spaces.

(c) **Algebraic Truncation**

Now we explain why the above discussion and Rankin-Selberg method have anything to do with our non-abelian zeta functions. For this, we introduce yet another truncation, the algebraic one.

So back to the moduli space of rank 2 lattices of volume 1 over $\mathbb{Q}$. Then classical reduction theory gives a natural map from this moduli space to the fundamental domain D above: For any lattice Λ, fix $\mathbf{x}_1 \in \Lambda$ such that its length gives the first Minkowski minimum λ_1 of Λ. Then via rotation, we may assume that $\mathbf{x}_1 = (\lambda_1, 0)$. Further, from the reduction theory $\frac{1}{\lambda_1}\Lambda$ may be viewed as the lattice of the volume $\lambda_1^{-2} = y_0$ which is generated by $(1,0)$ and $\omega = x_0 + iy_0 \in D$. That is to say, the points in D_T are in one-to-one correspondence with the rank two lattices of volume one whose first Minkowski minimum $\lambda_1^{-2} \leq T$, i.e, $\lambda_1 \geq T^{-\frac{1}{2}}$. Set $\mathcal{M}_{\mathbb{Q},2}^{\leq \frac{1}{2}\log T}[1]$ be the moduli space of rank 2 lattices Λ of volume 1 over $\mathbb{Q}$ whose sublattices Λ_1 of rank 1 have degrees $\leq \frac{1}{2}\log T$. As a direct consequence, we have the following

Proposition. (Geometric Truncation = Algebraic Truncation) *There is a natural one-to-one, onto morphism*

$$\mathcal{M}_{\mathbb{Q},2}^{\leq \frac{1}{2}\log T}[1] \simeq D_T.$$

In particular,

$$\mathcal{M}_{\mathbb{Q},2}^{\leq 0}[1] = \mathcal{M}_{\mathbb{Q},2}[1] \simeq D_1.$$

Consequently, we have the following

Example in Rank 2. $\xi_{\mathbb{Q},2}(s) = \frac{\xi(2s)}{s-1} - \frac{\xi(2s-1)}{s}$.

Remark. This explicit expression proves to be very useful.
(i) Using it, Suzuki and Lagarias show that all zeros of $\xi_{\mathbb{Q},2}(s)$ lie on the critical line $\mathrm{Re}(s) = 1/2$; (For details, see C.4.)
(ii) Special values $\zeta(2n)$ and $\zeta(2n-1)$ of the Riemann zeta function are naturally related by our rank two zeta functions. As Eisenstein series may be studied independently, our formula indicates that non-abelian zetas could be used to understand abelian zetas; and
(iii) The volume of D_T may be evaluated from this formula via a residue argument; Moreover, the volume of D^1 also have an interpretation in terms of the special values of the Riemann zeta.

B.3. Non-Abelian L-Functions for Number Fields

B.3.1. Automorphic Forms and Eisenstein Series

To facilitate our ensuing discussion, we make the following preparation. For details, see e.g., [MW] and [We8].

Fix a connected reduction group G defined over F, denote by Z_G its center. Fix a minimal parabolic subgroup P_0 of G. Then $P_0 = M_0U_0$, where as usual we fix once and for all the Levi M_0 and the unipotent radical U_0. A parabolic subgroup P is G is called standard if $P \supset P_0$. For such groups write $P = MU$ with $M_0 \subset M$ the standard Levi and U the unipotent radical. Denote by $\mathrm{Rat}(M)$ the group of rational characters of M, i.e, the morphism $M \to \mathbb{G}_m$ where $\mathbb{G}_m$ denotes the multiplicative group. Set

$$\mathfrak{a}_M^* := \mathrm{Rat}(M) \otimes_{\mathbb{Z}} \mathbb{C}, \qquad \mathfrak{a}_M := \mathrm{Hom}_{\mathbb{Z}}(\mathrm{Rat}(M), \mathbb{C}),$$

and

$$\mathrm{Re}\mathfrak{a}_M^* := \mathrm{Rat}(M) \otimes_{\mathbb{Z}} \mathbb{R}, \qquad \mathrm{Re}\mathfrak{a}_M := \mathrm{Hom}_{\mathbb{Z}}(\mathrm{Rat}(M), \mathbb{R}).$$

For any $\chi \in \mathrm{Rat}(M)$, we obtain a (real) character $|\chi| : M(\mathbb{A}) \to \mathbb{R}^*$ defined by $m = (m_v) \mapsto m^{|\chi|} := \prod_{v \in S} |m_v|_v^{\chi_v}$ with $|\cdot|_v$ the v-absolute values. Set then $M(\mathbb{A})^1 := \cap_{\chi \in \mathrm{Rat}(M)} \mathrm{Ker}|\chi|$, which is a normal subgroup of $M(\mathbb{A})$. Set X_M to be the group of complex characters which are trivial on $M(\mathbb{A})^1$. Denote by $H_M := \log_M : M(\mathbb{A}) \to \mathfrak{a}_M$ the map such that $\forall \chi \in \mathrm{Rat}(M) \subset \mathfrak{a}_M^*, \langle \chi, \log_M(m) \rangle := \log(m^{|\chi|})$. Clearly,

$$M(\mathbb{A})^1 = \mathrm{Ker}(\log_M); \qquad \log_M(M(\mathbb{A})/M(\mathbb{A})^1) \simeq \mathrm{Re}\mathfrak{a}_M.$$

Hence in particular there is a natural isomorphism $\kappa : \mathfrak{a}_M^* \simeq X_M$. Set

$$\mathrm{Re}X_M := \kappa(\mathrm{Re}\mathfrak{a}_M^*), \qquad \mathrm{Im}X_M := \kappa(i \cdot \mathrm{Re}\mathfrak{a}_M^*).$$

Moreover define our working space X_M^G to be the subgroup of X_M consisting of complex characters of $M(\mathbb{A})/M(\mathbb{A})^1$ which are trivial on $Z_{G(\mathbb{A})}$.

Fix a maximal compact subgroup $\mathbb{K}$ such that for all standard parabolic subgroups $P = MU$ as above, $P(\mathbb{A}) \cap \mathbb{K} = M(\mathbb{A}) \cap \mathbb{K} \cdot U(\mathbb{A}) \cap \mathbb{K}$. Hence we get the Langlands decomposition $G(\mathbb{A}) = M(\mathbb{A}) \cdot U(\mathbb{A}) \cdot \mathbb{K}$. Denote by $m_P : G(\mathbb{A}) \to M(\mathbb{A})/M(\mathbb{A})^1$ the map $g = m \cdot n \cdot k \mapsto M(\mathbb{A})^1 \cdot m$ where $g \in G(\mathbb{A}), m \in M(\mathbb{A}), n \in U(\mathbb{A})$ and $k \in \mathbb{K}$.

Fix Haar measures on $M_0(\mathbb{A}), U_0(\mathbb{A}), \mathbb{K}$ respectively such that
(1) the induced measure on $M(F)$ is the counting measure and the volume of the induced measure on $M(F) \backslash M(\mathbb{A})^1$ is 1. (Recall that it is a fundamental fact that $M(F) \backslash M(\mathbb{A})^1$ is compact.)
(2) the induced measure on $U_0(F)$ is the counting measure and the volume of $U(F) \backslash U_0(\mathbb{A})$ is 1. (Recall that being unipotent radical, $U(F) \backslash U_0(\mathbb{A})$ is compact.)
(3) the volume of $\mathbb{K}$ is 1.

Such measures then also induce Haar measures via $\log_M$ to the spaces $\mathfrak{a}_{M_0}, \mathfrak{a}_{M_0}^*$, etc. Furthermore, if we denote by ρ_0 the half of the sum of the positive roots of the maximal split torus T_0 of the central

Z_{M_0} of M_0, then

$$f \mapsto \int_{M_0(\mathbb{A})\cdot U_0(\mathbb{A})\cdot \mathbb{K}} f(mnk)\,dk\,dn\,m^{-2\rho_0}dm$$

defined for continuous functions with compact supports on $G(\mathbb{A})$ defines a Haar measure dg on $G(\mathbb{A})$. This in turn gives measures on $M(\mathbb{A}), U(\mathbb{A})$ and hence on $\mathfrak{a}_M, \mathfrak{a}_M^*$, $P(\mathbb{A})$, etc, for all parabolic subgroups P. In particular, one checks that the following compatibility condition holds

$$\begin{aligned}&\int_{M_0(\mathbb{A})\cdot U_0(\mathbb{A})\cdot \mathbb{K}} f(mnk)\,dk\,dn\,m^{-2\rho_0}dm\\ =&\int_{M(\mathbb{A})\cdot U(\mathbb{A})\cdot \mathbb{K}} f(mnk)\,dk\,dn\,m^{-2\rho_P}dm\end{aligned}$$

for all continuous functions f with compact supports on $G(\mathbb{A})$, where ρ_P denotes the half of the sum of the positive roots of the maximal split torus T_P of the central Z_M of M. For later use, denote also by Δ_P the set of positive roots determined by (P, T_P) and $\Delta_0 = \Delta_{P_0}$.

Fix an isomorphism $T_0 \simeq \mathbb{G}_m^R$. Embed $\mathbb{R}_+^*$ by the map $t \mapsto (1;t)$. Then we obtain a natural injection $(\mathbb{R}_+^*)^R \hookrightarrow T_0(\mathbb{A})$ which splits. Denote by $A_{M_0(\mathbb{A})}$ the unique connected subgroup of $T_0(\mathbb{A})$ which projects onto $(\mathbb{R}_+^*)^R$. More generally, for a standard parabolic subgroup $P = MU$, set $A_{M(\mathbb{A})} := A_{M_0(\mathbb{A})} \cap Z_{M(\mathbb{A})}$ where as used above Z_* denotes the center of the group $*$. Clearly, $M(\mathbb{A}) = A_{M(\mathbb{A})}\cdot M(\mathbb{A})^1$. For later use, set also $A_{M(\mathbb{A})}^G := \{a \in A_{M(\mathbb{A})} : \log_G a = 0\}$. Then $A_{M(\mathbb{A})} = A_{G(\mathbb{A})} \oplus A_{M(\mathbb{A})}^G$.

Note that $\mathbb{K}$, $M(F)\backslash M(\mathbb{A})^1$ and $U(F)\backslash U(\mathbb{A})$ are all compact, thus with the Langlands decomposition $G(\mathbb{A}) = U(\mathbb{A})M(\mathbb{A})\mathbb{K}$ in mind, the reduction theory for $G(F)\backslash G(\mathbb{A})$ or more generally $P(F)\backslash G(\mathbb{A})$ is reduced to that for $A_{M(\mathbb{A})}$ since $Z_G(F) \cap Z_{G(\mathbb{A})}\backslash Z_{G(\mathbb{A})} \cap G(\mathbb{A})^1$ is compact as well. As such for $t_0 \in M_0(\mathbb{A})$ set

$$A_{M_0(\mathbb{A})}(t_0) := \{a \in A_{M_0(\mathbb{A})} : a^\alpha > t_0^\alpha \forall \alpha \in \Delta_0\}.$$

Then, for a fixed compact subset $\omega \subset P_0(\mathbb{A})$, we have the corresponding Siegel set

$$S(\omega; t_0) := \{p \cdot a \cdot k : p \in \omega, a \in A_{M_0(\mathbb{A})}(t_0), k \in \mathbb{K}\}.$$

In particular, for big enough ω and small enough t_0, i.e, t_0^α is very close to 0 for all $\alpha \in \Delta_0$, the classical reduction theory may be restated as $G(\mathbb{A}) = G(F) \cdot S(\omega; t_0)$. More generally set

$$A^P_{M_0(\mathbb{A})}(t_0) := \{a \in A_{M_0(\mathbb{A})} : a^\alpha > t_0^\alpha \forall \alpha \in \Delta_0^P\},$$

and

$$S^P(\omega; t_0) := \{p \cdot a \cdot k : p \in \omega, a \in A^P_{M_0(\mathbb{A})}(t_0), k \in \mathbb{K}\}.$$

Then similarly as above for big enough ω and small enough t_0, $G(\mathbb{A}) = P(F) \cdot S^P(\omega; t_0)$. (Here Δ_0^P denotes the set of positive roots for $(P_0 \cap M, T_0)$.)

Fix an embedding $i_G : G \hookrightarrow SL_n$ sending g to (g_{ij}). Introducing a height function on $G(\mathbb{A})$ by setting $\|g\| := \prod_{v \in S} \sup\{|g_{ij}|_v : \forall i, j\}$. It is well known that up to $O(1)$, height functions are unique. This implies that the following growth conditions do not depend on the height function we choose.

A function $f : G(\mathbb{A}) \to \mathbb{C}$ is said to have moderate growth if there exist $c, r \in \mathbb{R}$ such that $|f(g)| \leq c \cdot \|g\|^r$ for all $g \in G(\mathbb{A})$. Similarly, for a standard parabolic subgroup $P = MU$, a function $f : U(\mathbb{A})M(F)\backslash G(\mathbb{A}) \to \mathbb{C}$ is said to have moderate growth if there exist $c, r \in \mathbb{R}, \lambda \in \mathrm{Re}X_{M_0}$ such that for any $a \in A_{M(\mathbb{A})}, k \in \mathbb{K}, m \in M(\mathbb{A})^1 \cap S^P(\omega; t_0)$,

$$|f(amk)| \leq c \cdot \|a\|^r \cdot m_{P_0}(m)^\lambda.$$

Also a function $f : G(\mathbb{A}) \to \mathbb{C}$ is said to be smooth if for any $g = g_f \cdot g_\infty \in G(\mathbb{A}_f) \times G(\mathbb{A}_\infty)$, there exist open neighborhoods V_* of g_* in $G(\mathbb{A})$ and a C^∞-function $f' : V_\infty \to \mathbb{C}$ such that $f(g'_f \cdot g'_\infty) = f'(g'_\infty)$ for all $g'_f \in V_f$ and $g'_\infty \in V_\infty$.

By contrast, a function $f : S(\omega; t_0) \to \mathbb{C}$ is said to be rapidly decreasing if there exists $r > 0$ and for all $\lambda \in \mathrm{Re}X_{M_0}$ there exists $c > 0$ such that for $a \in A_{M(\mathbb{A})}, g \in G(\mathbb{A})^1 \cap S(\omega; t_0)$, $|\phi(ag)| \leq c \cdot \|a\| \cdot m_{P_0}(g)^\lambda$. And a function $f : G(F)\backslash G(\mathbb{A}) \to \mathbb{C}$ is said to be rapidly decreasing if $f|_{S(\omega;t_0)}$ is so.

By definition, a function $\phi : U(\mathbb{A})M(F)\backslash G(\mathbb{A}) \to \mathbb{C}$ is called automorphic if

(i) ϕ has moderate growth;
(ii) ϕ is smooth;
(iii) ϕ is $\mathbb{K}$-finite, i.e, the $\mathbb{C}$-span of all $\phi(k_1 \cdot * \cdot k_2)$ parametrized by $(k_1, k_2) \in \mathbb{K} \times \mathbb{K}$ is finite dimensional; and
(iv) ϕ is $\mathfrak{z}$-finite, i.e, the $\mathbb{C}$-span of all $\delta(X)\phi$ parametrized by all $X \in \mathfrak{z}$ is finite dimensional. Here $\mathfrak{z}$ denotes the center of the universal enveloping algebra $\mathfrak{u} := \mathfrak{U}(\mathrm{Lie}G(\mathbb{A}_\infty))$ of the Lie algebra of $G(\mathbb{A}_\infty)$ and $\delta(X)$ denotes the derivative of ϕ along X.

For such a function ϕ, set $\phi_k : M(F)\backslash M(\mathbb{A}) \to \mathbb{C}$ by $m \mapsto m^{-\rho_P}\phi(mk)$ for all $k \in \mathbb{K}$. Then one checks that ϕ_k is an automorphic form in the usual sense. Set $A(U(\mathbb{A})M(F)\backslash G(\mathbb{A}))$ be the space of automorphic forms on $U(\mathbb{A})M(F)\backslash G(\mathbb{A})$.

For a measurable locally L^1-function $f : U(F)\backslash G(\mathbb{A}) \to \mathbb{C}$, define its constant term along with the standard parabolic subgroup $P = UM$ to be $f_P : U(\mathbb{A})\backslash G(\mathbb{A}) \to \mathbb{C}$ given by $g \to \int_{U(F)\backslash G(\mathbb{A})} f(ng)dn$. Then an automorphic form $\phi \in A(U(\mathbb{A})M(F)\backslash G(\mathbb{A}))$ is called a cusp form if for any standard parabolic subgroup P' properly contained in P, $\phi_{P'} \equiv 0$. Denote by $A_0(U(\mathbb{A})M(F)\backslash G(\mathbb{A}))$ the space of cusp forms on $U(\mathbb{A})M(F)\backslash G(\mathbb{A})$. One checks easily that
(i) all cusp forms are rapidly decreasing; and hence
(ii) there is a natural pairing

$$\langle \cdot, \cdot \rangle : A_0(U(\mathbb{A})M(F)\backslash G(\mathbb{A})) \times A(U(\mathbb{A})M(F)\backslash G(\mathbb{A})) \to \mathbb{C}$$

defined by $\langle \psi, \phi \rangle := \int_{Z_{M(\mathbb{A})}U(\mathbb{A})M(F)\backslash G(\mathbb{A})} \psi(g)\bar{\phi}(g)\, dg$.

Moreover, for a (complex) character $\xi : Z_{M(\mathbb{A})} \to \mathbb{C}^*$ of $Z_{M(\mathbb{A})}$ set

$$\begin{aligned} A(U(\mathbb{A})M(F)\backslash G(\mathbb{A}))_\xi := \{\phi \in A(U(\mathbb{A})M(F)\backslash G(\mathbb{A})) :\\ \phi(zg) = z^{\rho_P} \cdot \xi(z) \cdot \phi(g), \forall z \in Z_{M(\mathbb{A})}, g \in G(\mathbb{A})\} \end{aligned}$$

and

$$\begin{aligned} &A_0(U(\mathbb{A})M(F)\backslash G(\mathbb{A}))_\xi \\ &:= A_0(U(\mathbb{A})M(F)\backslash G(\mathbb{A})) \cap A(U(\mathbb{A})M(F)\backslash G(\mathbb{A}))_\xi. \end{aligned}$$

Set now

$$A(U(\mathbb{A})M(F)\backslash G(\mathbb{A}))_Z := \sum_{\xi \in \mathrm{Hom}(Z_{M(\mathbb{A})}, \mathbb{C}^*)} A(U(\mathbb{A})M(F)\backslash G(\mathbb{A}))_\xi$$

and

$$A_0(U(\mathbb{A})M(F)\backslash G(\mathbb{A}))_Z := \sum_{\xi\in \mathrm{Hom}(Z_{M(\mathbb{A})},\mathbb{C}^*)} A_0(U(\mathbb{A})M(F)\backslash G(\mathbb{A}))_\xi.$$

One checks that the natural morphism

$$\mathbb{C}[\mathrm{Re}\mathfrak{a}_M]\otimes A(U(\mathbb{A})M(F)\backslash G(\mathbb{A}))_Z \to A(U(\mathbb{A})M(F)\backslash G(\mathbb{A}))$$

defined by $(Q,\phi)\mapsto \big(g\mapsto Q(\log_M(m_P(g)))\big)\cdot\phi(g)$ is an isomorphism, using the special structure of $A_{M(\mathbb{A})}$-finite functions and the Fourier analysis over the compact space $A_{M(\mathbb{A})}\backslash Z_{M(\mathbb{A})}$. Consequently, we also obtain a natural isomorphism

$$\mathbb{C}[\mathrm{Re}\mathfrak{a}_M]\otimes A_0(U(\mathbb{A})M(F)\backslash G(\mathbb{A}))_Z \to A_0(U(\mathbb{A})M(F)\backslash G(\mathbb{A}))_\xi.$$

Set also $\Pi_0(M(\mathbb{A}))_\xi$ be isomorphism classes of irreducible representations of $M(\mathbb{A})$ occurring in the space $A_0(M(F)\backslash M(\mathbb{A}))_\xi$, and

$$\Pi_0(M(\mathbb{A}) := \cup_{\xi\in\mathrm{Hom}(Z_{M(\mathbb{A})},\mathbb{C}^*)}\Pi_0(M(\mathbb{A}))_\xi.$$

(More precisely, we should use $M(\mathbb{A}_f)\times(M(\mathbb{A})\cap\mathbb{K},\mathrm{Lie}(M(\mathbb{A}_\infty))\otimes_\mathbb{R}\mathbb{C}))$ instead of $M(\mathbb{A})$.) For any $\pi\in\Pi_0(M(\mathbb{A}))_\xi$ set $A_0(M(F)\backslash M(\mathbb{A})_\pi$ to be the isotypic component of type π of $A_0(M(F)\backslash M(\mathbb{A})_\xi$, i.e, the set of cusp forms of $M(\mathbb{A})$ generating a semi-simple isotypic $M(\mathbb{A}_f)\times(M(\mathbb{A})\cap\mathbb{K},\mathrm{Lie}(M(\mathbb{A}_\infty))\otimes_\mathbb{R}\mathbb{C}))$-module of type π. Set

$$\begin{aligned}&A_0(U(\mathbb{A})M(F)\backslash G(\mathbb{A}))_\pi\\ &:=\{\phi\in A_0(U(\mathbb{A})M(F)\backslash G(\mathbb{A})) : \phi_k\in A_0(M(F)\backslash M(\mathbb{A}))_\pi, \forall k\in\mathbb{K}\}.\end{aligned}$$

Clearly

$$A_0(U(\mathbb{A})M(F)\backslash G(\mathbb{A}))_\xi = \oplus_{\pi\in\Pi_0(M(\mathbb{A}))_\xi}A_0(U(\mathbb{A})M(F)\backslash G(\mathbb{A}))_\pi.$$

More generally, let $V\subset A(M(F)\backslash M(\mathbb{A}))$ be an irreducible $M(\mathbb{A}_f)\times(M(\mathbb{A})\cap\mathbb{K},\mathrm{Lie}(M(\mathbb{A}_\infty))\otimes_\mathbb{R}\mathbb{C}))$-module with π_0 the induced representation of $M(\mathbb{A}_f)\times(M(\mathbb{A})\cap\mathbb{K},\mathrm{Lie}(M(\mathbb{A}_\infty))\otimes_\mathbb{R}\mathbb{C}))$. Then we call π_0 an automorphic representation of $M(\mathbb{A})$. Denote by $A(M(F)\backslash M(\mathbb{A})_{\pi_0}$

the isotypic subquotient module of type π_0 of $A(M(F)\backslash M(\mathbb{A})$. One checks that

$$\begin{aligned}&V\otimes \mathrm{Hom}_{M(\mathbb{A}_f)\times(M(\mathbb{A})\cap\mathbb{K},\mathrm{Lie}(M(\mathbb{A}_\infty))\otimes_{\mathbb{R}}\mathbb{C}))}(V,A(M(F)\backslash M(\mathbb{A})))\\ \simeq& A(M(F)\backslash M(\mathbb{A}))_{\pi_0}.\end{aligned}$$

Set

$$\begin{aligned}&A(U(\mathbb{A})M(F)\backslash G(\mathbb{A}))_{\pi_0}\\ :=&\{\phi\in A(U(\mathbb{A})M(F)\backslash G(\mathbb{A})):\phi_k\in A(M(F)\backslash M(\mathbb{A}))_{\pi_0},\forall k\in\mathbb{K}\}.\end{aligned}$$

Moreover if $A(M(F)\backslash M(\mathbb{A}))_{\pi_0}\subset A_0(M(F)\backslash M(\mathbb{A}))$, we call π_0 a cuspidal representation.

Two automorphic representations π and π_0 of $M(\mathbb{A})$ are said to be equivalent if there exists $\lambda\in X_M^G$ such that $\pi\simeq\pi_0\otimes\lambda$. This, in practice, means that $A(M(F)\backslash M(\mathbb{A}))_\pi=\lambda\cdot A(M(F)\backslash M(\mathbb{A}))_{\pi_0}$. That is for any $\phi_\pi\in A(M(F)\backslash M(\mathbb{A}))_\pi$ there exists a $\phi_{\pi_0}\in A(M(F)\backslash M(\mathbb{A}))_{\pi_0}$ such that $\phi_\pi(m)=m^\lambda\cdot\phi_{\pi_0}(m)$. Consequently,

$$A(U(\mathbb{A})M(F)\backslash G(\mathbb{A}))_\pi=(\lambda\circ m_P)\cdot A(U(\mathbb{A})M(F)\backslash G(\mathbb{A}))_{\pi_0}.$$

Denote by $\mathfrak{P}:=[\pi_0]$ the equivalence class of π_0. Then $\mathfrak{P}$ is an X_M^G-principal homogeneous space, hence admits a natural complex structure. Usually we call $(M,\mathfrak{P})$ a cuspidal datum of G if π_0 is cuspidal. Also for $\pi\in\mathfrak{P}$ set $\mathrm{Re}\pi:=\mathrm{Re}\chi_\pi=|\chi_\pi|\in\mathrm{Re}X_M$, where χ_π is the central character of π, and $\mathrm{Im}\pi:=\pi\otimes(-\mathrm{Re}\pi)$.

Now fix an irreducible automorphic representation π of $M(\mathbb{A})$ and an automorphic form $\phi\in A(U(\mathbb{A})M(F)\backslash G(\mathbb{A}))_\pi$, define the associated Eisenstein series $E(\phi,\pi):G(F)\backslash G(\mathbb{A})\to\mathbb{C}$ by

$$E(\phi,\pi)(g):=\sum_{\delta\in P(F)\backslash G(F)}\phi(\delta g).$$

Then one checks that there is an open cone $\mathcal{C}\subset\mathrm{Re}X_M^G$ such that if $\mathrm{Re}\pi\in\mathcal{C}$, $E(\lambda\cdot\phi,\pi\otimes\lambda)(g)$ converges uniformly for g in a compact subset of $G(\mathbb{A})$ and λ in an open neighborhood of 0 in X_M^G. For example, if $\mathfrak{P}=[\pi]$ is cuspidal, we may even take $\mathcal{C}$ to be the cone $\{\lambda\in\mathrm{Re}X_M^G:\langle\lambda-\rho_P,\alpha^\vee\rangle>0,\forall\alpha\in\Delta_P^G\}$. As a direct consequence, then $E(\phi,\pi)\in A(G(F)\backslash G(\mathbb{A}))$. That is, it is an automorphic form.

B.3.2. Non-Abelian L-Functions

Being automorphic forms, Eisenstein series are of moderate growth. Consequently, they are not integrable over $G(F)\backslash G(\mathbb{A})^1$. On the other hand, Eisenstein series are also smooth and hence integrable over compact subsets of $G(F)\backslash G(\mathbb{A})^1$. So it is very natural for us to search for compact domains which are intrinsically defined.

As such, let us now return to the group $G = GL_r$. Then form 2.3.2, we obtain compact moduli spaces

$$\mathcal{M}_{F,r}^{\leq p}[\Delta_F^{\frac{r}{2}}] := \{g \in GL_r(F)\backslash GL_r(\mathbb{A}) : \deg g = 0, \bar{p}^g \leq p\}$$

for a fixed convex polygon $p : [0,r] \to \mathbb{R}$. For example, $\mathcal{M}_{\mathbb{Q},r}^{\leq 0}[1] = \mathcal{M}_{\mathbb{Q},r}[1]$, (the adelic inverse image of) the moduli space of rank r semi-stable $\mathbb{Z}$-lattices of volume 1.

More generally, for the standard parabolic subgroup P of GL_r, we introduce the moduli spaces

$$\mathcal{M}_{F,r}^{P;\leq p}[\Delta_F^{\frac{r}{2}}] := \{g \in P(F)\backslash GL_r(\mathbb{A}) : \deg g = 0, \bar{p}_P^g \leq p, \bar{p}_P^g \geq -p\}.$$

By 2.3.2, these moduli spaces $\mathcal{M}_{F,r}^{P;\leq p}[\Delta_F^{\frac{r}{2}}]$ are all compact.

As usual, we fix the minimal parabolic subgroup P_0 corresponding to the partition $(1,\ldots,1)$ with M_0 consisting of diagonal matrices. Then $P = P_I = U_I M_I$ corresponds to a certain partition $I = (r_1,\ldots,r_{|P|})$ of r with M_I the standard Levi and U_I the unipotent radical.

Now for a fixed irreducible automorphic representation π of $M_I(\mathbb{A})$, choose

$$\begin{aligned}\phi &\in A(U_I(\mathbb{A})M_I(F)\backslash G(\mathbb{A}))_\pi \cap L^2(U_I(\mathbb{A})M_I(F)\backslash G(\mathbb{A}))\\ &:= A^2(U_I(\mathbb{A})M_I(F)\backslash G(\mathbb{A}))_\pi,\end{aligned}$$

with $L^2(U_I(\mathbb{A})M_I(F)\backslash G(\mathbb{A}))$ the space of L^2-functions on the quotient space $Z_{G(\mathbb{A})}U_I(\mathbb{A})M_I(F)\backslash G(\mathbb{A})$. Denote the associated Eisenstein series by $E(\phi,\pi) \in A(G(F)\backslash G(\mathbb{A}))$. Then, we define the rank r non-abelian L-function $L_{F,r}^{\leq p}(\phi,\pi)$ for the number field F associated to the L^2-automorphic form $\phi \in A^2(U_I(\mathbb{A})M_I(F)\backslash G(\mathbb{A}))_\pi$ to be the integration

$$L_{F,r}^{\leq p}(\phi, \pi) := \int_{\mathcal{M}_{F,r}^{\leq p}[\Delta_F^{\frac{r}{2}}]} E(\phi, \pi)(g)\, dg, \qquad \mathrm{Re}\,\pi \in \mathcal{C}.$$

More generally, for any standard parabolic subgroup $P_J = U_J M_J \supset P_I$ (so that the partition J is a refinement of I), we obtain a relative Eisenstein series

$$E_I^J(\phi, \pi)(g) := \sum_{\delta \in P_I(F)\backslash P_J(F)} \phi(\delta g), \qquad \forall g \in P_J(F)\backslash G(\mathbb{A}).$$

There is an open cone $\mathcal{C}_I^J$ in $\mathrm{Re}X_{M_I}^{P_J}$ s.t. if $\mathrm{Re}\pi \in \mathcal{C}_I^J$, then $E_I^J(\phi, \pi) \in A(P_J(F)\backslash G(\mathbb{A}))$, where $X_{M_I}^{P_J}$ is defined similarly as X_M^G with G replaced by P_J. As such, we are able to define the associated non-abelian L-function by

$$L_{F,r}^{P_J;\leq p}(\phi, \pi) := \int_{\mathcal{M}_{F,r}^{P_J;\leq p}[\Delta_F^{\frac{r}{2}}]} E_I^J(\phi, \pi)(g)\, dg, \qquad \mathrm{Re}\pi \in \mathcal{C}_I^J.$$

Remark. Here when defining non-abelian L-functions we assume that ϕ comes from a single irreducible automorphic representations. But this restriction is rather artificial and can be removed easily: such a restriction only serves the purpose of giving the constructions and results in a very neat form.

We end this section by pointing out that the discussion for rank r non-abelian L-functions $L_{F,r}^{\leq p}$ holds for general non-abelian L-functions $L_{F,r}^{P_J;\leq p}$ as well. So from now on, we will leave the discussion on general $L_{F,r}^{P_J;\leq p}$ to the reader, while we concentrate ourselves on the special case $L_{F,r}^{\leq p}$.

B.3.3. Meromorphic Extension and Functional Equations

With the same notation as above, set $\mathfrak{P} = [\pi]$. For $w \in W$ the Weyl group of $G = GL_r$, fix once and for all representative $w \in G(F)$ of w. Set $M' := wMw^{-1}$ and denote the associated parabolic subgroup by $P' = U'M'$. W acts naturally on the automorphic representations, from which we obtain an equivalence classes $w\mathfrak{P}$ of automorphic representations of $M'(\mathbb{A})$. As usual, define the associated intertwining operator $M(w,\pi)$ by

$$(M(w,\pi)\phi)(g) := \int_{U'(F)\cap wU(F)w^{-1}\backslash U'(\mathbb{A})} \phi(w^{-1}n'g)dn', \ \forall g \in G(\mathbb{A}).$$

One checks that if $\langle \mathrm{Re}\pi, \alpha^\vee\rangle \gg 0, \forall \alpha \in \Delta_P^G$,
(i) for a fixed automorphic ϕ, $M(w,\pi)\phi$ depends only on the double coset $M'(F)wM(F)$. So $M(w,\pi)\phi$ is well-defined for $w \in W$;
(ii) the above integral converges absolutely and uniformly for g varying in a compact subset of $G(\mathbb{A})$;
(iii) $M(w,\pi)\phi \in A(U'(\mathbb{A})M'(F)\backslash G(\mathbb{A}))_{w\pi}$; and if ϕ is L^2, which from now on we always assume, so is $M(w,\pi)\phi$.

Basic Facts of Non-Abelian L-Functions. ([We6])
(I) (Meromorphic Continuation) $L_{F,r}^{\leq p}(\phi,\pi)$ *for* $\mathrm{Re}\pi \in \mathcal{C}$ *is well-defined and admits a unique meromorphic continuation to the whole space* $\mathfrak{P}$;
(II) (Functional Equation) *As meromorphic functions on* $\mathfrak{P}$,

$$L_{F,r}^{\leq p}(\phi,\pi) = L_{F,r}^{\leq p}(M(w,\pi)\phi, w\pi), \qquad \forall w \in W.$$

This is a direct consequence of the fundamental results of Langlands on Eisenstein series and spectrum decompositions and explains why only L^2-automorphic forms are used in the definition of non-abelian Ls. (See e.g, [Ar1], [La1], [MW] and/or [We8]).

B.3.4. Holomorphicity and Singularities

Let $\pi \in \mathfrak{P}$ and $\alpha \in \Delta_M^G$. Define the function $h : \mathfrak{P} \to \mathbb{C}$ by $\pi \otimes \lambda \mapsto \langle \lambda, \alpha^\vee\rangle, \forall \lambda \in X_M^G \simeq \mathfrak{a}_M^G$. Here as usual, $\alpha^\vee$ denotes the coroot associated to α. Set $H := \{\pi' \in \mathfrak{P} : h(\pi') = 0\}$ and call it a root hyperplane. Clearly the function h is determined by H, hence we also denote h by h_H. Note also that root hyperplanes depend on the base point π we choose.

Let D be a set of root hyperplanes. Then
(i) the singularities of a meromorphic function f on $\mathfrak{P}$ is said to be carried out by D if for all $\pi \in \mathfrak{P}$, there exist $n_\pi : D \to \mathbb{Z}_{\geq 0}$ zero almost everywhere such that $\pi' \mapsto \left(\Pi_{H\in D} h_H(\pi')^{n_\pi(H)}\right) \cdot f(\pi')$ is holomorphic at π';

(ii) the singularities of f are said to be without multiplicity at π if $n_\pi \in \{0,1\}$;
(iii) D is said to be locally finite, if for any compact subset $C \subset \mathfrak{P}$, $\{H \in D : H \cap C \neq \emptyset\}$ is finite.

Basic Facts of Non-Abelian L-Functions. ([We6])
(III) (Holomorphicity) (i) *When* $\mathrm{Re}\pi \in \mathcal{C}$, $L_{F,r}^{\leq p}(\phi,\pi)$ *is holomorphic;*
(ii) $L_{F,r}^{\leq p}(\phi,\pi)$ *is holomorphic at* π *where* $\mathrm{Re}\pi = 0$*;*
(IV) (Singularities) *Assume further that* ϕ *is a cusp form. Then*
(i) *There is a locally finite set of root hyperplanes* D *such that the singularities of* $L_{F,r}^{\leq p}(\phi,\pi)$ *are carried out by* D*;*
(ii) *The singularities of* $L_{F,r}^{\leq p}(\phi,\pi)$ *are without multiplicities at* π *if* $\langle \mathrm{Re}\pi, \alpha^\vee\rangle \geq 0, \forall \alpha \in \Delta_M^G$*;*
(iii) *There are only finitely many of singular hyperplanes of* $L_{F,r}^{\leq p}(\phi,\pi)$ *which intersect* $\{\pi \in \mathfrak{P} : \langle \mathrm{Re}\pi, \alpha^\vee\rangle \geq 0, \forall \alpha \in \Delta_M\}$.
As above, this is a direct consequence of the fundamental results of Langlands on Eisenstein series and spectrum decompositions. (See e.g, [Ar1], [La1], [MW] and/or [We8]).

B.4. Geometric and Analytic Truncations: A Bridge

B.4.1. Geometric Truncation: Revised

B.4.1.1. Slopes, Canonical Filtrations

Following Lafforgue [Laf], we call an abelian category $\mathcal{A}$ together with two additive morphisms

$$\mathrm{rk} : \mathcal{A} \to \mathbb{N}, \qquad \deg : \mathcal{A} \to \mathbb{R}$$

a *category with slope structure.* In particular, for non-zero $A \in \mathcal{A}$,

(**1**) define the *slope* of A by $\mu(A) := \frac{\deg(A)}{\mathrm{rk} A}$;

(**2**) If $0 = A_0 \subset A_1 \subset \cdots \subset A_l = A$ is a filtration of A in $\mathcal{A}$ with $\mathrm{rk}(A_0) < \mathrm{rk}(A_1) < \cdots < \mathrm{rk}(A_l)$, define the *associated polygon* to be the function $[0, \mathrm{rk}A] \to \mathbb{R}$ such that
(i) its values at 0 and $\mathrm{rk}(A)$ are 0;

(ii) it is affine on the intervals $[\mathrm{rk}(A_{i-1}), \mathrm{rk}(A_i)]$ with slope $\mu(A_i/A_{i-1}) -\mu(A)$;

(**3**) If $\mathfrak{a}$ is a collection of subobjects of A in $\mathcal{A}$, then $\mathfrak{a}$ is said to be *nice* if
(i) $\mathfrak{a}$ is stable under intersection and finite summation;
(ii) $\mathfrak{a}$ is Noetherian, i.e., every increasing chain of elements in $\mathfrak{a}$ has a maximal element in $\mathfrak{a}$;
(iii) if $A_1 \in \mathfrak{a}$ then $A_1 \neq 0$ if and only if $\mathrm{rk}(A_1) \neq 0$; and
(iv) for $A_1, A_2 \in \mathfrak{a}$ with $\mathrm{rk}(A_1) = \mathrm{rk}(A_2)$. Then $A_1 \subset A_2$ is proper implies that $\deg(A_1) < \deg(A_2)$;

(**4**) For any nice $\mathfrak{a}$, set

$$\mu^+(A) := \sup\Big\{\mu(A_1) : A_1 \in \mathfrak{a}, \mathrm{rk}(A_1) \geq 1\Big\},$$
$$\mu^-(A) := \inf\Big\{\mu(A/A_1) : A_1 \in \mathfrak{a}, \mathrm{rk}(A_1) < \mathrm{rk}(A)\Big\}.$$

Then we say $(A, \mathfrak{a})$ is *semi-stable* if $\mu^+(A) = \mu(A) = \mu^-(A)$. Moreover if $\mathrm{rk}(A) = 0$, set also $\mu^+(A) = -\infty$ and $\mu^-(A) = +\infty$.

Proposition 1. ([Laf]) *Let $\mathcal{A}$ be a category with slope structure, A an object in $\mathcal{A}$ and $\mathfrak{a}$ a nice family of subobjects of A in $\mathcal{A}$. Then*
(1) (**Canonical Filtration**) *A admits a unique filtration $0 = \overline{A}_0 \subset \overline{A}_1 \subset \cdots \subset \overline{A}_l = A$ with elements in $\mathfrak{a}$ such that*
(i) *$\overline{A}_i, 0 \leq i \leq k$ are maximal in $\mathfrak{a}$;*
(ii) *$\overline{A}_i/\overline{A}_{i-1}$ are semi-stable; and*
(iii) *$\mu(\overline{A}_1/\overline{A}_0) > \mu(\overline{A}_2/\overline{A}_1 > \cdots > \mu(\overline{A}_k/\overline{A}_{k-1})$;*
(2) (**Boundness**) *All polygons of filtrations of A with elements in $\mathfrak{a}$ are bounded from above by $\overline{p}$, where $\overline{p} := \overline{p}^A$ is the associated polygon for the canonical filtration in (1);*
(3) *For any $A_1 \in \mathfrak{a}, \mathrm{rk}(A_1) \geq 1$ implies $\mu(A_1) \leq \mu(A) + \frac{\overline{p}(\mathrm{rk}(A_1))}{\mathrm{rk}(A_1)}$;*
(4) *The polygon $\overline{p}$ is convex with maximal slope $\mu^+(A) - \mu(A)$ and minimal slope $\mu^-(A) - \mu(A)$;*
(5) *If $(A', \mathfrak{a}')$ is another pair, and $u : A \to A'$ is a homomorphism such that $\mathrm{Ker}(u) \in \mathfrak{a}$ and $\mathrm{Im}(u) \in \mathfrak{a}'$. Then $\mu^-(A) \geq \mu^+(A')$ implies that $u = 0$.*

This results from a Harder-Narasimhan type filtration consideration. A detailed proof may be found at pp. 87-88 in [Laf].

As an example, we have the following

Proposition 2. ([We6]) *Let F be a number field. Then*
(1) *the abelian category of hermitian vector sheaves on* $\mathrm{Spec}\,\mathcal{O}_F$ *together with the natural rank and the Arakelov degree is a category with slopes;*
(2) *For any hermitian vector sheaf (E,ρ), $\mathfrak{a}$ consisting of pairs (E_1,ρ_1) with E_1 sub vector sheaves of E and ρ_1 the restrictions of ρ, forms a nice family.*

Indeed, (1) is obvious, while (2) is a direct consequence of the following standard facts:
(i) For a fixed (E,ρ), $\Big\{\deg(E_1,\rho_1) : (E_1,\rho_1)\in\mathfrak{a}\Big\}$ is discrete subset of $\mathbb{R}$; and
(ii) for any two sublattices Λ_1, Λ_2 of Λ,

$$\mathrm{Vol}\Big(\Lambda_1/(\Lambda_1\cap\Lambda_2)\Big)\geq\mathrm{Vol}\Big((\Lambda_1+\Lambda_2)/\Lambda_2\Big).$$

Thus we get the canonical filtration of Harder-Narasimhan type for hermitian vector sheaves over $\mathrm{Spec}\,\mathcal{O}_F$. Recall that hermitian vector sheaves over $\mathrm{Spec}\,\mathcal{O}_F$ are $\mathcal{O}_F$-lattices in $(\mathbb{R}^{r_1}\times\mathbb{C}^{r_2})^{r=\mathrm{rk}(E)}$ in the language of Arakelov theory: Say, corresponding $\mathcal{O}_F$-lattices are induced from their H^0 via the natural embedding $F^r\hookrightarrow(\mathbb{R}^{r_1}\times\mathbb{C}^{r_2})^r$ where r_1 (resp. r_2) denotes the real (resp. complex) embeddings of F.

B.4.1.2. Micro-Global Relation for Geo-Ari Truncations

In algebraic geometry, or better in Geometric Invariant Theory, a fundamental principle, the Micro-Global Principle, claims that if a point is not GIT stable then there exists a parabolic subgroup which destroys the corresponding stability.

Here even we do not have a proper definition of GIT stability for lattices, in terms of intersection stability, an analogue of the Micro-Global Principle holds. (Recall that in geometry, for vector bundles, intersection stability and GIT stability are equivalent. But we do not know whether this is true say in geometry for manifolds and in arithmetic for lattices. It is our belief that intersection stability in application should be equally important. Say in the case for manifolds it claims that if a manifold is stable in terms of intersection, then certain kinds of submanifolds should not exist, a far better statement than merely claiming that the Hilbert or Chow points of the manifold in question are GIT stability. This is why we when working in geometry always try to convince the colleagues there to find a proper definition for intersection stability. Surely, we also understand that for theoretical purpose such as building up the moduli GIT stability plays a central role.)

Let $\Lambda = \Lambda^g$ be a rank r lattice associated to $g \in \mathrm{GL}_r(\mathbb{A})$ and P a parabolic subgroup. Denote the sublattices filtration associated to P by

$$0 = \Lambda_0 \subset \Lambda_1 \subset \Lambda_2 \subset \cdots \subset \Lambda_{|P|} = \Lambda.$$

Assume that P corresponds to the partition $I = (d_1, d_2, \ldots, d_{n=:|P|})$. Consequently, we have

$$\mathrm{rk}(\Lambda_i) = r_i := d_1 + d_2 + \cdots + d_i, \qquad \text{for } i = 1, 2, \ldots, |P|.$$

Let $p, q : [0, r] \to \mathbb{R}$ be two polygons such that $p(0) = q(0) = p(r) = q(r) = 0$. Then following Lafforgue, we say q is *bigger than p with respect to P* and denote it by $q >_P p$, if $q(r_i) - p(r_i) > 0$ for all $i = 1, \ldots, |P|-1$. Introduce also the characteristic function $\mathbf{1}(\overline{p}^* \leq p)$ by

$$\mathbf{1}(\overline{p}^g \leq p) = \begin{cases} 1, & \text{if } \overline{p}^g \leq p; \\ 0, & \text{otherwise.} \end{cases}$$

Recall that for a parabolic subgroup P, p_P^g denotes the polygon induced by P for (the lattice corresponding to) the element $g \in G(\mathbb{A})$.

Fundamental Relation. ([We9]) *For a fixed convex polygon $p : [0, r] \to \mathbb{R}$ such that $p(0) = p(r) = 0$, we have*

$$\mathbf{1}(\overline{p}^g \leq p) = \sum_{P:\,\text{standard parabolic}} (-1)^{|P|-1} \sum_{\delta \in P(F)\backslash G(F)} \mathbf{1}(p_P^{\delta g} >_P p).$$

This is an analogue for lattices of a result of Lafforgue [La] for vector bundles.

B.4.2. Arthur's Analytic Truncation

B.4.2.1. Parabolic Subgroups

Let F be a number field with $\mathbb{A} = \mathbb{A}_F$ the ring of adeles. Let G be a connected reductive group defined over F. Recall that a subgroup P of G is called *parabolic* if G/P is a complete algebraic variety. Fix a minimal F-parabolic subgroup P_0 of G with its unipotent radical $N_0 = N_{P_0}$ and fix a F-Levi subgroup $M_0 = M_{P_0}$ of P_0 so as to have a Levi decomposition $P_0 = M_0 N_0$. An F-parabolic subgroup P is called *standard* if it contains P_0. For such parabolic subgroups P, there exists a unique Levi subgroup $M = M_P$ containing M_0 which we call the *standard Levi subgroup* of P. Let $N = N_P$ be the unipotent radical. Let us agree to use the term parabolic subgroups and Levi subgroups to denote standard F-parabolic subgroups and standard Levi subgroups respectively, unless otherwise is stated.

Let P be a parabolic subgroup of G. Write T_P for the maximal split torus in the center of M_P and T_P' for the maximal quotient split torus of M_P. Set $\tilde{\mathfrak{a}}_P := X_*(T_P) \otimes \mathbb{R}$ and denote its real dimension by $d(P)$, where $X_*(T)$ is the lattice of 1-parameter subgroups in the torus T. Then it is known that $\tilde{\mathfrak{a}}_P = X_*(T_P') \otimes \mathbb{R}$ as well. The two descriptions of $\tilde{\mathfrak{a}}_P$ show that if $Q \subset P$ is a parabolic subgroup, then there is a canonical injection $\tilde{\mathfrak{a}}_P \hookrightarrow \tilde{\mathfrak{a}}_Q$ and a natural surjection $\tilde{\mathfrak{a}}_Q \twoheadrightarrow \tilde{\mathfrak{a}}_P$. We thus obtain a canonical decomposition $\tilde{\mathfrak{a}}_Q = \tilde{\mathfrak{a}}_Q^P \oplus \tilde{\mathfrak{a}}_P$ for a certain subspace $\tilde{\mathfrak{a}}_Q^P$ of $\tilde{\mathfrak{a}}_Q$. In particular, $\tilde{\mathfrak{a}}_G$ is a summand of $\tilde{\mathfrak{a}} = \tilde{\mathfrak{a}}_P$ for all P. Set $\mathfrak{a}_P := \tilde{\mathfrak{a}}_P / \tilde{\mathfrak{a}}_G$ and $\mathfrak{a}_Q^P := \tilde{\mathfrak{a}}_Q^P / \tilde{\mathfrak{a}}_G$. Then we have

$$\mathfrak{a}_Q = \mathfrak{a}_Q^P \oplus \mathfrak{a}_P$$

and $\mathfrak{a}_P$ is canonically identified as a subspace of $\mathfrak{a}_Q$. Set $\mathfrak{a}_0 := \mathfrak{a}_{P_0}$ and $\mathfrak{a}_0^P = \mathfrak{a}_{P_0}^P$ then we also have $\mathfrak{a}_0 = \mathfrak{a}_0^P \oplus \mathfrak{a}_P$ for all P.

B.4.2.2. Logarithmic Map

For a real vector space V, write V^* its dual space over $\mathbb{R}$. Then dually we have the spaces $\mathfrak{a}_0^*, \mathfrak{a}_P^*, \left(\mathfrak{a}_0^P\right)^*$ and hence the decompositions

$$\mathfrak{a}_0^* = \left(\mathfrak{a}_0^Q\right)^* \oplus \left(\mathfrak{a}_Q^P\right)^* \oplus \mathfrak{a}_P^*.$$

So $\mathfrak{a}_P^* = X(M_P)\otimes\mathbb{R}$ with $X(M_P)$ the group $\mathrm{Hom}_F\Big(M_P, GL(1)\Big)$ i.e., collection of characters on M_P. It is known that $\mathfrak{a}_P^* = X(A_P)\otimes\mathbb{R}$ where A_P denotes the split component of the center of M_P. Clearly, if $Q \subset P$, then $M_Q \subset M_P$ while $A_P \subset A_Q$. Thus via restriction, the above two expressions of $\mathfrak{a}_P^*$ also naturally induce an injection $\mathfrak{a}_P^* \hookrightarrow \mathfrak{a}_Q^*$ and a surjection $\mathfrak{a}_Q^* \twoheadrightarrow \mathfrak{a}_P^*$, compactible with the decomposition $\mathfrak{a}_Q^* = \left(\mathfrak{a}_Q^P\right)^* \oplus \mathfrak{a}_P^*$.

Every $\chi = \sum s_i\chi_i$ in $\mathfrak{a}_{P,\mathbb{C}}^* := \mathfrak{a}_P^* \otimes \mathbb{C}$ determines a morphism $P(\mathbb{A}) \to \mathbb{C}^*$ by $p \mapsto p^\chi := \prod |\chi_i(p)|^{s_i}$. Consequently, we have a natural logarithmic map $H_P : P(\mathbb{A}) \to \mathfrak{a}_P$ defined by

$$\langle H_P(p), \chi\rangle = p^\chi, \qquad \forall \chi \in \mathfrak{a}_P^*.$$

The kernel of H_P is denoted by $P(\mathbb{A})^1$ and we set $M_P(\mathbb{A})^1 := P(\mathbb{A})^1 \cap M_P(\mathbb{A})$.

Let also A_+ be the set of $a \in A_P(\mathbb{A})$ such that

(1) $a_v = 1$ for all finite places v of F; and

(2) $\chi(a_\sigma)$ is a positive number independent of infinite places σ of F for all $\chi \in X(M_P)$.

Then $M(\mathbb{A}) = A_+ \cdot M(\mathbb{A})^1$.

B.4.2.3. Roots, Coroots, Weights and Coweights

We now introduce standard bases for above spaces and their duals. Let Δ_0 and $\widehat{\Delta}_0$ be the subsets of simple roots and simple weights in $\mathfrak{a}_0^*$ respectively. (Recall that elements of $\widehat{\Delta}_0$ are non-negative linear combinations of elements in Δ_0.) Write $\Delta_0^\vee$ (resp. $\widehat{\Delta}_0^\vee$) for the basis of $\mathfrak{a}_0$ dual to $\widehat{\Delta}_0$ (resp. Δ_0). Being the dual of the collection of simple weights (resp. of simple roots), $\Delta_0^\vee$ (resp. $\widehat{\Delta}_0^\vee$) is the set of coroots (resp. coweights).

For every P, let $\Delta_P \subset \mathfrak{a}_0^*$ be the set of non-trivial *restrictions* of elements of Δ_0 to $\mathfrak{a}_P$. Denote the dual basis of Δ_P by $\widehat{\Delta}_P^\vee$. For each $\alpha \in \Delta_P$, let $\alpha^\vee$ be the projection of $\beta^\vee$ to $\mathfrak{a}_P$, where β is the root in Δ_0 whose restriction to $\mathfrak{a}_P$ is α. Set $\Delta_P^\vee := \Big\{\alpha^\vee : \alpha \in \Delta_P\Big\}$, and define the dual basis of $\Delta_P^\vee$ by $\widehat{\Delta}_P$.

More generally, if $Q \subset P$, write Δ_Q^P to denote the *subset* $\alpha \in \Delta_Q$ appearing in the action of T_Q in the unipotent radical of $Q \cap M_P$. (Indeed, $M_P \cap Q$ is a parabolic subgroup of M_P with nilpotent radical $N_Q^P := N_Q \cap M_P$. Thus Δ_Q^P is simply the set of roots of the parabolic subgroup $(M_P \cap Q, A_Q)$. And one checks that the map $P \mapsto \Delta_Q^P$ gives a natural bijection between parabolic subgroups P containing Q and subsets of Δ_Q.) Then $\mathfrak{a}_P$ is the subspace of $\mathfrak{a}_Q$ annihilated by Δ_Q^P. Denote by $(\widehat{\Delta}^\vee)_Q^P$ the dual of Δ_Q^P. Let $(\Delta_Q^P)^\vee := \left\{\alpha^\vee : \alpha \in \Delta_Q^P\right\}$ and denote by $\widehat{\Delta}_Q^P$ the dual of $(\Delta_Q^P)^\vee$.

B.4.2.4. Positive Cone and Positive Chamber

Let $Q \subset P$ be two parabolic subgroups of G. We extend the linear functionals in Δ_Q^P and $\widehat{\Delta}_Q^P$ to elements of the dual space $\mathfrak{a}_0^*$ by means of the canonical projection from $\mathfrak{a}_0$ to $\mathfrak{a}_Q^P$ given by the decomposition $\mathfrak{a}_0 = \mathfrak{a}_0^Q \oplus \mathfrak{a}_Q^P \oplus \mathfrak{a}_P$. Let τ_Q^P be the characteristic function of the *positive chamber*

$$\begin{aligned}&\left\{H \in \mathfrak{a}_0 : \langle \alpha, H\rangle > 0 \ \forall \alpha \in \Delta_Q^P\right\}\\ =&\,\mathfrak{a}_0^Q \oplus \left\{H \in \mathfrak{a}_Q^P : \langle \alpha, H\rangle > 0 \text{ for all } \alpha \in \Delta_Q^P\right\} \oplus \mathfrak{a}_P\end{aligned}$$

and let $\widehat{\tau}_Q^P$ be the characteristic function of the *positive cone*

$$\begin{aligned}&\left\{H \in \mathfrak{a}_0 : \langle \varpi, H\rangle > 0 \ \forall\ \varpi \in \widehat{\Delta}_Q^P\right\}\\ =&\,\mathfrak{a}_0^Q \oplus \left\{H \in \mathfrak{a}_Q^P : \langle \varpi, H\rangle > 0 \text{ for all } \varpi \in \widehat{\Delta}_Q^P\right\} \oplus \mathfrak{a}_P.\end{aligned}$$

Note that elements in $\widehat{\Delta}_Q^P$ are non-negative linear combinations of elements in Δ_Q^P, we have

$$\widehat{\tau}_Q^P \geq \tau_Q^P.$$

B.4.2.5. Partial Truncation and First Estimations

Denote τ_P^G and $\widehat{\tau}_P^G$ simply by τ_P and $\widehat{\tau}_P$.

Basic Estimation. (Arthur) *Suppose that we are given a parabolic subgroup P, and a Euclidean norm $\|\cdot\|$ on $\mathfrak{a}_P$. Then there are* *constants* c *and* N *such that for all* $x \in G(\mathbb{A})^1$ *and* $X \in \mathfrak{a}_P$,

$$\sum_{\delta \in P(F)\backslash G(F)} \widehat{\tau}_P\Big(H(\delta x) - X\Big) \leq c\Big(\|x\|e^{\|X\|}\Big)^N.$$

Moreover, *the sum is finite.*

As a direct consequence, we have the following

Corollary. Suppose that $T \in \mathfrak{a}_0$ and $N \geq 0$. Then *there exist constants* c' *and* N' *such that for any function* ϕ *on* $P(F)\backslash G(\mathbb{A})^1$, *and* $x, y \in G(\mathbb{A})^1$,

$$\sum_{\delta \in P(F)\backslash G(F)} \Big|\phi(\delta x)\Big| \cdot \widehat{\tau}_P\Big(H(\delta x) - H(y) - X\Big)$$

is bounded by

$$c'\|x\|^{N'} \cdot \|y\|^{N'} \cdot \sup_{u \in G(\mathbb{A})^1} \Big(|\phi(u)| \cdot \|u\|^{-N}\Big).$$

B.4.2.6. Langlands' Combinatorial Lemma

In this section, following Arthur, we recall Langlands' combinatorial lemma.

If $P_1 \subset P_2$, following Arthur [Ar2], set

$$\sigma_1^2(H) := \sigma_{P_1}^{P_2} := \sum_{P_3 : P_2 \supset P_2} (-1)^{\dim(A_3/A_2)} \tau_1^3(H) \cdot \hat{\tau}_3(H),$$

for $H \in \mathfrak{a}_0$. Then we have

Lemma 1. If $P_1 \subset P_2$, σ_1^2 *is a characteristic function of the subset of* $H \in \mathfrak{a}_1$ such that

(i) $\alpha(H) > 0$ for all $\alpha \in \Delta_1^2$;

(ii) $\sigma(H) \leq 0$ for all $\sigma \in \Delta_1 \backslash \Delta_1^2$; and

(iii) $\varpi(H) > 0$ for all $\varpi \in \hat{\Delta}_2$.

As a special case, with $P_1 = P_2$, we get the following important consequence:

Langlands' Combinatorial Lemma. *If $Q \subset P$ are parabolic subgroups, then for all $H \in \mathfrak{a}_0$,*

$$\sum_{R:Q\subset R\subset P} (-1)^{\dim(A_R/A_P)}\tau_Q^R(H)\hat{\tau}_R^P(H) = \delta_{QP};$$
$$\sum_{R:Q\subset R\subset P} (-1)^{\dim(A_Q/A_R)}\hat{\tau}_Q^R(H)\tau_R^P(H) = \delta_{QP}.$$

We next give an application of this combinatorial lemma, which will be used in 4.2.8. Suppose that $Q \subset P$ are parabolic subgroups. Fix a vector $\Lambda \in \mathfrak{a}_0^*$. Let

$$\varepsilon_Q^P(\Lambda) := (-1)^{\#\{\alpha\in\Delta_Q^P:\Lambda(\alpha^\vee)\leq 0\}},$$

and let

$$\phi_Q^P(\Lambda, H), \qquad H \in \mathfrak{a}_0,$$

be the characteristic function of the set

$$\Big\{H \in \mathfrak{a}_0 : \begin{matrix} \varpi(H) > 0, & \text{if } \Lambda(\alpha^\vee) \leq 0 \\ \varpi(H) \leq 0, & \text{if } \Lambda(\alpha^\vee) > 0 \end{matrix}, \forall \alpha \in \Delta_Q^P\Big\}.$$

Lemma 2. With the same notation as above,

$$\sum_{R:Q\subset R\subset P} \varepsilon_Q^R(\Lambda)\cdot\phi_Q^R(\Lambda,H)\cdot\tau_R^P(H) = \begin{cases} 0, & \text{if } \Lambda(\alpha^\vee) \leq 0,\ \exists \alpha \in \Delta_Q^P \\ 1, & \text{otherwise} \end{cases}.$$

B.4.2.7. Langlands-Arthur's Partition: Reduction Theory

Our aim is to derive Langlands-Arthur's partition of $G(F)\backslash G(\mathbb{A})$ into disjoint subsets, one for each (standard) parabolic subgroup.

To start with, suppose that ω is a compact subset of $N_0(\mathbb{A})M_0(\mathbb{A})^1$ and that $T_0 \in -\mathfrak{a}_0^+$. For any parabolic subgroup P_1, introduce the associated Siegel set $\mathfrak{s}^{P_1}(T_0,\omega)$ as the collection of

$$pak, \qquad p \in \omega,\ a \in A_0(\mathbb{R})^0,\ k \in K,$$

where $\alpha\Big(H_0(a) - T_0\Big)$ is positive for each $\alpha \in \Delta_0^1$. Then from classical reduction theory, we conclude that *for sufficiently big ω and sufficiently small T_0, $G(\mathbb{A}) = P_1(F)\mathfrak{s}^{P_1}(T_0,\omega)$.*

Suppose now that P_1 is given. Let $\mathfrak{s}^{P_1}(T_0,T,\omega)$ be the set of x in $\mathfrak{s}^{P_1}(T_0,\omega)$ such that $\varpi\Big(H_0(x)-T\Big)\leq 0$ for each $\varpi\in\hat{\Delta}_0^1$. Let $F^{P_1}(x,T):=F^1(x,T)$ be the characteristic function of the set of $x\in G(\mathbb{A})$ such that δx belongs to $\mathfrak{s}^{P_1}(T_0,T,\omega)$ for some $\delta\in P_1(F)$.

As such, $F^1(x,T)$ is left $A_1(\mathbb{R})^0N_1(\mathbb{A})M_1(F)$-invariant, and can be regarded as the characteristic function of the projection of the space $\mathfrak{s}^{P_1}(T_0,T,\omega)$ onto the space $A_1(\mathbb{R})^0N_1(\mathbb{A})M_1(F)\backslash G(\mathbb{A})$, a compact subset of $A_1(\mathbb{R})^0N_1(\mathbb{A})M_1(F)\backslash G(\mathbb{A})$.

For example, $F(x,T):=F^G(x,T)$ admits the following more direct description which will play a key role in our study of Arthur's periods:

If $P_1\subset P_2$ are (standard) parabolic subgroups, we write $A_1^\infty:=A_{P_1}^\infty$ for $A_{P_1}(\mathbb{A})^0$, the identity component of $A_{P_1}(\mathbb{R})$, and

$$A_{1,2}^\infty:=A_{P_1,P_2}^\infty:=A_{P_1}\cap M_{P_2}(\mathbb{A})^1.$$

Then the logarithmic map H_{P_1} maps $A_{1,2}^\infty$ isomorphically onto $\mathfrak{a}_1^2$, the orthogonal complement of $\mathfrak{a}_2$ in $\mathfrak{a}_1$. If T_0 and T are points in $\mathfrak{a}_0$, set $A_{1,2}^\infty(T_0,T)$ to be the set

$$\Big\{a\in A_{1,2}^\infty:\alpha\Big(H_1(a)-T\Big)>0,\ \alpha\in\Delta_1^2;\ \varpi\Big(H_1(a)-T\Big)<0,\ \varpi\in\hat{\Delta}_1^2\Big\},$$

where $\Delta_1^2:=\Delta_{P_1\cap M_2}$ and $\hat{\Delta}_1^2:=\hat{\Delta}_{P_1\cap M_2}$. In particular, for T_0 such that $-T_0$ is suitably regular, $F(x,T)$ *is the characteristic function of the compact subset of* $G(F)\backslash G(\mathbb{A})^1$ *obtained by projecting*

$$N_0(\mathbb{A})\cdot M_0(\mathbb{A})^1\cdot A_{P_0,G}^\infty(T_0,T)\cdot K$$

onto $G(F)\backslash G(\mathbb{A})^1$.

All in all, by Lemma 2 of 4.2.7, we arrive at the following

Arthur's Partition. ([We9]) *Fix P and let T be any suitably point in $T_0+\mathfrak{a}_0^+$. Then*

$$\sum_{P_1:P_0\subset P_1\subset P}\ \sum_{\delta\in P_1(F)\backslash G(F)}F^1(\delta x)\cdot\tau_1^P\Big(H_0(\delta x)-T\Big)=1\qquad\forall x\in G(\mathbb{A}).$$

B.4.2.8. Arthur's Analytic Truncation

Definition. Fix a suitably regular point $T \in \mathfrak{a}_0^+$. If ϕ is a continuous function on $G(F)\backslash G(\mathbb{A})^1$, define Arthur's analytic trunction $\Big(\Lambda^T\phi\Big)(x)$ to be the function

$$\Big(\Lambda^T\phi\Big)(x) := \sum_P (-1)^{\dim(A/Z)} \sum_{\delta\in P(F)\backslash G(F)} \phi_P(\delta x)\cdot\hat{\tau}_P\Big(H(\delta x)-T\Big),$$

where

$$\phi_P(x) := \int_{N(F)\backslash N(\mathbb{A})} \phi(nx)\,dn$$

denotes the constant term of ϕ along P, and the sum is over all (standard) parabolic subgroups.

The main purpose for introducing analytic truncation is to give a natural way to construct integrable functions: even from the example of GL_2, we know that automorphic forms are generally not integrable over the total fundamental domain $G(F)\backslash G(\mathbb{A})^1$ mainly due to the fact that in the Fourier expansions of such functions, constant terms are only of moderate growth (hence not integrable). Thus in order to naturally obtain integrable functions, we should truncate the original function along the cuspidal regions by removing constant terms. Simply put, Arthur's analytic truncation is a well-designed device in which constant terms are tackled in such a way that different levels of parabolic subgroups are suitably counted at the corresponding cuspidal region so that the whole truncation will not be overdone while there will be no parabolic subgroups left untackled.

Note that all parabolic subgroups of G can be obtained from standard parabolic subgroups by taking conjugations with elements from $P(F)\backslash G(F)$. So we have:

(a) $\Big(\Lambda^T\phi\Big)(x) = \sum_P (-1)^{\dim(A/Z)}\phi_P(x)\cdot\hat{\tau}_P\Big(H(x)-T\Big)$, *where the sum is over all, both standard and non-standard, parabolic subgroups*;

(b) *If ϕ is a cusp form, then $\Lambda^T\phi = \phi$*;

This is because by definition, all constant terms along proper $P : P \neq G$ are zero. Moreover, it is a direct consequence of the Basic

Estimation for partial truncation, (see e.g. the Corollary there), we have

(c) *If ϕ is of moderate growth* in the sense that there exist some constants C, N such that $|\phi(x)| \leq c\|x\|^N$ for all $x \in G(\mathbb{A})$, *then so is* $\Lambda^T\phi$.

Recall that an element $T \in \mathfrak{a}_0^+$ is called *sufficiently regular*, if for any $\alpha \in \Delta_0$, $\alpha(T) \gg 0$. Fundamental properties of Arthur's analytic truncation may be summarized as follows:

Theorem. ([Ar2,3]) *For sufficiently regular T in $\mathfrak{a}_0$,*
(1) *Let $\phi : G(F)\backslash G(\mathbb{A}) \to \mathbb{C}$ be a locally L^1 function. Then*

$$\Lambda^T\Lambda^T\phi(g) = \Lambda^T\phi(g)$$

for almost all g. If ϕ is also locally bounded, then the above is true for all g;
(2) *Let ϕ_1, ϕ_2 be two locally L^1 functions on $G(F)\backslash G(\mathbb{A})$. Suppose that ϕ_1 is of moderate growth and ϕ_2 is rapidly decreasing. Then*

$$\int_{Z_{G(\mathbb{A})}G(F)\backslash G(\mathbb{A})} \overline{\Lambda^T\phi_1(g)}\cdot\phi_2(g)\, dg = \int_{Z_{G(\mathbb{A})}G(F)\backslash G(\mathbb{A})} \overline{\phi_1(g)}\cdot\Lambda^T\phi_2(g)\, dg;$$

(3) *Let K_f be an open compact subgroup of $G(\mathbb{A}_f)$, and r, r' are two positive real numbers. Then there exists a finite subset $\Big\{X_i : i = 1, 2, \ldots, N\Big\} \subset \mathcal{U}$, the universal enveloping algebra of $\mathfrak{g}_\infty$, such that the following is satisfied: Let ϕ be a smooth function on $G(F)\backslash G(\mathbb{A})$, right invariant under K_f and let $a \in A_{G(\mathbb{A})}$, $g \in G(\mathbb{A})^1 \cap S$. Then*

$$\Big|\Lambda^T\phi(ag)\Big| \leq \|g\|^{-r} \sum_{i=1}^{N} \sup\Big\{|\delta(X_i)\phi(ag')|\,\|g'\|^{-r'} : g' \in G(\mathbb{A})^1\Big\},$$

where S is a Siegel domain with respect to $G(F)\backslash G(\mathbb{A})$; and
(4) *The function $\Lambda^T\mathbf{1}$ is a characteristic function of a compact subset of $G(F)\backslash G(\mathbb{A})^1$.*

The key to the proof is that nilpotent groups have simple structure. Say, for (1-3), while a bit complicated, it is based on the discussion in 4.2.6-8. As for (4), which has been considered to be only secondary, please see 4.3 where it plays a central role.

B.4.3. Arthur's Periods

B.4.3.1. Arthur's Period

Fix a sufficiently regular $T \in \mathfrak{a}_0$ and let ϕ be an automorphic form of G. Then by Theorem 4.2.9, Λ^T is rapidly decreasing, and hence integrable. Hence, the integration

$$A(\phi;T) := \int_{G(F)\backslash G(\mathbb{A})} \Lambda^T \phi(g)\, dg$$

makes sense. We claim that $A(\phi;T)$ can be written as an integration of the original automorphic form ϕ over a certain compact subset.

To start with, note that for Arthur's analytic truncation Λ^T, we have $\Lambda^T \circ \Lambda^T = \Lambda^T$. Hence,

$$\begin{aligned} A(\phi;T) &= \int_{Z_{G(\mathbb{A})}G(F)\backslash G(\mathbb{A})} \Lambda^T \phi\; d\mu(g) \\ &= \int_{Z_{G(\mathbb{A})}G(F)\backslash G(\mathbb{A})} \Lambda^T \Big(\Lambda^T \phi\Big)(g)\; d\mu(g). \end{aligned}$$

Moreover, by the self-adjoint property, for the constant function $\mathbf{1}$ on $G(\mathbb{A})$,

$$\begin{aligned} &\int_{Z_{G(\mathbb{A})}G(F)\backslash G(\mathbb{A})} \mathbf{1}(g)\cdot \Lambda^T \Big(\Lambda^T \phi\Big)(g)\; d\mu(g) \\ &= \int_{Z_{G(\mathbb{A})}G(F)\backslash G(\mathbb{A})} \Big(\Lambda^T \mathbf{1}\Big)(g)\cdot \Big(\Lambda^T \phi\Big)(g)\; d\mu(g) \\ &= \int_{Z_{G(\mathbb{A})}G(F)\backslash G(\mathbb{A})} \Lambda^T \Big(\Lambda^T \mathbf{1}\Big)(g)\cdot \phi(g)\; d\mu(g), \end{aligned}$$

since $\Lambda^T \phi$ and $\Lambda^T \mathbf{1}$ are rapidly decreasing. Therefore, using $\Lambda^T \circ \Lambda^T = \Lambda^T$ again, we arrive at

$$A(\phi;T) = \int_{Z_{G(\mathbb{A})}G(F)\backslash G(\mathbb{A})} \Lambda^T \mathbf{1}(g)\cdot \phi(g)\; d\mu(g). \qquad (*)$$

To go further, let us give a much more detailed study of Authur's analytic truncation for the constant function $\mathbf{1}$. Introduce the truncated subset $\Sigma(T) := \Big(Z_{G(\mathbb{A})}G(F)\backslash G(\mathbb{A})\Big)_T$ of the space $G(F)\backslash G(\mathbb{A})^1$ by $\Sigma(T) :=$

$$\Big(Z_{G(\mathbb{A})}G(F)\backslash G(\mathbb{A})\Big)_T := \Big\{g \in Z_{G(\mathbb{A})}G(F)\backslash G(\mathbb{A}) : \Lambda^T\mathbf{1}(g) = 1\Big\}.$$

We claim that $\Sigma(T)$ or the same $\Big(Z_{G(\mathbb{A})}G(F)\backslash G(\mathbb{A})\Big)_T$, is compact. In fact, much stronger result is correct. Namely, we have the following

Basic Fact. ([Ar4]) *For sufficiently regular* $T \in \mathfrak{a}_0^+$, $\Lambda^T\mathbf{1}(x) = F(x,T)$. *That is to say,* $\Lambda^1\mathbf{1}$ *is the characteristic function of the compact subset* $\Sigma(T)$ *of* $G(F)\backslash G(\mathbb{A})^1$ *obtained by projecting* $N_0(\mathbb{A}) \cdot M_0(\mathbb{A})^1 \cdot A_{P_0,G}^{\infty}(T_0,T) \cdot K$ *onto* $G(F)\backslash G(\mathbb{A})^1$.

Thus, by the equation $(*)$ and the Basic Fact above,

$$\begin{aligned}&\int_{Z_{G(\mathbb{A})}G(F)\backslash G(\mathbb{A})} \Lambda^T\phi(g)\, d\mu(g)\\ &= \int_{Z_{G(\mathbb{A})}G(F)\backslash G(\mathbb{A})} \Lambda^T\mathbf{1}(g)\cdot\phi(g)\, d\mu(g)\\ &= \int_{\Sigma(T)} \phi(g)\, d\mu(g).\end{aligned}$$

That is to say, we have obtained the following very beautiful relation:

Arthur's Periods. *For a sufficiently regular* $T \in \mathfrak{a}_0$, *and an automorphic form* ϕ *on* $G(F)\backslash G(\mathbb{A})$,

$$\int_{\Sigma(T)} \phi(g)\, d\mu(g) = \int_{G(F)\backslash G(\mathbb{A})^1} \Lambda^T\phi(g)\, d\mu(g).$$

The importance of this fundamental relation can hardly be overestimated. Say, when taking ϕ to be Eisenstein series associated with an L^2-automorphic forms,

(i) (*Analytic Evaluation for RHS*) The right hand side can be evaluated using certain analytic methods. For example, in the case for

Eisenstein series associated to cusp forms, as done in the next chapter, the RHS can be evaluated in terms of integrations of cusp forms twisted by intertwining operators, which normally are close related with L-functions;
(ii) (*Geometric Evaluation for LHS*) The left hand, by taking the residue (for suitable Eisenstein series), can be evaluated using geometric methods totally independently.
A good example is given in [KW], see also the end of B.4 where certain new yet fundamental relations among special values of zeta functions are exposed.

B.4.3.2. $\Lambda^1 1$ is a Characteristic Function

The proof of the Basic Fact is based on Arthur's partition for $G(F)\backslash G(\mathbb{A})$ and an inversion formula, which itself is a direct consequence of Langlands' Combinatorial lemma.

Recall that for $T \in \mathfrak{a}_0$, level *P-Arthur's analytic truncation* of ϕ is defined by the formula

$$\Lambda^{T,P}\phi(g) := \sum_{R:R\subset P} (-1)^{d(R)-d(P)} \sum_{\delta\in R(F)\backslash P(F)} \phi_R(\delta g)\cdot\hat{\tau}_R^P\Big(H(\delta g)-T\Big).$$

Then Λ^T stands simply for $\Lambda^{T,G}$. Thus, together with Langlands' combinatorial lemma, we have the following:

(a) $\displaystyle \Lambda^T\phi(g) := \sum_R (-1)^{d(R)-d(G)} \sum_{\delta\in R(F)\backslash G(F)} \phi_R(\delta g)\cdot\hat{\tau}_R\Big(H(\delta g)-T\Big);$

and

(b) (**Inversion Formula**) *For a $G(F)$-invariant function ϕ,*

$$\phi(g) = \sum_P \sum_{\delta\in P(F)\backslash G(F)} \Lambda^{T,P}\phi(\delta g)\cdot\tau_P\Big(H(\delta g)-T\Big).$$

(c) *Assume that $T \in \mathcal{C}_P$ and ϕ is automorphic on $N(\mathbb{A})M(F)\backslash G(\mathbb{A})$. Then $\Lambda^{T,P}\phi(m)$ is rapidly decreasing on $M(F)\backslash M(\mathbb{A})^1$. In particular, the integration $\int_{M(F)\backslash M(\mathbb{A})^1}\Lambda_{T,P}\phi(m)\,dm$ is well-defined.*

Proof of the Basic Fact. ([Ar4]) By Arthur's partition for $G(F)\backslash G(\mathbb{A})$, we have

$$\sum_{P}\sum_{\delta\in P(F)\backslash G(F)} F^P\Big(\delta x, T\Big)\cdot\tau_P\Big(H_P(\delta x)-T\Big)=1$$

where τ_P is the characteristic function of

$$\Big\{H\in\mathfrak{a}_0:\alpha(H)>0,\ \alpha\in\Delta_P\Big\}$$

and

$$F^P\Big(nmk, T\Big)=F^{M_P}\Big(m,T\Big),\qquad n\in N_P(\mathbb{A}), m\in M_P(\mathbb{A}), k\in K.$$

On the other hand, by the inversion formula, applying to the constant function $\mathbf{1}$, we have

$$\sum_{P}\sum_{\delta\in P(F)\backslash G(F)} \Big(\Lambda^{T,P}\mathbf{1}\Big)(\delta x)\cdot\tau_P\Big(H_P(\delta x)-T\Big)=1,$$

where $\Lambda^{T,P}$ is the partial analytic truncation operator defined above. With this, the desired result is immediately obtained by induction.

B.4.4. Geometric and Analytic Truncations: A Bridge

B.4.4.1. A Micro Bridge

For simplicity, we in this subsection work only with the field of rationals $\mathbb{Q}$ and use mixed languages of adeles and lattices. Also, without loss of generality, we assume that $\mathbb{Z}$-lattices are of volume one. Accordingly, set $G=SL_r$.

For a rank r lattice Λ of volume one, denote the sublattices filtration associated to a parabolic subgroup P by

$$0=\Lambda_0\subset\Lambda_1\subset\Lambda_2\subset\cdots\subset\Lambda_{|P|}=\Lambda.$$

Assume that P corresponds to the partition $I=(d_1,d_2,\ldots,d_{|P|})$. A polygon $p:[0,r]\to\mathbb{R}$ is called *normalized* if $p(0)=p(r)=0$. For a (normalized) polygon $p:[0,r]\to\mathbb{R}$, define the associated (real) character $T=T(p)$ of M_0 by the condition that

$$\alpha_i(T)=\Big[p(i)-p(i-1)\Big]-\Big[p(i+1)-p(i)\Big]$$

for all $i = 1, 2, \ldots, r-1$. Then one checks that $T(p) =$

$$\Big(p(1), p(2)-p(1), \ldots, p(i)-p(i-1), \ldots, p(r-1)-p(r-2), -p(r-1)\Big).$$

Now take $g = g(\Lambda) \in G(\mathbb{A})$. Denote its lattice by Λ^g, and its induced filtration from P by

$$0 = \Lambda_0^{g,P} \subset \Lambda_1^{g,P} \subset \cdots \subset \Lambda_{|P|}^{g,P} = \Lambda^g.$$

Consequently, the polygon $p_P^g = p_P^{\Lambda^g} : [0, r] \to \mathbb{R}$ is characterized by
(1) $p_P^g(0) = p_P^g(r) = 0$;
(2) p_P^g is affine on $[r_i, r_{i+1}]$, $i = 1, 2, \ldots, |P|-1$; and
(3) $p_P^g(r_i) = \deg\big(\Lambda_i^{g,P}\big) - r_i \cdot \frac{\deg(\Lambda^g)}{r}$, $i = 1, 2, \ldots, |P|-1$.
Note that the volume of Λ is assumed to be one, therefore (3) is equivalent to
(3)′ $p_P^g(r_i) = \deg\big(\Lambda_i^{g,P}\big)$, $i = 1, 2, \ldots, |P|-1$.

The advantage of partially using adelic language is that the values of p_P^g may be written down precisely. Indeed, using Langlands decomposition $g = n \cdot m \cdot a(g) \cdot k$ with $n \in N_P(\mathbb{A}), m \in M_P(\mathbb{A})^1, a \in A_+$ and $k \in K := \prod_p SL(\mathcal{O}_{\mathbb{Q}_p}) \times SO(r)$. Write

$$a = a(g) = \mathrm{diag}\Big(a_1 I_{d_1}, a_2 I_{d_2}, \ldots, a_{|P|} I_{d_{|P|}}\Big)$$

where $r = d_1 + d_2 + \cdots + d_{|P|}$ is the partition corresponding to P. Then it is a standard fact that

$$\deg\Big(\Lambda_i^{g,P}\Big) = -\log\Big(\prod_{j=1}^{i} a_j^{d_j}\Big) = -\sum_{j=1}^{i} d_j \log a_j, \qquad i = 1, \ldots, |P|.$$

Set now $\mathbf{1}(p_P^* >_P p)$ to be the characteristic function of the subset of g's such that $p_P^g >_P p$.

Micro Bridge. *For a fixed convex normalized polygon $p : [0, r] \to \mathbb{R}$, and $g \in SL_r(\mathbb{A})$, with respect to any parabolic subgroup P, we have*

$$\hat{\tau}_P\Big(-H_0(g) - T(p)\Big) = \mathbf{1}\Big(p_P^g >_P p\Big).$$

B.4.4.2. Global Bridge between Geometric and Analytic Truncations

With the micro bridge above, now we are ready to expose a beautiful intrinsic relation between algebraic and analytic truncations.

Global Bridge. *For a fixed normalized convex polygon $p : [0, r] \to \mathbb{R}$, let $T(p) :=$*

$$\Big(p(1), p(2)-p(1), \dots, p(i)-p(i-1), \dots, p(r-1)-p(r-2), -p(r-1)\Big)$$

be the associated vector in $\mathfrak{a}_0$. If $T(p)$ is sufficiently positive, then

$$\mathbf{1}(\bar{p}^g \leq p) = \Big(\Lambda^{T(p)}\mathbf{1}\Big)(g).$$

B.5. Rankin-Selberg Method

Global bridge between geometric and analytic truncations in B.4 shows that at least for the polygons p whose associated $T(p) \in \mathfrak{a}_0$ are sufficiently positive, the corresponding non-abelian L-functions may be viewed as a special kinds of Arthur's periods. Therefore, to understand our non-abelian Ls, it is quite helpful to understand the structure of Arthur's periods.

However, to write down them precisely, Arthur's periods for automorphic forms seems to be too general. It is for this purpose that we then narrow our class by considering Eisenstein periods: By definition, Eisenstein periods are integrations of Eisenstein series over truncated compact domains obtained for a sufficiently regular $T \in \mathfrak{a}_0$. Simply put, the aim of this chapter is to show that a special kind of Eisenstein periods and hence also certain classes of non-abelian Ls can be evaluated via an advanced version of Rankin-Selberg method.

B.5.1. Regularized Integration: Integration over Cones

To begin with, we here give a review of Jacquet-Lapid-Rogawski's beautiful yet elementary treatment of integrations over cones.

Let V be a real r dimensional vector space. Let V^* be the space of complex linear forms on V. Denote by $S(V^*)$ the symmetric algebra of V^*, which may be regarded as the space of polynomial functions on V as well. By definition, an *exponential polynomial* function on V is a function of the form

$$f(x) = \sum_{i=1}^{r} e^{\langle \lambda_i, x \rangle} P_i(x)$$

where the λ_i are distinct elements of V^* and the $P_i(x)$ are non-zero elements of $S(V^*)$. One checks easily that
(1) *The λ_i are uniquely determined and called the exponents of f.*

By a *cone* in V we shall mean a closed subset of the form

$$\mathcal{C} := \Big\{ x \in V : \langle \mu_i, x \rangle \geq 0 \ \forall i \Big\}$$

where $\{\mu_i\}$ is a basis of $\mathrm{Re}(V^*)$, the space of *real* linear forms on V. Let $\{e_j\}$ be the dual basis of V. We shall say that $\lambda \in V^*$ is *negative* (resp. *non-degenerate*) with respect to $\mathcal{C}$ if $\mathrm{Re}\,\langle \lambda, e_j \rangle < 0$ (resp. $\langle \lambda, e_j \rangle \neq 0$) for each $j = 1, \ldots, r$.
(2) *The function $f(s) = \sum_{i=1}^{r} e^{\langle \lambda_i, x \rangle} P_i(x)$ is integrable over $\mathcal{C}$ if and only if λ_i are negative with respect to $\mathcal{C}$ for all i.*

To define regularized integrals over $\mathcal{C}$, we study the integral

$$I_{\mathcal{C}}\Big(f; \lambda\Big) := \int_{\mathcal{C}} f(x) e^{\langle \lambda, x \rangle} dx.$$

It converges absolutely for λ in the open set

$$\Big\{ \lambda \in V^* : \mathrm{Re}\langle \lambda_i - \lambda, e_j \rangle < 0 \ \forall 1 \leq i \leq n, 1 \leq j \leq r \Big\},$$

and can be analytically continued as follows:
In the case when $\mathrm{Re}(\lambda) > 0$ and P is an polynomial of one variable,

$$\int_0^\infty e^{-\lambda x} P(x) dx = \sum_{m \geq 0} \frac{\Big(D^m P\Big)(0)}{\lambda^{m+1}}.$$

We then can use the right hand side to obtain the continuation.

To deal with more general case, fix an index k and let $\mathcal{C}_k$ be the intersection of $\mathcal{C}$ with the hyperplane $V_k := \Big\{x : \langle \mu_k, x\rangle = 0\Big\}$. Then, we can write $x = \langle \mu_k, x\rangle e_k + y$ with $y \in V_k$. Thus, for a suitable choice of Haar measures,

$$I_{\mathcal{C}}\Big(f;\lambda\Big) = \sum_{i=1}^{n}\sum_{m\geq 0} \frac{1}{\Big(\langle\lambda - \lambda_i, e_k\rangle\Big)^{m+1}} \int_{\mathcal{C}_k} e^{-\langle\lambda-\lambda_i, y\rangle}\Big(D_{e_k}^m\Big)P_i(y)\,dy.$$

This formula then gives the analytic continuation of $I_{\mathcal{C}}\Big(f;\lambda\Big)$ to the tube domain defined by $\mathrm{Re}\langle\lambda_i - \lambda, e_k\rangle > 0$ for $j \neq k$, $1 \leq i \leq n$ with singularities on hyperplanes $H_{k,i} := \Big\{\lambda : \langle\lambda, e_k\rangle = \langle\lambda_i, e_k\rangle\Big\}, 1 \leq i \leq n$. Moreover, since this is true for all $k = 1, 2, \ldots, r$, the function $I_{\mathcal{C}}\Big(f;\lambda\Big)$ has analytic continuation to V^* by Hartogs' lemma with (only) hyperplane singularities along $H_{i,k}$, $1 \leq k, \leq r, 1 \leq i \leq r$. In particular, from such an induction, one checkes

(3) *Suppose that f is absolute integrable over $\mathcal{C}$. Then $I_{\mathcal{C}}\Big(f;\lambda\Big)$ is holomorphic at 0 and $I_{\mathcal{C}}\Big(f;0\Big) = \int_{\mathcal{C}} f(x)\,dx$;* and

(4) *The function $I_{\mathcal{C}}\Big(f;\lambda\Big)$ is holomorphic at 0 if and only if for all i, λ_i are non-degenerate with respect to $\mathcal{C}$, i.e.,$\langle\lambda_i, e_k\rangle \neq 0$ for all pairs (i, k).*

Denote the characteristic function of a set Y by τ^Y. For $\lambda \in V^*$ such that $\lambda_i - \lambda$ are negative with respect to $\mathcal{C}$ for all $i = 1, 2, \ldots, n$, set

$$\begin{aligned}\widehat{F}\Big(\lambda;\mathcal{C},T\Big) &:= \int_V f(x)\cdot\tau^{\mathcal{C}}(x-T)\cdot e^{-\langle\lambda,x\rangle}dx\\ &= \sum_{i=1}^{n}\int_V P_i(x)\tau^{\mathcal{C}}(x-T)e^{-\langle\lambda-\lambda_i,x\rangle}dx\\ &= \sum_{i=1}^{n} e^{-\langle\lambda_i-\lambda,T\rangle} I_{\mathcal{C}}\Big(P_i(*+T)e^{\langle\lambda_i,*\rangle};\lambda\Big).\end{aligned}$$

The integrals are absolutely convergent and $\widehat{F}\Big(\lambda;\mathcal{C},T\Big)$ extends to a meromorphic function on V^*. With this, following [JLR], we call the function $f(x) \cdot \tau^{\mathcal{C}}(x-T)$ *#-integrable* if $\widehat{F}\Big(\lambda; \mathcal{C}, T\Big)$ is holomorphic at $\lambda = 0$. In this case, set

$$\int_V^{\#} f(x) \cdot \tau^{\mathcal{C}}(x-T) dx := \widehat{F}\Big(0; \mathcal{C}, T\Big).$$

By (4), the #-integral exists if and only if each exponent λ_j is non-degenerate with respect to $\mathcal{C}$.

To go further, suppose that $V = W_1 \oplus W_2$ is a decomposition of V as a direct sum, and let $\mathcal{C}_j$ be a cone in W_j. Write $T = T_1 + T_2$ and $x = x_1 + x_2$ relative to this decomposition. If the exponents λ_j of f are non-degenerate with respect to $\mathcal{C} = \mathcal{C}_1 + \mathcal{C}_2$, then the function

$$w_2 \mapsto \int_{W_1}^{\#} f(w_1 + w_2) \tau_1^{\mathcal{C}_1}(w_1 - T_1)\, dw_1$$

is defined and is an exponential polynomial. And it follows by analytic continuation that

$$(5) \int_V^{\#} f(x) \tau^{\mathcal{C}}(x-T)\, dx$$

$$= \int_{W_2}^{\#} \left(\int_{W_1}^{\#} f(w_1 + w_2) \tau_1^{\mathcal{C}_1}(w_1 - T_1)\, dw_1 \right) \cdot \tau_2^{\mathcal{C}_2}(w_2 - T_2)\, dw_2;$$

and as a function of T, it is an exponential polynomial with the same exponents as f.

More generally, let $g(x)$ be a compactly supported function on W_1 and consider functions of the form $g(w_1 - t_1)\tau_{\mathcal{C}}(w_2 - T_2)$, which we call functions of *type (C)*. It is clear that the integral

$$\widehat{F}(\lambda) := \int_V f(x) \cdot g(w_1 - t_1) \tau_{\mathcal{C}}(w_2 - T_2) \cdot e^{-\langle \lambda, x \rangle}\, dx$$

converges absolutely for an open set of λ whose restriction to W_2 is negative with respect to $\mathcal{C}_2$. Furthermore, $\widehat{F}(\lambda)$ has a meromorphic continuation to V^*. Following JLR, the function $f(x)g(w_1 -$

$t_1)\tau_{\mathcal{C}}(w_2-T_2)$ is called *#-integrable* if $\widehat{F}(\lambda)$ is holomorphic at $\lambda=0$. If so we define

$$\int_V^{\#} f(x)\cdot g(w_1-t_1)\tau_{\mathcal{C}}(w_2-T_2)\,dx := \widehat{F}(0).$$

Note that as a function of w_2, $\int_{W_1} f(w_1+w_2)g(w_1-T_1)\,dw_1$ is an exponential polynomial on W_2, hence

$$\begin{aligned}&\int_V^{\#} f(x)\,g(w_1-t_1)\tau_{\mathcal{C}}(w_2-T_2)\,dx\\ &=\int_{W_2}^{\#}\left(\int_{W_1} f(w_1+w_2)g(w_1-T_1)\,dw_1\right)\tau_2^{\mathcal{C}}(w_2-T_2)\,dw_2.\end{aligned}$$

We end this discussion by the following explicit formulas:

$$(6)\ \int_V^{\#} e^{\langle\lambda,x\rangle}\tau^{\mathcal{C}}(x-T)\,dx = (-1)^r\mathrm{Vol}\Big(e_1,e_2,\dots,e_r\Big)\cdot\frac{e^{\langle\lambda,T\rangle}}{\prod_{j=1}^r\langle\lambda,e_j\rangle},$$

where $\mathcal{C}:=\Big\{\sum_{j=1}^r a_je_j\,:\,a_j\geq 0\Big\}$ *and* $\mathrm{Vol}\Big(e_1,e_2,\dots,e_r\Big)$ *is the volume of the associated parallelepiped* $\Big\{\sum_{j=1}^r a_je_j : a_j\in[0,1]\Big\}$.

In particular, if $V=\mathbb{R}$, we have $\displaystyle\int_T^{\infty} e^{\lambda t}dt = -\frac{e^{\lambda T}}{\lambda}$ and hence for any cone $\mathcal{C}\subset\mathbb{R}$,

$$\int_V^{\#} e^{\langle\lambda,x\rangle}\Big(1-\tau^{\mathcal{C}}\big(x-T\big)\Big)\,dx = -\int_V^{\#} e^{\langle\lambda,x\rangle}\tau^{\mathcal{C}}(x-T)\,dx$$

since $1-\tau^{\mathcal{C}}$ is the characteristic function of the cone $-\mathcal{C}$.

B.5.2. JLR Periods of Automorphic Forms

Now let us follow [JLR] to apply the above discussion on integration over cones to the theory of periods of automorphic forms.

Let $\tau_k(X)$ be a function of type (C) on $\mathfrak{a}_P$ that depends continuously on $k\in K$, i.e., we assume that there is a decomposition $\mathfrak{a}_P=W_1\oplus W_2$ such that $\widehat{\tau}_k$ has the form

$$g_k(w_1-T_1)\tau_{\mathcal{C}_{2k}}(w_2-T_2)$$

where the compactly supported function g_k varies continuously in the L^1-norm and linear inequalities defining the cone $\mathcal{C}_{2k}$ vary continuously.

Also with the application to automorphic forms in mind, let f be a function on $M(F)N(\mathbb{A})\backslash G(\mathbb{A})$ of the form

$$f(namk) = \sum_{j=1}^{s} \phi_j(m,k) \cdot \alpha_j\Big(H_P(a), k\Big) e^{\langle \lambda_j + \rho_P, H_P(a)\rangle}$$

for $n \in N(\mathbb{A}), a \in A_P, m \in M(\mathbb{A})^1$ and $k \in K$, where, for all j,
(a) $\phi_j(m,k)$ is absolutely integrable on $M(F)\backslash M(\mathbb{A})^1 \times K$; and
(b) $\lambda_j \in \mathfrak{a}_P^*$ and $\alpha_j(X,k)$ is a continuous family of polynomials on $\mathfrak{a}_P$ such that, for all $k \in K$, $\alpha_J(X,k)e^{\langle \lambda_j, X\rangle}\tau_k(X)$ is #-integrable.

As such, following [JLR], define the *#-integral*

$$\int_{P(F)\backslash G(\mathbb{A})}^{\#} f(g) \cdot \tau_k\Big(H_P(g)\Big)\, dg :=$$
$$\sum_{j=1}^{s} \int_K \left(\int_{M(F)\backslash M(\mathbb{A})^1} \phi_j(mk)\, dm \right) \cdot \left(\int_{\mathfrak{a}_P}^{\#} \alpha_j(X,k) \tau_k(X)\, dX \right) dk.$$

Remark. The reader should notice that in this definition there is no integration involving $\int_{N(F)\backslash N(\mathbb{A})} dn$: In practice, f is supposed to be the constant term along P, so is $N(\mathbb{A})$-invariant.

Clearly, the set $\mathcal{E}_P(f)$ of distinct exponents $\{\lambda_j\}$ of f is uniquely determined by f, but the functions ϕ_j and α_j are not. So we need to show that the #-integral is independent of the choices of these functions. Clearly, the function

$$X \mapsto \int_{M(F)\backslash M(\mathbb{A})^1} f(e^X mk)\, dm$$

is an exponential polynomial on $\mathfrak{a}_P$ and the above #-integral is equal to

$$\int_K \int_{\mathfrak{a}_P}^{\#} \left(\int_{M(F)\backslash M(\mathbb{A})^1} f(e^X mk)\, dm \right) e^{-\langle \rho_P, X\rangle} \tau_k(X)\, dX\, dk.$$

This shows in particular that the original #-integral is independent of the decomposition of f as used in the definition. Moreover, if each of the exponents λ_j is negative with respect to $\mathcal{C}_{2k}$ for all $k \in K$, then the ordinary integral

$$\int_{P(F)\backslash G(\mathbb{A})^1} f(g) \cdot \tau_k(H_P(g))\, dg$$

is absolutely convergent and equals to the #-integral by Basic Property (3) of integration over cones. That is to say, we have the following

Lemma. (1) *The above #-integral* $\int_{P(F)\backslash G(\mathbb{A})}^{\#} f(g) \cdot \tau_k\Big(H_P(g)\Big)\, dg$ *is well-defined;* and
(2) *If each of the exponents λ_j of f is negative with respect to $\mathcal{C}_{2k}$ for all $k \in K$, then the ordinary integral*

$$\int_{P(F)\backslash G(\mathbb{A})^1} f(g) \cdot \tau_k\Big(H_P(g)\Big)\, dg$$

is absolutely convergent and equals to the #-integral.

Fix a sufficiently regular element $T \in \mathfrak{a}_0^+$. Then the above construction applies to Arthur's analytic truncation $\Lambda^{T,P}\Psi(g)$ and the **characteristic function** $\tau_P\Big(H_P(g) - T\Big)$ where $\Psi \in \mathcal{A}_P(G)$, the space of level P automorphic forms. Indeed, with

$$\Psi(namk) = \sum_{j=1}^{s} Q_j\Big(H_P(a)\Big) \cdot \psi_j(amk)$$

for $n \in N(\mathbb{A}), a \in A_P, m \in M(\mathbb{A})^1$ and $k \in K$, where Q_j are polynomials and $\psi_j \in \mathcal{A}_P(G)$ satisfies

$$\psi_j(ag) = e^{\langle \lambda_j + \rho_P, H_P(a)\rangle} \psi_j(g)$$

for some $\lambda_j \in \mathfrak{a}_P^*$ and all $a \in A_P$, $\int_{M(F)\backslash M(\mathbb{A})^1 \times K} \Lambda^{T,P}\psi_j(mk) dm\, dk$ is well-defined, since the function $m \mapsto \Lambda^{T,P}\psi_j(mk)$ is rapidly decreasing on $M(F)\backslash M(\mathbb{A})^1 \times K$. Now τ_P is the characteristic function of the cone spanned by the coweights $\widehat{\Delta}_P^\vee$, so

$$\int_{P(F)\backslash G(\mathbb{A})^1}^{\#} \Lambda^{T,P}\Psi(g)\cdot\tau_P\Big(H_P(g)-T\Big)\,dg$$

exists if and only if

$$\langle\lambda_j,\varpi^\vee\rangle\neq 0\qquad \forall\varpi\in\widehat{\Delta}_P^\vee\quad\text{and}\quad\lambda_j\in\mathcal{E}_P(\Psi).$$

The same is true for $\Lambda^{T,P}\Psi(g)\cdot\tau_P(H_P(gx)-T)$ for $x\in G(\mathbb{A})$. Indeed, for $x\in G(\mathbb{A})$, let $K(g)\in K$ be an element such that $gK(g)^{-1}\in P_0(\mathbb{A})$, then $H_P(gx)=H_P(g)+H_P(K(g)x)$ and hence $\tau_P\Big(X-T-H_P(K(g)x)\Big)$ is the characteristic function of a cone depending continuously on g. Set then

$$I_G^{P,T}(\Psi):=\int_{P(F)\backslash G(\mathbb{A})^1}^{\#} \Lambda^{T,P}\Psi(g)\cdot\tau_P\Big(H_P(g)-T\Big)\,dg.$$

For any automorphic form $\phi\in\mathcal{A}(G)$, write $\mathcal{E}_P(\phi)$ for the set of exponents $\mathcal{E}_P(\phi_P)$. Set $\mathcal{A}(G)^*:=$

$$\Big\{\phi\in\mathcal{A}(G):\langle\lambda,\varpi^\vee\rangle\neq 0\qquad\forall\varpi^\vee\in\widehat{\Delta}_P^\vee,\ \lambda\in\mathcal{E}_P(\phi),\ P\neq G\Big\}.$$

If $\phi\in\mathcal{A}(G)^*$, then $I_{P,T}^G(\phi_P)$ exists for all P; and define JLR's *regularized period* by

$$\int_{G(F)\backslash G(\mathbb{A})}^{*}\phi(g)\,dg=\sum_P I_G^{P,T}(\phi_P):=I_G^T(\phi).$$

In the follows, let us expose some of the basic properties of JLR's regularized periods. For this, we make the following preparation. As above, for $x\in G(\mathbb{A}_f)$, let $\rho(x)$ denote right translation by x: $\rho(x)\phi(g)=\phi(gx)$. The space $\mathcal{A}(G)$ is stable under right translation by $G(\mathbb{A}_f)$. Furthermore, $\rho(x)\phi$ has the same set of exponents as ϕ. Consequently, the space $\mathcal{A}(G)_*$ is invariant under right translation by $G(\mathbb{A}_f)$. Indeed, for $k \in K$, write the Iwasawa decomposition of kx as $kx = n'a'm'K(kx)$ and write ϕ_P in the form

$$\phi_P\Big(namk\Big) = \sum_j Q_j\Big(H_P(a)\Big) \cdot e^{\langle \lambda_j + \rho_P, H(a)\rangle} \phi_j(mk).$$

Since $amkx = n^*aa'mm'K(kx)$ for some $n^* \in N(\mathbb{A})$, we have

$$\begin{aligned} &\phi_P\Big(namk \cdot x\Big) \\ =&\sum_j Q_j\Big(H_P(a) + H_P(a')\Big) \cdot e^{\langle \lambda_j + \rho_P, H(a) + H(a')\rangle} \phi_j\Big(mm'K(kx)\Big). \end{aligned}$$

This shows that $\mathcal{E}_P\Big(\rho(x)\phi\Big) = \mathcal{E}_P(\phi)$.

Theorem. ([JLR]) (1) *I_G^T defines a $G(\mathbb{A}_f)^1$-invariant linear functional on $\mathcal{A}(G)_*$;*
(2) *$I_G^T(\phi) := \int_{G(F)\backslash G(\mathbb{A})^1}^* \phi(g)\,dg$ is independent of the choice of T;* and
(3) *If $\phi \in \mathcal{A}(G)$ is integrable over $G(F)\backslash G(\mathbb{A})^1$, then $\phi \in \mathcal{A}(G)_*$ and*

$$\int_{G(F)\backslash G(\mathbb{A})^1}^* \phi(g)\,dg = \int_{G(F)\backslash G(\mathbb{A})^1} \phi(g)\,dg.$$

B.5.3. Arthur's Periods and JLR's Periods

For $f \in \mathcal{A}_P(G)$ with

$$f\Big(namk\Big) = \sum_{j=1}^{l} \phi_j(mk) \cdot \alpha_j\Big(H(a)\Big) e^{\langle \lambda_j + \rho_P, H(a)\rangle}$$

where $n \in N(\mathbb{A})$, $a \in A_P$, $m \in M(\mathbb{A})^1$, $k \in K$, and for all j, $\alpha_j(X)$ is a polynomial, and $\phi_j(g)$ is an automorphic form in $\mathcal{A}_P(G)$ such that $\phi_j(ag) = \phi_j(g)$ for $a \in A_P$, following 5.2, define

$$\begin{aligned}&\int_{P(F)\backslash G(\mathbb{A})^1}^{*} f(g) \cdot \tau_P\Big(H(g) - T\Big)\, dg \\ :=&\sum_{j=1}^{l} \int_K \int_{M(F)\backslash M(\mathbb{A})^1}^{*} \phi_j(mk)\, dm\, dk \\ &\times \left(\int_{\mathfrak{a}_P}^{\#} \alpha_j(X) e^{\langle \lambda_j, X\rangle} \tau_P(X - T)\, dX \right).\end{aligned}$$

This is well-defined provided that the following two conditions are satisfied:
(i*) $\langle \mu, \varpi^\vee \rangle \neq 0 \qquad \forall Q \subset P,\ \varpi^\vee \in (\widehat{\Delta}^\vee)_Q^P,\ \mu \in \mathcal{E}_Q(\phi)$; and
(ii*) $\langle \lambda, \alpha^\vee \rangle \neq 0 \qquad \forall \alpha \in \Delta_P,\ \lambda \in \mathcal{E}_P(\phi)$.

Let $\mathcal{A}(G)^{**}$ be the space of $\phi \in \mathcal{A}(G)$ such that (i*) and hence also (ii*) are satisfied for any P. As a direct consequence of the theorem in 5.2, we have the following

Theorem. ([JLR]) *For* $\phi \in \mathcal{A}(G)^{**}$,

$$\begin{aligned}&\int_{G(F)\backslash G(\mathbb{A})^1} \Lambda^T \phi(g)\, dg \\ =&\sum_P (-1)^{d(P)-d(G)} \int_{P(F)\backslash G(\mathbb{A})^1}^{*} \phi_P(g) \cdot \widehat{\tau}_P\Big(H(g) - T\Big)\, dg.\end{aligned}$$

B.5.4. Bernstein Principle

With above, to understand non-abelian L-functions at least for $T \gg 0$, i.e., for $T \in \mathfrak{a}_0^+$ sufficiently positive, we need to study Eisenstein periods. For this, the following result plays a key role.

Bernstein's Principle. ([JLR]) *Let* $P = MN$ *be a proper parabolic subgroup and let* σ *be an irreducible cuspidal representation in the space* $L^2\Big(M(F)\backslash M(\mathbb{A})^1\Big)$. *Let* $E(g, \phi, \lambda)$ *be an Eisenstein series associated to* $\phi \in \mathcal{A}_P(G)_\sigma$. *Then*

$$I_G^T\Big(E(g, \phi, \lambda)\Big) = 0$$

for all λ *such that* $E(g, \phi, \lambda)$ *and* $I_G^T\big(E(g, \phi, \lambda)\big)$ *are defined.*

We believe that this principle is the main reason behind special yet very important relations among multiple and ordinary zeta functions in literature.

B.5.5. Eisenstein Periods Associated to Cusp Forms

As a direct consequence of Bernstein principle, with the explicit formula for constant terms of Eisenstein series associated to cusp forms, (see e.g., [We8],) we have the following closed formula for Eisenstein periods associated with cusp forms:

Theorem. ([JLR]) *Fix a sufficiently positive $T \in \mathfrak{a}_0^+$. Let $P = MN$ be a parabolic subgroup and σ an irreducible* **cuspidal** *representation in $L^2\big(M(F)\backslash M(\mathbb{A})^1\big)$. Let $E(g,\phi,\lambda)$ be an Eisenstein series associated to $\phi \in \mathcal{A}_P(G)_\sigma$. Then the integration $\int_{G(F)\backslash G(\mathbb{A})} \Lambda^T E(g,\phi,\lambda)\, dg$ is equal to*

(1) *0 if $P \neq P_0$ is not minimal;* and

(2) $\mathrm{Vol}\Big(\Big\{ \sum_{\alpha\in\Delta_0} a_\alpha \alpha^\vee : a_\alpha \in [0,1)\Big\}\Big) \cdot \sum_{w\in\Omega} \frac{e^{\langle w\lambda-\rho,T\rangle}}{\prod_{\alpha\in\Delta_0}\langle w\lambda-\rho,\alpha^\vee\rangle}$ $\cdot \int_{M_0(F)\backslash M_0(\mathbb{A})^1\times K} \Big(M(w,\lambda)\phi\Big)(mk)\, dm\, dk,$ *if $P = P_0 = M_0N_0$ is minimal.*

Consequently, we have the following

Theorem$'$. ([We9]) *Fix a convex polygon p such that $T(p) \in \mathfrak{a}_0$ is sufficiently positive. Then for a P-level cusp form ϕ, $L_{F,r}^{\leq p}(\phi,\lambda)$ is equal to*

(1) 0 *if P is not minimal;*

(2) $\mathrm{Vol}\Big(\Big\{ \sum_{\alpha\in\Delta_0} a_\alpha \alpha^\vee : a_\alpha \in [0,1)\Big\}\Big) \cdot \sum_{w\in\Omega} \frac{e^{\langle w\lambda-\rho,T\rangle}}{\prod_{\alpha\in\Delta_0}\langle w\lambda-\rho,\alpha^\vee\rangle}$ $\cdot \int_{M_0(F)\backslash M_0(\mathbb{A})^1\times K} \Big(M(w,\lambda)\phi\Big)(mk)\, dm\, dk,$ *if $P = P_0 = M_0N_0$ is minimal.*

We remind the reader that this result is merely the beginning of our quantitive study about non-abelian L-functions. Say,

(1) When the polygon p satisfies $T(p) \gg 0$, we still need to understand $L_{F,r}^{P';\leq p}(\phi,\lambda)$ where $P' \supset P$ is a parabolic subgroup of G;

(2) We have to weak the restriction that $T(p) \gg 0$; and

(3) We need also to study non-abelian L-functions for principal lattices, i.e., that for general reductive groups.

In fact all this proves to be very interesting and quite fruitful: In (1), Artin L-functions naturally appears; to (2), the above discussion is valid as long as $T(p) \in \mathfrak{a}_0^+$; and for (3), no essentially difficulties appear despite its generality. For more details, see [We9].

As an application, we here recall a recent result of H. Kim and the author.

Let then G be a split, semi-simple group of rank r. For any $w \in \Omega$, let $J_w := \{\alpha \in \Delta_0 : w\alpha \in \Delta_0\}$ and P_w be the standard parabolic subgroup corresponding to J_w. Accordingly, let $r_w := \{\alpha \in \Delta_0 : w\alpha < 0\}$, and for $i > 1$,

$$\begin{aligned} n_i &:= \#\{\alpha > 0 : \langle \rho, \alpha^\vee \rangle = i\} - \#\{\alpha > 0, : \langle \rho, \alpha^\vee \rangle = i-1\}, \\ n_{i,w} &:= \#\{\alpha > 0, w\alpha < 0 : \langle \rho, \alpha^\vee \rangle = i\} \\ &\quad - \#\{\alpha > 0, w\alpha < 0 : \langle \rho, \alpha^\vee \rangle = i-1\}. \end{aligned}$$

As usual, denote by Δ_F the discriminant of F, r_2 the number of complex embeddings of F, and $\xi_F(1)$ the residue of the completed Dedekind zeta function $\xi_F(s)$ at $s = 1$. Denote also by v the volume of $\{\sum_{\alpha \in \Delta_0} a_\alpha \alpha^\vee : a_\alpha \in [0,1)\}$.

Theorem. ([KW]) *For a sufficiently positive $T \in \mathfrak{a}_0^+$ the volume of $\Big(G(F)\backslash G(\mathbb{A})^1\Big)_T$ is given by*

$$\begin{aligned} & v \cdot (2\pi)^{r_2 r} \Delta_F^{-\frac{r}{2}} \prod_{i>1} \xi_F(i)^{-n_i} \\ & \cdot \sum_{w \in W_0} (-1)^{\mathrm{rank} P_w} \frac{\xi_F(1)^{r_w} \prod_{i>1} \xi_F(i)^{n_{i,w}}}{\prod_{\alpha \in \Delta_0 - wJ_w} \big(1 - \langle w\rho, \alpha^\vee \rangle\big)} \cdot e^{\langle w\rho - \rho, T \rangle}. \end{aligned}$$

Note that as $T \to \infty$, only the term corresponding to $w = 1$ survives and the limit is exactly the volume of the fundamental domain given by Langlands [La2], namely,

Corollary. ([La2]) *If G is split and semi-stable, then the volume of a fundamental domain $\mathfrak{F}$ for $G(F)\backslash G(\mathbb{A})^1$ is given by*

$$\mathrm{Vol}(\mathfrak{F}) = v \cdot (2\pi)^{r_2 r} \Delta_F^{-\frac{r}{2}} \cdot \prod_{i>1} \xi_F(i)^{-n_i}.$$

In particular, this formula suggests that the volume of cuspidal regions can be expressed in terms of special values of Dedekind zeta functions.

B.5.6. Fundamental Relations between Special Values of Zeta Functions

By strengthening the results in 5.5, namely working with all $T \in \mathfrak{a}_0^+$ instead of $T \gg 0$, we then can obtain the following result which reveals certain fundamental relations among special values of abelian and non-abelian zeta functions ([We9]). To state them, let $\xi_{F,r}(1) := \mathrm{Res}_{s=1}\xi_{F,r}(s)$.

Level 1. $\xi_{F,1}(1) = \xi_F(1)$;

Level 2. $\xi_{F,2}(1) = \xi_F(2)\xi_F(1) - \xi_F(1)\xi_F(1)$;

Level 3.
$$\xi_{F,3}(1) = \xi_F(3)\xi_F(2)\xi_F(1) -\frac{1}{3}\xi_F(2)\xi_F(1)^2 - \frac{1}{3}\xi_F(2)\xi_F(1)^2 + \frac{1}{4}\xi_F(1)^3;$$

Level 4.
$$\begin{aligned}\xi_{F,4}(1) = {} & \xi_F(4)\xi_F(3)\xi_F(2)\xi_F(1)\\ & - \frac{1}{4}\xi_F(3)\xi_F(2)\xi_F(1)^2 - \frac{1}{4}\xi_F(2)^2\xi_F(1)^2 - \frac{1}{4}\xi_F(3)\xi_F(2)\xi_F(1)^2\\ & + \frac{1}{6}\xi_F(2)^2\xi_F(1)^2 + \frac{1}{9}\xi_F(2)\xi_F(1)^2 + \frac{1}{6}\xi_F(2)\xi_F(1)^3\\ & - \frac{1}{8}\xi_F(1)^4;\end{aligned}$$

Level r. The r-level relation can be described as follows.
(1) $\xi_{F,r}(1)$ is a degree homogenous polynomial in $\xi_F(i)$, $i = 1, 2, \ldots, r$;
(2) there is a one-to-one and onto correspondence between the terms of this polynomial and partitions of r, or better, standard parabolic subgroups. For example, in the case $r = r_1 + r_2 + \cdots + r_N$, the corresponding term consists of three parts, that is,
(i) the sign $(-1)^{N-1}$;
(ii) the product of zeta values: If $r_i \geq 2$, the product $\xi_F(r_i)\xi_F(r_i - 1)\cdots\xi_F(2)$ does appear. Moreover, these subfactors together with certain powers of $\xi_F(1)$ then gives all the product of zeta values (under the constrain that the final product is of degree r); and

(iii) The coefficient $\frac{1}{(r_1+r_2)(r_2+r_3)\cdots(r_{N-1}+r_N)}$.

It is well known that following a result of Siegel, in order to read the special value $\xi_F(n)$, we have to go to rank n lattices. The above relations may be viewed as a refined version of this basic result. Roughly put,
(i) among all rank n-lattices, only stable lattices are essential;
(ii) other rank n lattices can be built up from stable lattices of lower ranks;
(iii) corresponding modular points of lattices in (ii) are sitting in cuspidal regions; and
(iv) volumes of cupidal regions can be expressed in terms of $\xi_F(m)$'s, $m < n$.

We end this discussion by pointing out that our result above is closely related with the result of Harder and Narasimhan ([HN]) on counting stable bundles over curves on finite fields: Their method, an arithmetic geometry one, is algebraic while ours, a geometric arithmetic one, is analytic.

C. Explicit Formula, Functional Equation and Geo-Ari Intersection

C.1. The Riemann Hypothesis for Curves

C.1.1. Weil's Explicit Formula: The Reciprocity Law

Let C be a projective irreducible reduced regular curve of genus g defined over a finite field $k := \mathbf{F}_q$. Denote by $\zeta_C(s)$ the associated Artin zeta function. Set $t = q^{-s}$ and $Z_C(t) := \zeta_C(s)$. Then by the rationality, there exists a polynomial $P_C(t)$ of degree $2g$ such that

$$Z_C(t) = \frac{P_C(t)}{(1-t)(1-qt)}.$$

Now let C_n be the curve obtained from C by extending the field of constants from $\mathbf{F}_q$ to $\mathbf{F}_{q^n}$. Then by a discussion on covering of curves, we obtain the following

Reciprocity Law. $Z_{C^n}(t^n) = \prod_\zeta Z_C(\zeta t)$ *where the product runs over all n-th roots of 1.*

Moreover, by the Euler product,

$$t\frac{Z_C'}{Z_C}(t) = \sum_{n=1}^\infty \sum_P d(P) t^{nd(P)} = \sum_{m=1}^\infty \Big(\sum_P d(P)\Big) t^m$$

where $\sum_P$ is taken over those closed points rational over $\mathbf{F}_q$ whose degree divides m. Hence,

$$Z_C(t) = \exp\Big(\sum_{m=1}^\infty N_m \frac{t^m}{m}\Big)$$

where $N_m = \sum_{P,d(P)|m} d(P)$. Clearly when $m = 1$, the sum $\sum_{P,d(P)|1} 1$ simply counts the number of closed points of C rational over $\mathbf{F}_q$. Thus, by the reciprocity law above,

$$Z_{C^n}(t^n) = \exp\Big(\sum_{m=1}^\infty M_m \frac{t^{nm}}{m}\Big) = \exp\Big(\sum_{m=1}^\infty N_m \frac{t^m}{m}\Big(\sum_\zeta \zeta^m\Big)\Big).$$

This implies that $N_n = N_1(C_n)$, i.e., N_n is the number of $\mathbf{F}_{q^n}$-rational points on C.

Weil's Explicit Formula. $N_n = q^n + 1 - \sum_{\zeta_C(\rho)=0} \rho^n$.

C.1.2. Geometric Version of Explicit Formula

Let C be an algebraic curve defined over $\mathbf{F}_p$. Over $C \times C$, for $n \in \mathbf{Z}$, introduce (micro) divisors A_n (via algebraic correspondence) as follows:

$$A_n := \begin{cases} \Big\{(x, x^{p^n}) : x \in C\Big\}, & \text{if } n \geq 0; \\ \Big\{(x^{p^{-n}}, x) : x \in C\Big\}, & \text{if } n \leq 0. \end{cases}$$

Then we have the following relations for the intersections among A_n's.

(i) If $n \geq m \geq 0$, $\langle A_n, A_m \rangle = p^m \langle A_{n-m}, A_0 \rangle$;

(ii) If $m \geq n \geq 0$, $\langle A_n, A_m \rangle = p^n \langle A_{m-n}, A_0 \rangle$;

(iii) If $m \geq 0 \geq n$, $\langle A_n, A_m \rangle = \langle A_{n-m}, A_0 \rangle$;

(iv) If $n \leq m \leq 0$, $p^m \langle A_n, A_m \rangle = \langle A_{m-n}, A_0 \rangle$;

(v) If $m \leq n \leq 0$, $p^n \langle A_n, A_m \rangle = \langle A_{n-m}, A_0 \rangle$;

(vi) If $n \geq 0 \geq m$, $\langle A_n, A_m \rangle = \langle A_{n-m}, A_0 \rangle$.

Hence, we have the following

Lemma 1. *(1) For all $m, n \in \mathbf{Z}$, $\langle A_n, A_m \rangle = \langle A_m, A_n \rangle$;*
(2) For all $m, n \in \mathbf{Z}$, $\langle A_{-n}, A_{-m} \rangle = \langle A_m, A_n \rangle$;
(3) For all $n \geq m \geq 0$ in $\mathbf{Z}$, $\langle A_n, A_m \rangle = p^m \langle A_{n-m}, A_0 \rangle$;
(4) For all $n \geq 0 \geq m$ in $\mathbf{Z}$, $\langle A_n, A_m \rangle = \langle A_{n-m}, A_0 \rangle$.

Obviously, (1) $\sim$ (4) are equivalent to (i) $\sim$ (vi) above. Therefore, in order to understand $\langle A_n, A_m \rangle$ for all $n, m \in \mathbf{Z}$, we only need to know $\langle A_n, A_0 \rangle$ for $n \geq 0$.

For this latest purpose, first, note that A_0 is simply the diagonal. Hence, by definition, for $n > 0$, $\langle A_n, A_0 \rangle$ is $N_n(C)$, the number of $\mathbf{F}_{p^n}$-rational points of C;

Secondly, by the functional equation of $\zeta_C(s)$, (and the rationality of $\zeta_C(s)$), we have the following

Lemma 2. (**Explicit Formula of Weil**) *Denote by F_1 and F_2 horizontal and vertical fibers in two directions of $C \times C$ respectively. For $n \in \mathbf{Z}_{\geq 0}$,*

$$\langle A_n, A_0 \rangle = \langle A_n, F_1 \rangle + \langle A_n, F_2 \rangle - \sum_{\zeta_C(s)=0} s^n.$$

C.1.3. Riemann Hypothesis for Function Fields

Following Weil again, (see e.g., [Ha1]), we now prove the (Artin-)Riemann hypothesis.

Let $f : p^{\mathbf{Z}} \to \mathbf{Z}$ be a function with finite supports. Define its Mellin transform via $\hat{f}(s) := \sum_n f(p^n)p^{ns}$. Clearly, if $f^*(p^n) := f(p^{-n})p^{-n}$, $\widehat{f^*}(s) = \hat{f}(1-s)$; moreover,

$$\text{if } (f * g)(p^n) := \sum_m f(p^m)g(p^{n-m}),\ \widehat{f * g}(s) = \hat{f}(s) \cdot \hat{g}(s).$$

In particular, $\widehat{f * g^*}(s) = \hat{f}(s) \cdot \hat{g}(1-s)$.

With f, we may use the divisors A_n's above to define a (global) $\mathbf{Q}$-divisor $D_{\hat{f}}$ on $C \times C$ as follows:

$$D_{\hat{f}} := \sum_{n>0} f(p^n)A_n + \sum_{n\geq 0} f(p^{-n})p^{-n}A_n.$$

Theorem. (Weil) *(1) (Relative Degrees)* $\langle D_{\hat{f}}, F_1 \rangle = \hat{f}(1)$; $\langle D_{\hat{f}}, F_2 \rangle = \hat{f}(0)$;
(2) (Fixed Points Formula) $\langle D_{\hat{f}}, D_{\hat{g}} \rangle = \langle D_{\widehat{f*g^*}}, \mathrm{Diag} \rangle$, *where Diag denotes the diagonal;*
(3) (Explicit Formula) $\hat{f}(0) + \hat{f}(1) - \sum_{\zeta(s)=0} \hat{f}(s) = \langle D_{\hat{f}}, \mathrm{Diag} \rangle$.
Indeed, by definition,

$$\langle D_{\hat{f}}, F_1 \rangle = \sum_{n>0} f(p^n) \cdot p^n + \sum_{n\geq 0} f(p^{-n})p^{-n} \cdot 1 = \hat{f}(1);$$

$$\langle D_{\hat{f}}, F_2 \rangle = \sum_{n>0} f(p^n) + \sum_{n\geq 0} f(p^{-n})p^{-n} \cdot p^n = \hat{f}(0).$$

This gives (1). (2) is a direct consequence of the relations (i)-(vi) for the intersections of A_n's, and hence comes from Lemma 1 in 2.1. Finally, (3) is simply Lemma 2 of 1.2 by definition.

Since $\langle F_1+F_2, D_{\hat{f}}-\hat{f}(1)F_2-\hat{f}(0)F_1\rangle = 0$, by the two-dimensional intersection theory, in particular, the Hodge Index Theorem,

$$\langle D_{\hat{f}} - \hat{f}(1)F_2 - \hat{f}(0)F_1, D_{\hat{f}} - \hat{f}(1)F_2 - \hat{f}(0)F_1\rangle \leq 0.$$

That is,

$$\hat{f}(0) \cdot \hat{f}(1) \geq \frac{1}{2}\langle D_{\hat{f}}, D_{\hat{f}}\rangle.$$

Thus by the theorem above, this last equality is equivalent to

$$\sum_{\zeta(s)=0} \hat{f}(s) \cdot \hat{f}(1-s) \geq 0.$$

From this, easily, we get the following

Riemann Hypothesis for Artin Zetas. (Hasse-Weil) *If* $\zeta(s) = 0$, *then* $\mathrm{Re}(s) = \frac{1}{2}$.

C.2. Geo-Ari Intersection: A Mathematics Model

C.2.1. Motivation from Cramér's Formula

In the above discussion, the summation $\sum_{\zeta_C(s)=0} s^n$ plays a key role in understanding Artin-Riemann Hypothesis, via the so-called micro explicit formula of Weil. So naturally, we want to know whether this approach works for number fields, and hence are led to study the formal summation

$$\sum_{\xi_{\mathbf{Q}}(\rho)=0} x^\rho, \qquad \text{for } x \in [1, \infty). \tag{$*$}$$

Here $\xi_{\mathbf{Q}}(s)$ denotes the completed Riemann zeta, that is to say, $\xi_{\mathbf{Q}}(s) := \pi^{-\frac{s}{2}}\Gamma(\frac{s}{2})\zeta_{\mathbf{Q}}(s)$ with $\Gamma(s)$ the standard gamma function and $\zeta_{\mathbf{Q}}(s)$ the Riemann zeta function.

The reader at this point certainly would reject $(*)$, since the summation does not make any sense. How could we write such a monster down?! Well, do not panic!!! After all, in the study of the prime distributions, conditional convergent summation $\sum_{\xi_{\mathbf{Q}}(\rho)=0} \frac{x^\rho}{\rho}$ does appear. Recall that we have the following Riemann-von Mangoldt formula

$$\sum_{p^n \leq x}{}' \log p = x - \sum_{\xi_{\mathbf{Q}}(\rho)=0} \frac{x^\rho}{\rho} - \frac{\zeta'}{\zeta}(0) - \frac{1}{2}\log(1-x^2),$$

which itself motivates and hence stands as a special form of the more general form of explicit formulas. See e.g., Jorgenson and Lang's lecture notes on Explicit Formulas. More generally, in various discussions about prime distributions, we do use the summations such as $\sum_{\xi_{\mathbf{Q}}(\rho)=0} \frac{x^{\rho\pm 1}}{\rho(\rho\pm 1)}$. In this sense, the problem is not whether we should introduce $(*)$, rather, it should be how to justify it.

Anyway, let us make a change of variables $x := e^t$ with $t \geq 0$. Then $(*)$ becomes

$$V(t) := \sum_{\xi_{\mathbf{Q}}(\rho)=0} e^{t\rho}. \qquad (**)$$

So, following Riemann, we may further view $V(t)$ as a function of complex variable z, i.e.,

$$V(z) := \sum_{\xi_{\mathbf{Q}}(\rho)=0} e^{z\rho}. \qquad (**')$$

Now, we claim that there is a nice way to regularize $(**')$ hence also $(**)$ and $(*)$.

To explain this in a simpler form, which in fact would not really make our life any easier, we instead consider the partial sum $V_+(z)$ defined by

$$V_+(z) := \sum_{\xi_{\mathbf{Q}}(\rho)=0, \mathrm{Re}(\rho)>0} e^{z\rho}.$$

Then we have the following result, dated in 1919.

Theorem. (Cramér) *The function $V_+(z)$ converges absolutely for* $\mathrm{Im}(z) > 0$. *Moreover,*

$$2\pi i V_+(z) - \frac{\log z}{1 - e^z}$$

has a meromorphic continuation to $\mathbf{C}$, *with simple poles at the points* $\pm\pi\sqrt{-1}n$ *for all integers* n, *and at the points* $\pm\log p^m$ *for all powers of primes.*

We claim that this theorem actually offers us a natural analytic way to normalize the formal summation $\sum_{\xi_{\mathbf{Q}}(\rho)=0} x^\rho$ for $x \in [1, \infty)$, by using [C], [JL1,2,3], and [DS]. In a sense, this is in a similar way as what we do when normalizing $\infty!$: by the Stirling formula

$$n! = \sqrt{2\pi} \cdot \sqrt{n} \cdot \left(\frac{n}{e}\right)^n \cdot \exp\left(\frac{\theta_n}{12}\right)$$

for n sufficient large with $|\theta_n| < 1$, we set $\infty! = \sqrt{2\pi}$.

However, in this article, we do it very differently – we are going to introduce a mathematics model to normalize this formal summation geometrically.

C.2.2. Micro Divisors

We will not use Grothendieck's scheme language. Instead, formally, we call (the set theoretical product) $S := \overline{\mathrm{Spec}(\mathbf{Z})} \times \overline{\mathrm{Spec}(\mathbf{Z})}$ a geometric arithmetic base (surface).

Similarly as in 1.2, associated to all $x \in [0, \infty]$ are symbols D_x which will be called *micro divisors*; and associated to any two micro divisors D_x, D_y is the intersection number $\langle D_x, D_y\rangle \in \mathbf{R}$. Assume that the following fundamental relations are satisfied by our micro intersections:

Axiom 1. (*Symmetry*) For any $x, y \in [0, \infty]$,

$$\langle D_x, D_y \rangle = \langle D_y, D_x \rangle;$$

Axiom 2. (*Mirrow Image*) For any $x, y \in [0, \infty]$,

$$\langle D_x, D_y \rangle = \langle D_{\frac{1}{x}}, D_{\frac{1}{y}} \rangle;$$

Axiom 3. (*Fixed Points 1*) If $0 \le x \le y \le 1$, then

$$\langle D_x, D_y \rangle = y \langle D_{\frac{x}{y}}, D_1 \rangle;$$

Axiom 4. (*Fixed Points 2*) If $0 \le x \le 1 \le y \le \infty$, then

$$\langle D_x, D_y \rangle = \langle D_{\frac{x}{y}}, D_1 \rangle.$$

Note that $[0, \infty] \times [0, \infty]$ is simply the union of $\{0 \le x \le y \le 1\}$, $\{0 \le y \le x \le 1\}$, $\{1 \le x \le y \le \infty\}$, $\{1 \le y \le x \le \infty\}$, $\{0 \le x \le 1, 1 \le y \le \infty\}$ and $\{0 \le y \le 1, 1 \le x \le \infty\}$. Thus by the above relations, if we know the pricise intersection $\langle D_x, D_1 \rangle$ for all $x \in [0, 1]$, then we obtain all the intersections $\langle D_x, D_y \rangle$ for all $x, y \in [0, \infty]$.

Axiom 5. (*Micro Explicit Formula*) Denote the completed Riemann zeta function by $\xi_{\mathbf{Q}}(s)$. Then, for all $x \in [0, 1]$,

$$\langle D_x, D_1 \rangle = \langle D_0, D_x \rangle + \langle D_\infty, D_x \rangle - \sum_{\xi_{\mathbf{Q}}(s)=0} x^s.$$

Here, as said in 2.1, we certainly encounter with the convergence problem of $\sum_{\xi_{\mathbf{Q}}(s)=0} x^s$. Instead of solving it, in our model, we simply assume that our micro intersection offers a natural normalization of $\sum_{\xi_{\mathbf{Q}}(s)=0} x^s$ via the (5). It is in this sense we say that our model gives a geometric way to normalize $\sum_{\xi_{\mathbf{Q}}(s)=0} x^s$.

The compatibility among these Axioms is guaranteed by the functional equation for Riemann zeta function. Moreover, if $x \ge 1$, by the Mirrow principle, we see that $\langle D_x, D_1\rangle = \langle D_{\frac{1}{x}}, D_1\rangle$. So, by using the Explicit Formula 1, i.e., Axiom 5 above, together with the Relations II below, we have

$$\langle D_{\frac{1}{x}}, D_1\rangle = 1 + \frac{1}{x} - \sum_{\xi_{\mathbf{Q}}(s)=0} x^{-s}.$$

Multiplying both sides by x, we get

$$x\langle D_{\frac{1}{x}}, D_1\rangle = 1 + x - \sum_{\xi_{\mathbf{Q}}(s)=0} x^{1-s} = 1 + x - \sum_{\xi_{\mathbf{Q}}(s)=0} x^{s},$$

by the functional equation. On the other hand, by the Fixed Point 1, i.e., Axiom 3 above, we get

$$\langle D_x, D_1\rangle = x\langle D_{\frac{1}{x}}, D_1\rangle.$$

That is to say, formally, for all x,

$$\langle D_x, D_1\rangle = 1 + x - \sum_{\xi_{\mathbf{Q}}(s)=0} x^{s}.$$

From Axioms 1-4, we may formally get the following relations.

Relations I. *(i)* $\langle D_0, D_0\rangle = \langle D_\infty, D_\infty\rangle$*;*
(ii) $\langle D_0, D_1\rangle = \langle D_\infty, D_1\rangle = \langle D_0, D_\infty\rangle$*.*

Note that D_0 and D_∞ are supposed to be the vertical and horizontal fibers of $\overline{\mathrm{Spec}(\mathbf{Z})} \times \overline{\mathrm{Spec}(\mathbf{Z})}$ over $\infty \in \overline{\mathrm{Spec}(\mathbf{Z})}$, so we normalize our intersection further by the following

Axiom 6. (*Normalization*) $\langle D_0, D_0\rangle = 0, \quad \langle D_0, D_1\rangle = 1.$

With this, formally we may further get the following

Relations II. *(i) For* $\langle D_0, D_x\rangle$*,*

$$\langle D_0, D_x\rangle = \begin{cases} x, & \text{if } x \in [0,1]; \\ 1, & \text{if } x \in [1,\infty]; \end{cases}$$

(ii) For $\langle D_\infty, D_x\rangle$,

$$\langle D_\infty, D_x\rangle = \begin{cases} 1, & \text{if } x \in [0,1]; \\ \frac{1}{x}, & \text{if } x \in [1,\infty]; \end{cases}$$

In particular, for $x \in [0,1]$,

$$\langle D_x, D_1\rangle = 1 + x - \sum_{\xi_{\mathbf{Q}}(s)=0} x^s.$$

Remark. In (iii) above, taking $x = 1$, we get

$$\langle D_1, D_1\rangle = 2 + \sum_{\xi_{\mathbf{Q}}(s)=0} 1.$$

In other words, $\sum_{\xi_{\mathbf{Q}}(s)=0} 1$ is supposed to be related to the self-intersection of the *diagonal divisor.*

C.2.3. Global Divisors and Their Intersections: Geometric Reciprocity Law

Motivated by C.1, i.e., the discussion about Artin-Riemann Hypothesis, we start with a standard construction in function theory. Let $f : \mathbf{R}^+ \to \mathbf{R}$ be a smooth, compactly supported function. Define its *Mellin transform* via

$$\hat{f}(s) := \int_0^\infty f(x) x^s \frac{dx}{x}.$$

Then, if $f^*(x) := f(\frac{1}{x}) \cdot \frac{1}{x}$,

$$\widehat{f^*}(s) = \hat{f}(1-s);$$

and if $(f * g)(x) := \int_0^\infty f(y) g(\frac{x}{y}) \frac{dy}{y}$ denotes the standard multiplicative convolution,

$$\widehat{f * g}(s) = \hat{f}(s) \cdot \hat{g}(s).$$

In particular,

$$\widehat{f * g^*}(s) = \hat{f}(s) \cdot \hat{g}(1-s).$$

Next, we give a parallel construction for our divisors. Standard wisdom says that we should use linear combinations of generalized divisors $D_x, x \in [0, \infty]$ to form new type of divisors. But as we clearly see that such a conventional way does not result sufficiently many divisors, we do it very differently.

Starting from divisors D_x, for any might-be-interesting function f, formally define the associated *global divisor* $D_{\hat{f}}$ by setting

$$D_{\hat{f}} := \int_0^1 f(x) \cdot D_x \cdot \frac{dx}{x} + \int_1^\infty f(x) \cdot x D_x \cdot \frac{dx}{x}.$$

Remark. In all formal discussions here, we pay no attention to the convergence problem. See however 2.5 below. In other words, we assume also in our model that global divisors do exist and their relations with micro divisors are given as above.

With this definition, we may extend the intersection in 2.2 to $D_{\hat{f}}$'s by linearity. For example, the intersection $\langle D_{\hat{f}}, D_1 \rangle$ is by definition given by

$$\langle D_{\hat{f}}, D_1 \rangle = \int_0^1 f(x) \cdot \langle D_x, D_1 \rangle \cdot \frac{dx}{x} + \int_1^\infty f(x) \cdot x \langle D_x, D_1 \rangle \cdot \frac{dx}{x}.$$

Then formally, we have the following

Key Relations. *(i) (Relative Degrees in Two Fiber Directions)*

$$\deg_1 D_{\hat{f}} := \langle D_0, D_{\hat{f}} \rangle = \hat{f}(1)$$

and

$$\deg_2 D_{\hat{f}} := \langle D_\infty, D_{\hat{f}} \rangle = \hat{f}(0).$$

(ii) (Fixed Point Formula)

$$\langle D_{\hat{f}}, D_{\hat{g}} \rangle = \langle D_{\widehat{f*g^*}}, D_1 \rangle;$$

(iii) (Global Explicit Formula)

$$\langle D_{\hat{f}}, D_1 \rangle = \hat{f}(0) + \hat{f}(1) - \sum_{\xi_{\mathbf{Q}}(s)=0} \hat{f}(s).$$

In fact, this may be formally checked as follows using axioms. First consider the Fixed Point Formula. Set $D'_x := D_{\frac{1}{x}}$. Then, by definition,

$$
\begin{aligned}
&\langle D_{\widehat{f*g^*}}, D_1\rangle \\
&= \left\langle \int_0^1 \widehat{f*g^*}(x) D_x \frac{dx}{x} + \int_0^1 \widehat{f*g^*}\left(\frac{1}{x}\right)\frac{1}{x} D'_x \frac{dx}{x}, D_1 \right\rangle \\
&= \left\langle \int_0^1 \int_0^\infty f(y) g^*\left(\frac{x}{y}\right)\frac{dy}{y} D_x \frac{dx}{x} \right. \\
&\qquad \left. + \int_0^1 \int_0^\infty f(y) g^*\left(\frac{1}{xy}\right) xy \frac{dy}{y}\frac{1}{x} D'_x \frac{dx}{x}, D_1 \right\rangle
\end{aligned}
$$

By changing variables $x' := \frac{y}{x}$, the latest quantity is simply

$$
= \left\langle \int_0^\infty f(y)\frac{dy}{y}\int_y^\infty g(x) x D_{\frac{y}{x}}\frac{dx}{x} + \int_0^\infty f(y)\frac{dy}{y}\int_0^y g(x) y D_{\frac{x}{y}}\frac{dx}{x}, D_1 \right\rangle.
$$

But for $x \geq y$, we may split the region into three parts, i.e.,

(1.1) $1 \geq x \geq 0, 1 \geq y \geq 0$ and $x \geq y$;

(1.2) $\infty \geq x \geq 1, \infty \geq y \geq 1$ and $x \geq y$;

(1.3) $\infty \geq x \geq 1$ and $1 \geq y \geq 0$.

Similarly, the region of $x < y$ may be split into three parts, i.e.,

(2.1) $1 \geq x \geq 0, 1 \geq y \geq 0$ and $x < y$;

(2.2) $\infty \geq x \geq 1, \infty \geq y \geq 1$ and $x < y$;

(2.3) $\infty \geq y \geq 1$ and $1 \geq x \geq 0$.

Hence the latest quantity is simply, by writing according to (2.1),(2.2) and (2.3) (resp. (1.1), (1.2) and (1.3)) for the first term (resp. second term),

$$\begin{aligned}&\Big\langle\Big(\Big(\int_0^1 f(y)\frac{dy}{y}\int_0^y g(x)yD_{\frac{x}{y}}\frac{dx}{x}+\int_1^\infty f(y)\frac{dy}{y}\int_1^y g(x)yD_{\frac{x}{y}}\frac{dx}{x}\\&+\int_1^\infty f(y)\frac{dy}{y}\int_0^1 g(x)yD_{\frac{x}{y}}\frac{dx}{x}\Big)+\Big(\int_0^1 f(y)\frac{dy}{y}\int_y^1 g(x)xD_{\frac{y}{x}}\frac{dx}{x}\\&+\int_1^\infty f(y)\frac{dy}{y}\int_y^\infty g(x)xD_{\frac{y}{x}}\frac{dx}{x}+\int_0^1 f(y)\frac{dy}{y}\int_1^\infty g(x)xD_{\frac{y}{x}}\frac{dx}{x}\Big),\\&\hspace{20em}D_1\Big\rangle\\&=\Big\langle\Big(\int_0^1 f(y)\frac{dy}{y}\Big[\int_0^y g(x)yD_{\frac{x}{y}}\frac{dx}{x}+\int_y^1 g(x)xD_{\frac{y}{x}}\frac{dx}{x}\Big]\\&+\int_1^\infty f(y)\frac{dy}{y}\Big[\int_1^y g(x)yD_{\frac{x}{y}}\frac{dx}{x}+\int_y^\infty g(x)xD_{\frac{y}{x}}\frac{dx}{x}\Big]\\&+\int_1^\infty f(y)\frac{dy}{y}\int_0^1 g(x)yD_{\frac{x}{y}}\frac{dx}{x}+\int_0^1 f(y)\frac{dy}{y}\int_1^\infty g(x)xD_{\frac{y}{x}}\frac{dx}{x}\Big),\\&\hspace{20em}D_1\Big\rangle\\&=\int_0^1 f(y)\frac{dy}{y}\Big[\int_0^y g(x)y\langle D_{\frac{x}{y}},D_1\rangle\frac{dx}{x}+\int_y^1 g(x)x\langle D_{\frac{y}{x}},D_1\rangle\frac{dx}{x}\Big]\\&+\int_1^\infty f(y)\frac{dy}{y}\Big[\int_1^y g(x)y\langle D_{\frac{x}{y}},D_1\rangle\frac{dx}{x}+\int_y^\infty g(x)x\langle D_{\frac{y}{x}},D_1\rangle\frac{dx}{x}\Big]\\&+\int_1^\infty f(y)\frac{dy}{y}\int_0^1 g(x)y\langle D_{\frac{x}{y}},D_1\rangle\frac{dx}{x}\\&+\int_0^1 f(y)\frac{dy}{y}\int_1^\infty g(x)x\langle D_{\frac{y}{x}},D_1\rangle\frac{dx}{x}.\end{aligned}$$

Now accordingly call each of the terms from the beginning as (3.1), (3.2), (4.1), (4.2), (5.1) and (5.2), we then see that for (3.1), (resp. (3.2), (4.1), (4.2), (5.1) and (5.2)) we have $0\leq x\leq y\leq 1$, (resp. $0\leq y\leq x\leq 1$, $1\leq x\leq y\leq\infty$ and $\infty\geq x\geq y\geq 1$,) hence by our axioms, for $0\leq x\leq y\leq 1$

$$y\langle D_{\frac{x}{y}},D_1\rangle=\langle D_y,D_x\rangle,$$

(resp. for $0 \leq y \leq x \leq 1$

$$x\langle D_{\frac{y}{x}}, D_1\rangle = \langle D_x, D_y\rangle,$$

for $1 \leq x \leq y \leq \infty$

$$\langle D_{\frac{x}{y}}, D_1\rangle = x\langle D_x, D_y\rangle,$$

for $\infty \geq x \geq y \geq 1$

$$\langle D_{\frac{y}{x}}, D_1\rangle = y\langle D_x, D_y\rangle,$$

for $0 \leq x \leq 1 \leq y \leq \infty$

$$y\langle D_{\frac{x}{y}}, D_1\rangle = y\langle D_y, D_x\rangle,$$

and, for $0 \leq y \leq 1 \leq x \leq \infty$

$$x\langle D_{\frac{y}{x}}, D_x\rangle = x\langle D_y, D_x\rangle).$$

We see that the latest combination is simply

$$\begin{aligned}&\int_0^1 f(y)\frac{dy}{y}\Big[\int_0^y g(x)\langle D_y, D_x\rangle\frac{dx}{x} + \int_y^1 g(x)\langle D_x, D_y\rangle\frac{dx}{x}\Big]\\ &+ \int_1^\infty f(y)\frac{dy}{y}\Big[\int_1^y g(x)(xy)\langle D_x, D_y\rangle\frac{dx}{x}\\ &+ \int_y^\infty g(x)(xy)\langle D_x, D_y\rangle\frac{dx}{x}\Big] + \int_1^\infty f(y)\frac{dy}{y}\int_0^1 g(x)y\langle D_y, D_x\rangle\frac{dx}{x}\\ &+ \int_0^1 f(y)\frac{dy}{y}\int_1^\infty g(x)x\langle D_y, D_x\rangle\frac{dx}{x}\end{aligned}$$

that is,

$$\begin{aligned}&\int_0^1 f(y)\frac{dy}{y}\int_0^1 g(x)\frac{dx}{x}\langle D_y, D_x\rangle\\ &+ \int_0^1 f(y)\frac{dy}{y}\int_1^\infty g(x)\frac{dx}{x}\cdot x\langle D_y, D_x\rangle\\ &+ \int_1^\infty f(y)\frac{dy}{y}\int_0^1 g(x)\frac{dx}{x}\cdot y\langle D_y, D_x\rangle\\ &+ \int_1^\infty f(y)\frac{dy}{y}\int_1^\infty g(x)\frac{dx}{x}\cdot xy\langle D_y, D_x\rangle.\end{aligned}$$

Clearly, by defintion, this latest combination is simply

$$\langle D_{\hat{f}}, D_{\hat{g}}\rangle.$$

This then completes the proof of the Fixed Point Formula.

To see the relative degree relation, we have the following formal arguments.

$$\begin{aligned}
&\langle D_{\hat{f}}, D_0\rangle\\
&= \int_0^1 f(x)\langle D_x, D_0\rangle \frac{dx}{x} + \int_1^\infty f(x)x\langle D_x, D_0\rangle \frac{dx}{x}\\
&= \int_0^1 f(x)x\frac{dx}{x} + \int_1^\infty f(x)x\frac{dx}{x} = \int_0^\infty f(x)x\frac{dx}{x} = \hat{f}(1)
\end{aligned}$$

and

$$\begin{aligned}
&\langle D_{\hat{f}}, D_\infty\rangle\\
&= \int_0^1 f(x)\langle D_x, D_\infty\rangle \frac{dx}{x} + \int_1^\infty f(x)x\langle D_x, D_\infty\rangle \frac{dx}{x}\\
&= \int_0^1 f(x)\frac{dx}{x} + \int_1^\infty f(x)\frac{dx}{x}\\
&= \int_0^\infty f(x)\frac{dx}{x} = \hat{f}(0).
\end{aligned}$$

(Here a standard regularization is needed. For details, see 2.5 below on Not So Serious Convergence Problems.)

Finally, let see how global explicit formula is established. Here the functional equation plays a key role as in function fields case.

Indeed, by definition,

$$\langle D_{\hat{f}}, D_1\rangle = \int_0^1 f(x)\langle D_x, D_1\rangle \frac{dx}{x} + \int_1^\infty f(x)x\langle D_x, D_1\rangle \frac{dx}{x}.$$

Now by the micro explicit formula, we have

$$\begin{aligned}
\langle D_{\hat{f}}, D_1\rangle = &\int_0^1 f(x)\Big(\langle D_x, D_0\rangle + \langle D_x, D_\infty\rangle - \sum_{\xi_{\mathbf{Q}}(s)=0} x^s\Big)\frac{dx}{x}\\
&+ \int_1^\infty f(x)x\Big(\langle D_x, D_0\rangle + \langle D_x, D_\infty\rangle - \sum_{\xi_{\mathbf{Q}}(s)=0} x^{s-1}\Big)\frac{dx}{x}.
\end{aligned}$$

Recall that by the micro explicit formula, for $x \in [0, \infty]$,

$$\langle D_x, D_1 \rangle = \langle D_x, D_0 \rangle + \langle D_x, D_\infty \rangle - \sum_{\xi_{\mathbf{Q}}(s)=0} x^s.$$

Consequently,

$$\begin{aligned}\langle D_{\hat{f}}, D_1 \rangle =& \langle D_{\hat{f}}, D_0 \rangle + \langle D_{\hat{f}}, D_\infty \rangle + \int_0^\infty f(x) \sum_{\xi_{\mathbf{Q}}(s)=0} x^s \frac{dx}{x} \\ =& \hat{f}(0) + \hat{f}(1) - \sum_{\xi_{\mathbf{Q}}(s)=0} \hat{f}(s).\end{aligned}$$

C.2.4. The Riemann Hypothesis

To finally relate our intersection with the Riemann Hypothesis, we should have a certain positivity. That is to say, we need an analog of the so-called Hodge Index Theorem.

A Weak Version of Hodge Index Theorem. *Self-intersection of the global divisor* $L_{\hat{f}} := D_{\hat{f}} - \hat{f}(1)D_\infty - \hat{f}(0)D_0$ *is non-positive, i.e.,*

$$\langle L_{\hat{f}}, L_{\hat{f}} \rangle \leq 0. \tag{$*$}$$

Remark. We call $(*)$ a weak version of Hodge Index Theorem, since

$$\langle L_{\hat{f}}, D_0 + D_\infty \rangle = 0.$$

From now on, let us assume that $(*)$ holds. Then by the Key Relations in the previous section, with a direct calculation, we certainly will arrive at

$$\sum_{\xi_{\mathbf{Q}}(s)=0} \hat{f}(s) \cdot \hat{f}(1-s) \geq 0.$$

This is very nice, since then, following Weil, we may get the Riemann Hypothesis from this latest inequality. (See e.g., page 342 of the second edition of Lang's Algebraic Number Theory for more details.)

C.2.5. Not So Serious Convergence Problem

We add some remarks on the formal calculation appeared in 2.3. Roughly speaking, to justify them, what we meet is a certain regularization process. This may be done as what Jorgenson and Lang do in their lecture notes on Basic Analysis of Regularized Series and Product. More precisely, motivated by the definitions of hyperbolic Green's functions (of Selberg, Hejhal, Gross and Zagier,) Ray-Singer's analytic torsions, we may first introduce *imaginary divisors* $D_{\hat{f},s}$ by setting

$$D_{\hat{f},s} = \frac{1}{\Gamma(s)}\left(\int_0^1 f(x)D_x x^s \frac{dx}{x} + \int_1^\infty f(x)xD_x x^s \frac{dx}{x}\right)$$

for certain type of suitable functions f and for s whose real parts are sufficiently large; then assume that in our model $D_{\hat{f},s}$ has a meromorphic continuation to the half space $\mathrm{Re}(s) \geq -\varepsilon$ with $\varepsilon > 0$, from which we could finally get a well-defined $D_{\hat{f}}$ after removing the singularity at $s = 0$.

C.2.6. Weil's Explicit Formula and Two-Dimensional Geo-Ari Intersections

We still have not explain why we say the above intersection is indeed an intersection over the geometric arithmetic surface $\overline{\mathrm{Spec}(\mathbf{Z})} \times \overline{\mathrm{Spec}(\mathbf{Z})}$. To understand this, we have to use yet another fundamental result of Weil, the Weil Explicit Formula.

Note that by the Key Relations, we obtain the following crucial formula:

$$\langle D_{\hat{f}}, D_1 \rangle = \hat{f}(0) + \hat{f}(1) - \sum_{\xi_{\mathbf{Q}}(s)=0} \hat{f}(s). \qquad (*)$$

However, as an intersection over $\overline{\mathrm{Spec}(\mathbf{Z})} \times \overline{\mathrm{Spec}(\mathbf{Z})}$, $\langle D_{\hat{f}}, D_1 \rangle$ should be counted locally over each point $(p, q) \in \overline{\mathrm{Spec}(\mathbf{Z})} \times \overline{\mathrm{Spec}(\mathbf{Z})}$, i.e., we should have the decomposition

$$\langle D_{\hat{f}}, D_1 \rangle = \sum_{p,q \leq \infty} \langle D_{\hat{f}}, D_1 \rangle_{(p,q)}.$$

Now note that D_1 is the diagonal, so besides at the points (p, p), $p \leq \infty$ on the diagonal, $\langle D_{\hat{f}}, D_1 \rangle_{(p,q)}$ is naturally zero. With this, then we would have

$$\langle D_{\hat{f}}, D_1 \rangle = \sum_{p \leq \infty} \langle D_{\hat{f}}, D_1 \rangle_{(p,p)}.$$

Clearly, this latest expression suggests that via $(*)$ above

$$\hat{f}(0) + \hat{f}(1) - \sum_{\xi_{\mathbf{Q}}(s)=0} \hat{f}(s) = \sum_{p \leq \infty} W_p(f), \qquad (**)$$

where for each places p of $\mathbf{Q}$, i.e., the primes p and the Archimedean place ∞. Without any mistake, it is then nothing but Weil's explicit formula.

This interpretation naturally leads to a question about the explicit formula for the micro intersection. Recall that one of the key assumption for our micro intersection is that, for $x \in [0, 1]$,

$$\langle D_x, D_1 \rangle = \langle D_x, D_0 \rangle + \langle D_x, D_\infty \rangle - \sum_{\xi_{\mathbf{Q}}(s)=0} x^s.$$

Therefore, if we believe that $\langle \cdot, \cdot \rangle$ is indeed a two-dimensional intersection on $\overline{\text{Spec}(\mathbf{Z})} \times \overline{\text{Spec}(\mathbf{Z})}$, then we similarly should have

$$1 + x - \sum_{\xi_{\mathbf{Q}}(s)=0} x^s = \sum_{p \leq \infty} w_p(x), \qquad \text{for } x \in [0, 1]. \qquad (***)$$

We will call $(***)$ the *micro explicit formula* of Cremér, motivated by Jorgenson and Lang's 'ladder principle'.

Remark. There are also fundamental works of Deninger and Connes on the Riemann Hypothesis, based on certain cohomological consideration. While these approaches appear quite different, they share one common part, the Weil Explicit Formula.

C.3. Towards a Geo-Ari Cohomology in Lower Dimensions

C.3.1. Classical Approach in Dimension One

It is clear that we should go beyond a geo-ari intersection: With our experiences from algebraic geometry and Arakelov theory, to establish an analog of Hodge index theorem, we need to develop a corresponding cohomology theory.

It is our belief that a general yet well-behaved cohomology theory is at the present time beyond our reach. However this does not mean that we cannot do anything about it. After all, what we need is a practical yet uniform cohomology (and intersection) in dimensions one and two, which satisfies duality, adjunction formula and Riemann-Roch.

Recall that in algebraic geometry, we do have a general cohomology theory, thanks to Master Grothendieck. While extremely powerful, this theory is still far short when dealing with global arithmetic problems. Say, at least, analysis is largely left behind. Thus, for global arithmetic, new approach has to be taken. For this, we suggest to go back to one generation before Grothendieck, i.e., that of Chevalley and Weil who once believed that adelic language will sevre us equally if not better.

To give the reader a good test, let us start with cohomology in dimension one i.e, that for curves. A good reference is Serre's GTM on Algebraic Groups and Class Fields.

Let C be a regular irreducible reduced projective curve of genus g defined over k, a field which may not be finite, with F its associated function fields. Fix a divisor D on C. Then main points of the corresponding cohomology theory may be summarized as follows.
(1) by definition, the 0-th cohomology group of (the divisor class associated to) D is given by $H^0(C, D) := \{f \in F : \mathrm{div}(f) + D \geq 0\}$;
(2) From the short exact sequence of sheaves

$$0 \to \mathcal{O}_C(D) \to \mathcal{F} \to \mathcal{F}/\mathcal{O}_C(D) \to 0,$$

where $\mathcal{F}$ denotes the constant sheaf on C associated to the function field F, we get a long exact sequence

$$0 \to H^0(C, \mathcal{O}_C(D)) \to F \to H^0(C, \mathcal{F}/\mathcal{O}_C(D)) \to H^1(C, \mathcal{O}_C(D)) \to 0.$$

Consequently, there are canonical isomorphisms

$$H^0(X, \mathcal{O}_C(D)) \simeq \mathbf{A}(D) \cap F, \text{ and } H^1(X, \mathcal{O}_C(D)) \simeq \mathbf{A}/(\mathbf{A}(D) + F),$$

where $\mathbf{A}$ denotes the adelic ring of F, and

$$\mathbf{A}(D) := \{(r_p) \in \mathbf{A} : \mathrm{ord}_p(r_p) + \mathrm{ord}_p(D) \geq 0\};$$

(3) For any point p, from the structural exact sequence of sheaves $0 \to \mathcal{O}_C(D) \to \mathcal{O}_C(D+p) \to \mathcal{O}_C(D)|_p \to 0$, we get a long exact sequence of cohomology

$$\begin{aligned} 0 \to H^0(C, \mathcal{O}_C(D)) \to H^0(C, \mathcal{O}_C(D+p)) \to H^0(C, \mathcal{O}_C(D)|_p) \\ \to H^1(C, \mathcal{O}_C(D)) \to H^1(C, \mathcal{O}_C(D+p)) \to 0; \end{aligned}$$

(4) By studying the residue pairing, we get a canonical isomorphism

$$\mathbf{A}/(\mathbf{A}(D) + F) \simeq (\mathbf{A}(K_C - D) \cap F)^\vee$$

which in particular implies the Serre duality;
(5) Cohomology groups H^0 and H^1 are all finite dimensional vector spaces. Thus in particular,

$$h^0(C, D) - h^1(C, D) = h^0(C, D+p) - h^1(C, D+p) - 1.$$

This then implies the Riemann-Roch

$$h^0(C, D) - h^1(C, D) = d(D) - (g-1).$$

As it stands, at least for the algebraic part, i.e., the part on cohomology groups, the above theory may be completely understood in terms of adelic language.

(1′) For any divisor D on C, define its associated cohomology groups by

$$H^0(X, \mathcal{O}_C(D)) = \mathbf{A}(D) \cap F, \qquad \text{and} \qquad H^1(C, D) := \mathbf{A}/\mathbf{A}(D) + F;$$

(2′) There is the following commutative 9-diagram $\Sigma(D)$, whose rows and colums are all exact:

$$\begin{array}{ccccccccc} & & 0 & & 0 & & 0 & & \\ & & \downarrow & & \downarrow & & \downarrow & & \\ 0 & \rightarrow & \mathbf{A}(D)\cap F & \rightarrow & \mathbf{A}(D) & \rightarrow & \mathbf{A}(D)/\mathbf{A}(D)\cap F & \rightarrow & 0 \\ & & \downarrow & & \downarrow & & \downarrow & & \\ 0 & \rightarrow & F & \rightarrow & \mathbf{A} & \rightarrow & \mathbf{A}/F & \rightarrow & 0 \\ & & \downarrow & & \downarrow & & \downarrow & & \\ 0 & \rightarrow & F/\mathbf{A}(D)\cap F & \rightarrow & \mathbf{A}/\mathbf{A}(D) & \rightarrow & \mathbf{A}/\mathbf{A}(D)+F & \rightarrow & 0 \\ & & \downarrow & & \downarrow & & \downarrow & & \\ & & 0 & & 0 & & 0. & & \end{array}$$

So, we get the exact sequences $0 \to H^0(C, \mathcal{O}_C(D)) \to F \to F/\mathbf{A}(D)\cap F \to 0$ and

$$0 \to F/\mathbf{A}(D)\cap F \to \mathbf{A}/\mathbf{A}(D) \to H^1(C, \mathcal{O}_C(D)) \to 0;$$

(3′) For any $p \in C$, there exists a natural morphism from $\Sigma(D)$ to $\Sigma(D+p)$. Thus, by the five-lemma, based on the fact that F and $\mathbf{A}$ are the same in these two 9-diagrams, we obtain the exact sequences

$$0 \to \mathbf{A}(D)\cap F \to \mathbf{A}(D+p)\cap F \to \mathbf{A}(D+p)\cap F/\mathbf{A}(D)\cap F \to 0$$

$$\mathbf{A}(D+p)\cap F/\mathbf{A}(D)\cap F \rightarrowtail \mathbf{A}(D+p)/\mathbf{A}(D) \twoheadrightarrow \mathbf{A}(D+p)+F/\mathbf{A}(D)+F$$

$$0 \to \mathbf{A}(D+p)+F/\mathbf{A}(D)+F \to \mathbf{A}/\mathbf{A}(D)+F \to \mathbf{A}/\mathbf{A}(D+p)+F \to 0;$$

(4′) Residue pairing works in terms of adelic language as well by the self-dual property of $\mathbf{A}$.

Thus, what is left is to count the cohomology groups in order to have a cohomology theory using only adelic language (which satisfies duality and the Riemann-Roch).

Generally speaking, to find a reasonable way to count is very difficult. In Grothendieck's pure algebraic approach, this is based on the fact that *all coherent sheaves are locally finitely generated.* With adelic language, in general dimensions, we do not know how to do it. However for certain special kinds of adelic ring $\mathbf{A}$, particular for curves, we do have a limited success. To explain it, as an example, consider the simplest case for which the constant field k is a finite field say $\mathbf{F}_q$.

Recall that with respect to the natural topology on $\mathbf{A}$, F is discrete and $\mathbf{A}(D)$ is compact. Thus in particular, $H^0(C, \mathcal{O}_C(D)) = \mathbf{A}(D) \cap F$ is finite. Similarly, since $\mathbf{A}(D)$ is compact and $\mathbf{A}/F$ is compact so $\mathbf{A}(D) + F/F$ is again compact. But $k = \mathbf{F}_q$ is a finite fields, so compactness implies finiteness, and the number of elements in a finite dimensional space is simply q to the power of the corresponding dimension. Thus, with repsect to the natural Haar measures on the associated groups induced from that on $\mathbf{A}$,

$$\mathrm{Vol}(\mathbf{A}(D)) = q^{h^0(C,\mathcal{O}_C(D))} \cdot \mathrm{Vol}(\mathbf{A}(D) + F/F)$$

and

$$\mathrm{Vol}(\mathbf{A}/F) = q^{h^1(C,\mathcal{O}_C(D))} \cdot \mathrm{Vol}(\mathbf{A}(D) + F/F).$$

Therefore,

$$q^{h^0(C,\mathcal{O}_C(D))-h^1(C,\mathcal{O}_C(D))} = \frac{\mathrm{Vol}\mathbf{A}(D)}{\mathrm{Vol}(\mathbf{A}/F)}.$$

Easily by definition,

$$\mathrm{Vol}\mathbf{A}(D) = q^{d(D)}, \quad \mathrm{Vol}(\mathbf{A}/F) = q^{g-1},$$

so we obtain an analytic proof of the Riemann-Roch theorem

$$h^0(C, \mathcal{O}_C(D)) - h^1(C, \mathcal{O}_C(D)) = d(D) - (g-1).$$

C.3.2. Chevalley's Linear Compacity

The above discussion for curves over finite fields is not valid for curves over general fields, since even the space is finite dimensional, it is not finite itself. To overcome this difficult, Chevalley introduced his notion of linear compacity. Next, we indicate how Chevalley's idea works. (Instead of recalling all the details, we indicate what are the essential points involved. For details, we recommend the reader to consult Iwasawa's Princeton lecture notes.)

(0) There exists the 9-diagram as above;

(1) (Additive Structure) Among subquotient groups of the adelic ring are topological spaces called discrete objects and linearly compact objects. Moreover, all groups used in our 9-diagram are supposed to be locally linearly compact, i.e., they are either extensions of discrete objects by linearly compact objects, or extensions of linearly compact objects by discrete objects, or simply generated by finitely many discrete objects and linearly compact objects. In particular, $\mathbf{A}$ is selfdual and locally linearly compact;
(2) Discrete objects and linearly compact objects are dual to each other. Thus, if an object is both locally linearly compact and discrete, it is then isomorphic to a finite dimensional k vector space, and hence whose dimension makes sense;
(3) (Multiplicative Structure) Under the multiplication, $\mathbf{A}$ is selfdual. As a direct consequence, we get the duality.
(4) The count may be proceeded as follows to offer the Riemann-Roch:
(i) By definition, $\dim(\mathbf{A}(0) \cap F) = 1$ and $\dim(\mathbf{A}/\mathbf{A}(0) + F) = g$;
(ii) By counting local contribution, $[\mathbf{A}(D+p) : \mathbf{A}(D)] = d(D+p) - d(D) = 1$, despite the fact that $\mathbf{A}(D)$ and $\mathbf{A}(D+p)$ themselves cannot be counted;
(iii) By definition,

$$[\mathbf{A}(D+p) : \mathbf{A}(D)] = \dim\big(\mathbf{A}(D+p)/\mathbf{A}(D)\big)$$

(iv) The fundamental theorem of isomorphisms for groups implies that

$$\begin{aligned}
&\dim\big(\mathbf{A}(D+p)/\mathbf{A}(D)\big)\\
&= \dim\big(\mathbf{A}(D+p)+F/\mathbf{A}(D)+F\big) + \dim\big(\mathbf{A}(D+p)\cap F/\mathbf{A}(D)\cap F\big)\\
&= \dim(\mathbf{A}/\mathbf{A}(D)+F) - \dim(\mathbf{A}/\mathbf{A}(D+p)+F)\\
&\quad + \dim(\mathbf{A}(D+p)\cap F) - \dim(\mathbf{A}(D)\cap F)\\
&= \big(h^0(C,D+p) - h^1(C,D+p)\big) - \big(h^0(C,D) - h^1(C,D)\big).
\end{aligned}$$

C.3.3. Adelic Approach in Geometric Dimension Two

Now let us consider the situation in dimension two. Here in Grothendieck's theory, with a proper definition of cohomology groups and a suitable way of count, we can prove a (weak) Riemann-Roch theorem and duality using the adjunction formula, a one-dimensional Riemann-Roch and a long exact sequence of cohomology groups resulting from a short exact sequence of sheaves

$$0 \to \mathcal{O}(D) \to \mathcal{O}(D+C) \to \mathcal{O}(D)|_C \to 0$$

for a divisor D and a regular curve C on the surface, and a residual discussion. (See e.g. [Har].)

On the other hand, just as for curves, there is a two-dimensional cohomology theory in adelic language. There are two parts:
(I) Algebraic construction of cohomology groups and some associated structural exact sequences;
(II) Analytic Count of cohomology groups in (I).

We start with algebraic construction. In algebraic geometry, this is essentially given by Parshin as a by-product of his study on residues. (We now also have its higher dimensional generalization of Beilinson.) As above, we here only indicate the main points. The interested reader may consult Parshin's original paper for the details.

So for a surface S with function field F, denote by $\mathbf{A}$ its ring of adeles. Inside $\mathbf{A}$ are two subrings, which we denote by $\mathbf{A}_0$ and $\mathbf{A}_1$, respectively. Similarly, for a divisor D on S, introduce its associated subgroup $\mathbf{A}(D)$ as in curve case. Then we obtain **three** 9-diagrams:
(a)

$$\begin{array}{ccccc}
0 & & 0 & & 0 \\
\downarrow & & \downarrow & & \downarrow \\
\mathbf{A}(D)\cap\ \mathbf{A}_0\cap\mathbf{A}_1 & \rightarrowtail & \mathbf{A}(D) & \twoheadrightarrow & \mathbf{A}(D)/\mathbf{A}(D)\cap\ \mathbf{A}_0\cap\mathbf{A}_1 \\
\downarrow & & \downarrow & & \downarrow \\
\mathbf{A}_0\cap\mathbf{A}_1 & \rightarrowtail & \mathbf{A} & \twoheadrightarrow & \mathbf{A}/\mathbf{A}_0\cap\mathbf{A}_1 \\
\downarrow & & \downarrow & & \downarrow \\
\mathbf{A}_0\cap\mathbf{A}_1/\mathbf{A}(D)\cap\mathbf{A}_0\cap\mathbf{A}_1 & \rightarrowtail & \mathbf{A}/\mathbf{A}(D) & \twoheadrightarrow & \mathbf{A}/\mathbf{A}(D)+\mathbf{A}_0\cap\mathbf{A}_1 \\
\downarrow & & \downarrow & & \downarrow \\
0 & & 0 & & 0
\end{array}$$

for which isomorphisms

$$\mathbf{A}(D)/\mathbf{A}(D)\cap\big(\mathbf{A}_0\cap\mathbf{A}_1\big) \simeq \mathbf{A}(D)+\mathbf{A}_0\cap\mathbf{A}_1/\mathbf{A}_0\cap\mathbf{A}_1$$

$$\mathbf{A}_0\cap\mathbf{A}_1/\mathbf{A}(D)\cap\mathbf{A}_0\cap\mathbf{A}_1 \simeq \mathbf{A}(D)+\big(\mathbf{A}_0\cap\mathbf{A}_1\big)/\mathbf{A}(D)$$

are used; (**b**)

$$\begin{array}{ccccc}
0 & & 0 & & 0 \\
\downarrow & & \downarrow & & \downarrow \\
\mathbf{A}(D)\cap\ \mathbf{A}_0\cap\mathbf{A}_1 & \rightarrowtail & \mathbf{A}(D)\cap\mathbf{A}_1 & \twoheadrightarrow & \frac{\mathbf{A}(D)\cap\mathbf{A}_1}{\mathbf{A}(D)\cap\ \mathbf{A}_0\cap\mathbf{A}_1} \\
\downarrow & & \downarrow & & \downarrow \\
\mathbf{A}(D)\cap\mathbf{A}_0 & \rightarrowtail & \mathbf{A}(D)\cap\ \mathbf{A}_0+\mathbf{A}_1 & \twoheadrightarrow & \frac{\mathbf{A}(\mathbf{D})\cap\ \mathbf{A}_0+\mathbf{A}_1}{\mathbf{A}(D)\cap\mathbf{A}_0} \\
\downarrow & & \downarrow & & \downarrow \\
\frac{\mathbf{A}(D)\cap\mathbf{A}_0}{\mathbf{A}(D)\cap\ \mathbf{A}_0\cap\mathbf{A}_1} & \rightarrowtail & \frac{\mathbf{A}(D)\cap\ \mathbf{A}_0+\mathbf{A}_1}{\mathbf{A}(D)\cap\mathbf{A}_1} & \twoheadrightarrow & \frac{\mathbf{A}(D)\cap\ \mathbf{A}_0+\mathbf{A}_1}{\mathbf{A}(D)\cap\mathbf{A}_0+\mathbf{A}(D)\cap\mathbf{A}_1} \\
\downarrow & & \downarrow & & \downarrow \\
0 & & 0 & & 0
\end{array}$$

for which isomorphisms

$$\mathbf{A}(D)\cap\mathbf{A}_1/\mathbf{A}(D)\cap\big(\mathbf{A}_0\cap\mathbf{A}_1\big)\simeq\mathbf{A}(D)\cap\mathbf{A}_0+\mathbf{A}(D)\cap\mathbf{A}_1/\mathbf{A}(D)\cap\mathbf{A}_0,$$

$$\mathbf{A}(D)\cap\mathbf{A}_0/\mathbf{A}(D)\cap\big(\mathbf{A}_0\cap\mathbf{A}_1\big)\simeq\mathbf{A}(D)\cap\mathbf{A}_0+\mathbf{A}(D)\cap\mathbf{A}_1/\mathbf{A}(D)\cap\mathbf{A}_1$$

are used; and (**c**)

$$\begin{array}{ccccc}
0 & & 0 & & 0 \\
\downarrow & & \downarrow & & \downarrow \\
\mathbf{A}(D)\cap\ \mathbf{A}_0+\mathbf{A}_1 & \rightarrowtail & \mathbf{A}(D) & \twoheadrightarrow & \mathbf{A}(D)/\mathbf{A}(D)\cap\ \mathbf{A}_0+\mathbf{A}_1 \\
\downarrow & & \downarrow & & \downarrow \\
\mathbf{A}_0+\mathbf{A}_1 & \rightarrowtail & \mathbf{A} & \twoheadrightarrow & \mathbf{A}/\ \mathbf{A}_0+\mathbf{A}_1 \\
\downarrow & & \downarrow & & \downarrow \\
\mathbf{A}_0+\mathbf{A}_1\ /\mathbf{A}(D)\cap\ \mathbf{A}_0+\mathbf{A}_1 & \rightarrowtail & \mathbf{A}/\mathbf{A}(D) & \twoheadrightarrow & \mathbf{A}/\mathbf{A}(D)+\ \mathbf{A}_0+\mathbf{A}_1 \\
\downarrow & & \downarrow & & \downarrow \\
0 & & 0 & & 0
\end{array}$$

for which isomorphisms

$$\mathbf{A}(D)/\mathbf{A}(D)\cap\big(\mathbf{A}_0+\mathbf{A}_1\big)\simeq\mathbf{A}(D)+\big(\mathbf{A}_0+\mathbf{A}_1\big)/\big(\mathbf{A}_0+\mathbf{A}_1\big)$$

$$\big(\mathbf{A}_0+\mathbf{A}_1\big)/\mathbf{A}(D)\cap\big(\mathbf{A}_0+\mathbf{A}_1\big)\simeq\mathbf{A}(D)+\big(\mathbf{A}_0+\mathbf{A}_1\big)/\mathbf{A}(D)$$

are used.

Now, define the cohomology groups H^i, $i=0,1,2$ by setting

$$H^0(S,D):=\mathbf{A}(D)\cap\big(\mathbf{A}_0\cap\mathbf{A}_1\big);$$

$$H^1(S,D):=\mathbf{A}(D)\cap\big(\mathbf{A}_0+\mathbf{A}_1\big)/\mathbf{A}(D)\cap\mathbf{A}_0+\mathbf{A}(D)\cap\mathbf{A}_1;$$

and

$$H^2(S,D) := \mathbf{A}/\mathbf{A}(D) + \big(\mathbf{A}_0 + \mathbf{A}_1\big).$$

Moreover, by working with $D+C$ for regular curve C on S indtead of D, we get another set of three 9-diagrams, to which there are natural morphisms from the three 9-diagrams for D above to those for $D+C$. As a direct consequence, by a snake chasing, we arrive at the following long exact sequence of cohomologies, which as stated above, plays a key role in the induction process:

$$\begin{aligned} 0 &\to H^0(S,D) \to H^0(S,D+C) \to H^0(C,D+C|_C) \\ &\to H^1(S,D) \to H^1(S,D+C) \to H^1(C,D+C|_C) \\ &\to H^2(S,D) \to H^2(S,D+C) \to 0. \end{aligned}$$

In this way, provided that a good counting process is available, we then are able to give an adelic approach to the weak Riemann-Roch in dimension two by using the Riemann-Roch for curves with the help of the adjunction formula.

Thus the problem left is how to count. However we fail badly at this point. Say, in Parshin-Beilinson, this problem is 'solved' by claiming that adelic cohomology groups are the same as these resulting from Gerothendieck's theory and consequently we can count them.

Motivated by Chevalley's theory on linear compacity for adelic rings of algebraic curves, we suggest to study the terms used in the above algebraic discussion. The key is the self-dual property of $\mathbf{A}$. In fact, we have the following canonical isomorphisms:
(a) $A_0^\perp \simeq \mathbf{A}_1, A_1^\perp \simeq \mathbf{A}_0$, and $\mathbf{A}(D)^\perp = \mathbf{A}(K_S - D)$, where K_S denotes a canonical divisor of S. In particular,

$$\big(\mathbf{A}_0 + \mathbf{A}_1\big)^\perp \simeq \mathbf{A}_1 \cap \mathbf{A}_2, \qquad \big(\mathbf{A}_1 \cap \mathbf{A}_2\big)^\perp \simeq \mathbf{A}_0 + \mathbf{A}_1.$$

(b) Duality between $H^0(S,D)$ and $H^2(S,K_S-D)$

$$\begin{aligned} &\big(\mathbf{A}(D) \cap \mathbf{A}_0 \cap \mathbf{A}_1\big)^\vee \simeq \mathbf{A}/\big(\mathbf{A}(D)\cap\mathbf{A}_0\cap\mathbf{A}_1\big)^\perp \\ \simeq &\,\mathbf{A}/\mathbf{A}(D)^\perp + \mathbf{A}_0^\perp + \mathbf{A}_1^\perp \simeq \mathbf{A}/\mathbf{A}(K_S - D) + \big(\mathbf{A}_0 + \mathbf{A}_1\big), \end{aligned}$$

(c) Note that $\mathbf{A}_0 \cap \mathbf{A}_1 = F$ is nothing but the function field of S. Moreover, algebrically

$$\begin{aligned}&\mathbf{A}(D) \cap (\mathbf{A}_0 + \mathbf{A}_1)/\mathbf{A}(D) \cap \mathbf{A}_0 + \mathbf{A}(D) \cap \mathbf{A}_1\\ \simeq\, &\mathbf{A}_0 \cap (\mathbf{A}(D) + \mathbf{A}_1)/(\mathbf{A}(D) \cap \mathbf{A}_0 + \mathbf{A}_0 \cap \mathbf{A}_1)\\ \simeq\, &\mathbf{A}_1 \cap (\mathbf{A}(D) + \mathbf{A}_0)/(\mathbf{A}(D) \cap \mathbf{A}_1 + \mathbf{A}_0 \cap \mathbf{A}_1);\end{aligned}$$

(d) Duality between $H^1(S, D)$ and $H^1(S, K_S - D)$:

$$\begin{aligned}&\Big(\mathbf{A}_0 \cap (\mathbf{A}(D) + \mathbf{A}_1)/(\mathbf{A}(D) \cap \mathbf{A}_0 + \mathbf{A}_0 \cap \mathbf{A}_1)\Big)^\vee\\ \simeq\, &(\mathbf{A}(D) \cap \mathbf{A}_0 + \mathbf{A}_0 \cap \mathbf{A}_1)^\perp/(\mathbf{A}_0 \cap (\mathbf{A}(D) + \mathbf{A}_1))^\perp\\ \simeq\, &(\mathbf{A}(K_S - D) + \mathbf{A}_1) \cap (\mathbf{A}_0 + \mathbf{A}_1)/(\mathbf{A}_1 + \mathbf{A}(K_S - D) \cap \mathbf{A}_1)\\ \simeq\, &\mathbf{A}(K_S - D) \cap (\mathbf{A}_0 + \mathbf{A}_1)/\mathbf{A}(K_S - D) \cap \mathbf{A}_0 + \mathbf{A}(D) \cap \mathbf{A}_1.\end{aligned}$$

Thus, to attack (II), the analytic structure aiming at a reasonable count, we should introduce a geo-ari theory for surfaces which is compactible with (a), (b), (c) and (d) above, similar to that of linear compactness of Chevalley used in proving the Riemann-Roch in dimension one. For examples, there should be a notion of geo-ari compactness, and if a space is both discrete and geo-ari compact, it should be of finite dimensional; moreover, the duality should transform discrete spaces to geo-ari compact speces and vice versa.

Remark. The above program proves to be very valuable. Say along with the proposal outlined above, we successfully use Tate's thesis to provide a beautiful geo-ari cohomology theory over 1+0 dimensional bases, i.e., over number fields. For details, please refer to B.2 of this program. Also we expect that an adelic cohomology for 1+1 dimensional bases will be established shortly, while we have confidence on the theory over 2+0 dimensional bases up to Riemann-Roch. (Keep in mind that the main part of the Program was created around 1999-2000 and that the present one is a revised version in which important new developments are added.)

C.4. Riemann Hypothesis in Rank Two

C.4.1. Non-Abelian Zeta Functions and Eisenstein Series

C.4.1.1. Non-Abelian Zeta Functions for Number Fields

Let K be an algebraic number field (of finite degree n) with Δ_K the absolute value of its discriminant. Denote by $\mathcal{O}_K$ the ring of integers. For a fixed positive integer $r \in \mathbb{N}$, denote by $\mathcal{M}_{F,r}$ the moduli space of semi-stable $\mathcal{O}_K$-lattices of rank r with $d\mu$ the associated (Tamagawa type) measure (induced from that on GL). Recall that for each $\Lambda \in \mathcal{M}_{K,r}$, the associated *0-th geo-arithmetical cohomology* $h^0(K,\Lambda)$ is

$$h^0(K,\Lambda) := \log\left(\sum_{x\in\Lambda} \exp\left(-\pi\sum_{\sigma:\mathbb{R}} \|x_\sigma\|^2_{\rho_\sigma} - 2\pi\sum_{\sigma:\mathbb{C}} \|x_\sigma\|^2_{\rho_\sigma}\right)\right)$$

where $x = (x_\sigma)_{\sigma\in S_\infty}$ and $(\rho_\sigma)_{\sigma\in S_\infty}$ denote the σ-component of the metric $\rho = \rho_\Lambda$ determined by the lattice Λ with S_∞ a collection Archimedean places of K; and that the *rank r (non-abelian) zeta function $\xi_{K,r}(s)$ of K* is the integration

$$\xi_{K,r}(s) := \int_{\Lambda\in\mathcal{M}_{K,r}} \left(e^{h^0(K,\Lambda)-1}\right)\cdot\left(e^{-s}\right)^{-\log \mathrm{Vol}(\Lambda)} d\mu(\Lambda),\ \mathrm{Re}(s) > 1,$$

or the same, $\xi_{K,r}(s)$

$$:= \left(\Delta_K^{\frac{r}{2}}\right)^s \cdot \int_{\Lambda\in\mathcal{M}_{K,r}} \left(e^{h^0(K,\Lambda)-1}\right)\cdot\left(e^{-s}\right)^{\deg(\Lambda)} d\mu(\Lambda),\quad \mathrm{Re}(s) > 1.$$

Theorem. ([We6]) (1) (**Meromorphic Continuation**) *$\xi_{K,r}(s)$ is well-defined when* $\mathrm{Re}(s) > 1$ *and admits a meromorphic continuation, denoted also by $\xi_{K,r}(s)$, to the whole complex s-plane;*
(2) (**Functional Equation**) $\xi_{K,r}(1-s) = \xi_{K,r}(s)$;
(3) (**Singularities & Residues**) *$\xi_{K,r}(s)$ has two singularities, all simple poles, at $s = 0, 1$, with the same residues* $\mathrm{Vol}\left(\mathcal{M}_{K,r}\left([\Delta_K^{\frac{r}{2}}]\right)\right)$, *where $\mathcal{M}_{K,r}\left([\Delta_K^{\frac{r}{2}}]\right)$ denotes the moduli space of rank r semi-stable $\mathcal{O}_K$-lattices whose volumes are fixed to be $\Delta_K^{\frac{r}{2}}$.*

C.4.1.2. Space of $\mathcal{O}_K$-Lattices via Special Linear Groups

Recall that an $\mathcal{O}_K$-lattice Λ consists of a underlying projective $\mathcal{O}_K$-module P and a metric structure on the space $V = \Lambda \otimes_{\mathbb{Z}} \mathbb{R} = \prod_{\sigma \in S_\infty} V_\sigma$. Moreover, in assuming that the $\mathcal{O}_K$-rank of P is r, we can identify P with one of the $P_i := P_{r;\mathfrak{a}_i} := \mathcal{O}_K^{(r-1)} \oplus \mathfrak{a}_i$, where $\mathfrak{a}_i, i = 1, \dots, h$, are chosen integral $\mathcal{O}_K$-ideals so that $\Big\{[\mathfrak{a}_1], [\mathfrak{a}_2], \dots, [\mathfrak{a}_h]\Big\} = CL(K)$ the class group of K.

With this said, following Minkowski, we obtain a natural embedding for P:

$$P := \mathcal{O}_K^{(r-1)} \oplus \mathfrak{a} \hookrightarrow K^{(r)} \hookrightarrow \Big(\mathbb{R}^{r_1} \times \mathbb{C}^{r_2}\Big)^r \cong \Big(\mathbb{R}^r\Big)^{r_1} \times \Big(\mathbb{C}^r\Big)^{r_2},$$

which is simply the space $V = \Lambda \otimes_{\mathbb{Z}} \mathbb{R}$ above. As a direct consequence, our lattice Λ then is determined by a metric structure on $V = \prod_{\sigma \in S_\infty} V_\sigma$, or the same, on $\big(\mathbb{R}^r\big)^{r_1} \times \big(\mathbb{C}^r\big)^{r_2}$. It is well known that all these metrices are parametrized by the space

$$\Big(GL(r, \mathbb{R}) \big/ O(r)\Big)^{r_1} \times \Big(GL(r, \mathbb{C}) \big/ U(r)\Big)^{r_2}.$$

Next let us shift our discussion from GL to SL. For this, let us start with a local discussion on $\mathcal{O}_K$-lattice structures: For complex places τ, by fixing a branch of the n-th root, we get

$$GL(r, \mathbb{C}) \big/ U(r) \cong \Big(SL(r, \mathbb{C}) \big/ SU(r)\Big) \times \mathbb{R}_+^*;$$

while for real places σ, we have

$$GL(r, \mathbb{R}) \big/ O(r) \cong \Big(SL(r, \mathbb{R}) \big/ SO(r)\Big) \times \mathbb{R}_+^*.$$

Consequently, metric structures on $V = \prod_{\sigma \in S_\infty} V_\sigma \simeq (\mathbb{R}^r)^{r_1} \times (\mathbb{C}^r)^{r_2}$ are parametrized by

$$\bigg(\Big(SL(r, \mathbb{R}) \big/ SO(r)\Big)^{r_1} \times \Big(SL(r, \mathbb{C}) \big/ SU(r)\Big)^{r_2}\bigg) \times (\mathbb{R}_+^*)^{r_1 + r_2}.$$

Furthermore, when we work with $\mathcal{O}_K$-lattice strucures on P, i.e., with the space $\mathbf{\Lambda} = \mathbf{\Lambda}(P)$, from above parametrized space of metric structures on $V = \prod_{\sigma\in S_\infty} V_\sigma$, we need to also factor out $GL(P)$, the automorphism group $\mathrm{Aut}_{\mathcal{O}_K}(\mathcal{O}_K^{(r-1)}\oplus\mathfrak{a})$ of $\mathcal{O}_K^{(r-1)}\oplus\mathfrak{a}$ as $\mathcal{O}_K$-modules, to get

$$GL(P)\Big\backslash \left(\left(\left(SL(r,\mathbb{R})\big/SO(r)\right)^{r_1}\times\left(SL(r,\mathbb{C})\big/SU(r)\right)^{r_2}\right)\times(\mathbb{R}_+^*)^{r_1+r_2}\right).$$

Hence we want
(a) to study the structure of the group $\mathrm{Aut}_{\mathcal{O}_K}(\mathcal{O}_K^{(r-1)}\oplus\mathfrak{a})$ in terms of SL and units; and
(b) to see how this group acts on the space of metric structures

$$\left(\left(SL(r,\mathbb{R})/SO(r)\right)^{r_1}\times\left(SL(r,\mathbb{C})/SU(r)\right)^{r_2}\right)\times(\mathbb{R}_+^*)^{r_1+r_2}.$$

View $\mathrm{Aut}_{\mathcal{O}_K}(\mathcal{O}_K^{(r-1)}\oplus\mathfrak{a})$ as a subgroup of $GL(r,K)$. Then, $\mathrm{Aut}_{\mathcal{O}_K}(\mathcal{O}_K^{(r-1)}\oplus\mathfrak{a})$

$$=\left\{A\in GL(r,K)\cap\begin{pmatrix} & & & \mathfrak{a}\\ & \mathcal{O}_K & & \vdots\\ & & & \mathfrak{a}\\ \mathfrak{a}^{-1} & \cdots & \mathfrak{a}^{-1} & \mathcal{O}_K\end{pmatrix}:\det A\in U_K\right\}.$$

Introduce subgroups $GL^+(r,\mathbb{R}) := \Big\{g\in GL(r,\mathbb{R}) : \det g > 0\Big\}$ and $O^+(r) := \Big\{A\in O(r,\mathbb{R}) : \det g > 0\Big\}$, and the subgroup $\mathrm{Aut}^+_{\mathcal{O}_K}(\mathcal{O}_K^{(r-1)}\oplus\mathfrak{a})$ of $\mathrm{Aut}_{\mathcal{O}_K}(\mathcal{O}_K^{(r-1)}\oplus\mathfrak{a})$ consisting of these elements whose local determinants at real places are all positive. Then, by checking directly, we obtain a natural identification of quotient spaces between

$$\mathrm{Aut}_{\mathcal{O}_K}(\mathcal{O}_K^{(r-1)}\oplus\mathfrak{a})\Big\backslash\Big(\big(GL(r,\mathbb{R})/O(r)\big)^{r_1}\times\big(GL(r,\mathbb{C})/U(r)\big)^{r_2}\Big)$$

and

$$\mathrm{Aut}^+_{\mathcal{O}_K}(\mathcal{O}_K^{(r-1)}\oplus\mathfrak{a})\Big\backslash\Big(\big(GL^+(r,\mathbb{R})/O^+(r)\big)^{r_1}\times\big(GL(r,\mathbb{C})/U(r)\big)^{r_2}\Big).$$

With all this, we are ready to shift further to the special linear group SL. It is here that Dirichlet's Unit Theorem plays a key role.

Recall that for a unit $\varepsilon \in U_K$,
(a) $\mathrm{diag}(\varepsilon, \ldots, \varepsilon) \in \mathrm{Aut}_{\mathcal{O}_K}(\mathcal{O}_K^{(r-1)} \oplus \mathfrak{a})$; and
(b) $\det \mathrm{diag}(\varepsilon, \ldots, \varepsilon) = \varepsilon^r \in U_F^r := \{\varepsilon^r : \varepsilon \in U_K\}$.
So to begin with, note that to pass from GL to SL over K, we need to use the intermediate subgroup GL^+. Accordingly, introduce a subgroup U_K^+ of U_K by setting

$$U_K^+ := \{\varepsilon \in U_K : \varepsilon_\sigma > 0, \forall \sigma \text{ real}\}$$

so as to get a well-controlled subgroup $U_K^{r,+} := U_K^+ \cap U_K^r$: by Dirichlet's Unit Theorem, the quotient group $U_K^+/(U_K^+ \cap U_K^r)$ is finite.

With this said, next we use $U_K^+ \cap U_K^r$ to decomposite the automorphism group $\mathrm{Aut}_{\mathcal{O}_K}(\mathcal{O}_K^{(r-1)} \oplus \mathfrak{a})$. Thus, choose elements

$$u_1, \ldots, u_{\mu(r,K)} \in U_K^+$$

such that $\Big\{[u_1], \ldots, [u_{\mu(r,K)}]\Big\}$ gives a complete representatives of the finite quotient group $U_K^+/\Big(U_K^+ \cap U_K^r\Big)$, where $\mu(r, K)$ denotes the cardinality of the group $U_K^+/(U_K^+ \cap U_K^r)$. Set also

$$SL(\mathcal{O}_K^{(r-1)} \oplus \mathfrak{a}) := SL(r, K) \cap GL(\mathcal{O}_K^{(r-1)} \oplus \mathfrak{a}).$$

Lemma. ([We7]) *There exist $A_1, \ldots, A_{\mu(r,K)}$ in $GL^+(\mathcal{O}_K^{(r-1)} \oplus \mathfrak{a})$ such that*
(i) $\det A_i = u_i$ for $i = 1, \ldots, \mu(r, K)$; and
(ii) $\Big\{A_1, \ldots, A_{\mu(r,K)}\Big\}$ *is a completed representative of the quotient of* $\mathrm{Aut}^+_{\mathcal{O}_K}\Big(\mathcal{O}_K^{(r-1)} \oplus \mathfrak{a}\Big)$ *modulo* $SL(\mathcal{O}_K^{(r-1)} \oplus \mathfrak{a}) \times \Big(U_K^{r,+} \cdot \mathrm{diag}(1, \ldots, 1)\Big)$.

That is to say, for automorphism groups,
(a) $\mathrm{Aut}^+_{\mathcal{O}_K}(\mathcal{O}_K^{(r-1)} \oplus \mathfrak{a})$ is naturally identified with the disjoint union

$$\cup_{i=1}^{\mu(r,K)} A_i \cdot \left(SL(\mathcal{O}_K^{(r-1)} \oplus \mathfrak{a}) \times \Big(U_K^{r,+} \cdot \mathrm{diag}(1, \ldots, 1)\Big)\right);$$

and, consequently,
(b) The $\mathcal{O}_K$-lattice structures $\mathbf{\Lambda}(P)$ on the projective $\mathcal{O}_K$-module $P=\mathcal{O}_K^{(r-1)}\oplus\mathfrak{a}$ are parametrized by the disjoint union

$$\cup_{i=1}^{\mu(r,K)}A_i\backslash\Bigg(\bigg(SL(\mathcal{O}_K^{(r-1)}\oplus\mathfrak{a})\backslash\Big(\big(SL(r,\mathbb{R})/SO(r)\big)^{r_1}\times\big(SL(r,\mathbb{C})/SU(r)\big)^{r_2}\Big)\bigg)\times\Big(|U_K^r\cap U_K^+|\backslash\big(\mathbb{R}_+^*\big)^{r_1+r_2}\Big)\Bigg).$$

Therefore, to understand the space of $\mathcal{O}_K$-lattice structures, beyond the spaces $SL(r,\mathbb{R})/SO(r)$ and $SL(r,\mathbb{C})/SU(r)$, we further need to study
(i) the quotient space $|U_K^r\cap U_K^+|\backslash\big(\mathbb{R}_+^*\big)^{r_1+r_2}$, which is more or less standard (see e.g. [Ne] or [We7]); and more importantly,
(ii) the quotient space

$$SL(\mathcal{O}_K^{(r-1)}\oplus\mathfrak{a})\backslash\Big(\big(SL(r,\mathbb{R})/SO(r)\big)^{r_1}\times\big(SL(r,\mathbb{C})/SU(r)\big)^{r_2}\Big).$$

C.4.1.3. Non-Abelian Zeta Functions and Epstein Zeta Functions

Choose integral $\mathcal{O}_K$-ideals $\mathfrak{a}_1=\mathcal{O}_K,\mathfrak{a}_2,\dots,\mathfrak{a}_h$ such that the ideal class group $CL(K)$ is given by $\Big\{[\mathfrak{a}_1],\dots,[\mathfrak{a}_h]\Big\}$, and that any rank r projective $\mathcal{O}_K$-module P is isomorphic to $P_{\mathfrak{a}_i}$ for a certain $i, 1\leq i\leq h$. Here, $P_\mathfrak{a}:=P_{r,\mathfrak{a}}:=\mathcal{O}_K^{(r-1)}\oplus\mathfrak{a}$ for a fractional $\mathcal{O}_K$-ideal $\mathfrak{a}$. Introduce the partial non-abelian zeta function $\xi_{K,r;\mathfrak{a}}(s)$ associated to $\mathfrak{a}$ by setting

$$\xi_{K,r;\mathfrak{a}}(s):=\int_{\mathcal{M}_{K,r}(\mathfrak{a})}\Big(e^{h^0(K,\Lambda)}-1\Big)\cdot\Big(e^{-s}\Big)^{-\log\mathrm{Vol}(\Lambda)}d\mu(\Lambda),\mathrm{Re}(s)>1$$

where $\mathcal{M}_{K,r}(\mathfrak{a})$ denotes the part of the moduli space of semi-stable $\mathcal{O}_K$-lattices whose corresponding points lie in

$$\Big(SL(\mathcal{O}_K^{(r-1)}\oplus\mathfrak{a})\Big\backslash\Big(\big(SL(r,\mathbb{R})/SO(r)\big)^{r_1}\times\big(SL(r,\mathbb{C})/SU(r)\big)^{r_2}\Big)\Big)\times\Big(|U_K^r\cap U_K^+|\backslash(\mathbb{R}_+^*)^{r_1+r_2}\Big).$$

Accordingly, for $\mathrm{Re}(s) > 1$, define the *completed Epstein zeta function* $\hat{E}_{K,r;\mathfrak{a}}(s)$ by

$$\hat{E}_{K,r;\mathfrak{a}}(s) := \left(\pi^{-\frac{rs}{2}}\Gamma\left(\frac{rs}{2}\right)\right)^{r_1} \cdot \left((2\pi)^{-rs}\Gamma(rs)\right)^{r_2} \cdot \left[\left(N(\mathfrak{a}) \cdot \Delta_K^{\frac{r}{2}}\right)^s \sum_{x\in(\Lambda\backslash\{0\})/U_{r,F}^+} \frac{1}{\|x\|_\Lambda^{rs}}\right].$$

With this, from the discussion in 4.1.2, by applying the Mellin transform and using the formula

$$\int_0^\infty e^{-At^B} t^s \frac{dt}{t} = \frac{1}{B} \cdot A^{-\frac{s}{B}} \cdot \Gamma\left(\frac{s}{B}\right)$$

(whenever both sides make sense,) we obtain the following

Proposition. ([We7]) (1) (**Decomposition**) *The rank r non-abelian zeta function of K admits a natural decomposition*

$$\xi_{K,r}(s) = \mu(r,K) \cdot \sum_{i=1}^h \xi_{K,r;\mathfrak{a}_i}(s);$$

(2) (**Non-Abelian Zeta = Integration of Epstein Zeta**) *The partial rank r non-abelian zeta function $\xi_{F,r;\mathfrak{a}}(s)$ of K associated to $\mathfrak{a}$ is given by an integration of a completed Epstein type zeta function:*

$$\xi_{F,r;\mathfrak{a}}(s) = \left(\frac{r}{2}\right)^{r_1+r_2} \cdot \int_{\mathcal{M}_{F,r;\mathfrak{a}}[N(\mathfrak{a})\cdot\Delta_K^{\frac{r}{2}}]} \hat{E}_{K,r;\mathfrak{a}}(s)\, d\mu, \qquad \mathrm{Re}(s) > 1.$$

C.4.2. Rank Two $\mathcal{O}_K$-Lattices: Stability and Distance to Cusps

C.4.2.1. Upper Half Plane

On the upper half plane $\mathcal{H} := \{z = x+iy \in \mathbb{C}, x \in \mathbb{R}, y \in \mathbb{R}_+^*\}$, the natural hyperbolic metric is given by the line element $ds^2 := \frac{dx^2+dy^2}{y^2}$ with volume form $d\mu := \frac{dx\wedge dy}{y^2}$. Consequently, the associated hyperbolic Laplace-Beltrami operator can be written as

$$\Delta := y^2\left(\frac{\partial^2}{\partial x^2} + \frac{\partial^2}{\partial y^2}\right).$$

The group $SL(2,\mathbb{R})$ naturally acts on $\mathcal{H}$ via:

$$M\,z := \frac{az+b}{cz+d}, \qquad \forall M = \begin{pmatrix} a & b \\ c & d \end{pmatrix} \in SL(2,\mathbb{R}), \quad z \in \mathcal{H}.$$

The stablizer of $i = (0,1) \in \mathcal{H}$ is equal to $SO(2) := \{A \in O(2) : \det A = 1\}$. Since this action is transitive, we can identify the quotient $SL(2,\mathbb{R})/SO(2)$ with $\mathcal{H}$ with the quotient map induced from $SL(2,\mathbb{R}) \to \mathcal{H}$, $g \mapsto g \cdot i$.

$\mathcal{H}$ admits the real line $\mathbb{R}$ as its boundary. Consequently, to compactify it, we add on it the real projective line $\mathbb{P}^1(\mathbb{R})$ with $\infty = \begin{bmatrix} 1 \\ 0 \end{bmatrix}$. Naturally, the above action of $SL(2,\mathbb{R})$ also extends to $\mathbb{P}^1(\mathbb{R})$ via

$$\begin{pmatrix} a & b \\ c & d \end{pmatrix} \begin{bmatrix} x \\ y \end{bmatrix} = \begin{bmatrix} ax+by \\ cx+dy \end{bmatrix}.$$

C.4.2.2. Upper Half Space

The 3-dimensional hyperbolic space may be written as

$$\begin{aligned} \mathbb{H} := \mathbb{C} \times]0,\infty[&= \Big\{(z,r) : z = x+iy \in \mathbb{C}, r \in \mathbb{R}_+^*\Big\} \\ &= \Big\{(x,y,r) : x,y \in \mathbb{R}, r \in \mathbb{R}_+^*\Big\}. \end{aligned}$$

We will think of $\mathbb{H}$ as a subset of Hamilton's quaternions with 1, i, j, k the standard $\mathbb{R}$-basis. Write points P in $\mathbb{H}$ as

$$P = (z,r) = (x,y,r) = z + rj \qquad \text{where } \; z = x+iy, \; j = (0,0,1).$$

We equip $\mathbb{H}$ with the hyperbolic metric coming from the line element $ds^2 := \frac{dx^2+dy^2+dr^2}{r^2}$ with hyperbolic volume form $d\mu := \frac{dx\wedge dy\wedge dr}{r^3}$. Consequently, the hyperbolic Laplace-Beltrami operator associated to the hyperbolic metric ds^2 is simply

$$\Delta := r^2\Big(\frac{\partial^2}{\partial x^2} + \frac{\partial^2}{\partial y^2} + \frac{\partial^2}{\partial r^2}\Big) - r\frac{\partial}{\partial r}.$$

The natural action of $SL(2,\mathbb{C})$ on $\mathbb{H}$ and on its boundary $\mathbb{P}^1(\mathbb{C})$ may be described as follows: We represent an element of $\mathbb{P}^1(\mathbb{C})$ by

$\begin{bmatrix} x \\ y \end{bmatrix}$ where $x, y \in \mathbb{C}$ with $(x, y) \neq (0,0)$. Then the action of the matrix $M = \begin{pmatrix} a & b \\ c & d \end{pmatrix} \in SL(2, \mathbb{C})$ on $\mathbb{P}^1(\mathbb{C})$ is defined to be

$$\begin{bmatrix} x \\ y \end{bmatrix} \mapsto \begin{pmatrix} a & b \\ c & d \end{pmatrix} \begin{bmatrix} x \\ y \end{bmatrix} := \begin{bmatrix} ax + by \\ cx + dy \end{bmatrix}.$$

Moreover, if we represent points $P \in \mathbb{H}$ as quaternions whose fourth component equals zero, then the action of M on $\mathbb{H}$ is defined to be

$$P \mapsto M\,P := (aP + b)(cP + d)^{-1},$$

where the inverse on the right is taken in the skew field of quaternions.

Furthermore, with this action, the stabilizer of $j = (0,0,1) \in \mathbb{H}$ in $SL(2, \mathbb{C})$ is equal to $SU(2) := \{A \in U(2) : \det A = 1\}$. Since the action of $SL(2, \mathbb{C})$ on $\mathbb{H}$ is transitive, we obtain also a natural identification $\mathbb{H} \simeq SL(2, \mathbb{C})/SU(2)$ via the quotient map induced from $SL(2, \mathbb{C}) \to \mathbb{H}, \quad g \mapsto g \cdot j$.

C.4.2.3. Rank Two $\mathcal{O}_K$-Lattices: Upper Half Space Model

Identifying $\mathcal{H}$ with $SL(2, \mathbb{R})/SO(2)$ and $\mathbb{H}$ with $SL(2, \mathbb{C})/SU(2)$, using the discussion in C.4.1, we conclude that

$$\mathcal{M}_{K,2;\mathfrak{a}}[N(\mathfrak{a}) \cdot \Delta_K] \simeq \Big(SL(\mathcal{O}_K \oplus \mathfrak{a}) \Big\backslash \big(\mathcal{H}^{r_1} \times \mathbb{H}^{r_2}\big)\Big)_{\mathrm{ss}},$$

where as usual ss means the subset consisting of points corresponding to rank two semi-stable $\mathcal{O}_K$-lattices in

$$SL(\mathcal{O}_K \oplus \mathfrak{a}) \Big\backslash \Big(\big(SL(2, \mathbb{R})/SO(2)\big)^{r_1} \times \big(SL(2, \mathbb{C})/SU(2)\big)^{r_2}\Big).$$

More precisely, if the metric on $\mathcal{O}_K \oplus \mathfrak{a}$ is given by $g = (g_\sigma)_{\sigma \in S_\infty}$ with $g_\sigma \in SL(2, K_\sigma)$, then the corresponding points on the right hand side is $g(\mathrm{ImJ})$ with $\mathrm{ImJ} := (i^{(r_1)}, j^{(r_2)})$, i.e., the point given by $(g_\sigma \tau_\sigma)_{\sigma \in S_\infty}$ where $\tau_\sigma = i_\sigma := (0,1)$ if σ is real and $\tau_\sigma = j_\sigma := (0,0,1)$ if σ is complex.

C.4.2.4. Cusps

As such, the working site becomes the space $SL(\mathcal{O}_K\oplus\mathfrak{a})\Big\backslash\Big(\mathcal{H}^{r_1}\times \mathbb{H}^{r_2}\Big)$. Here the action of $SL(\mathcal{O}_K\oplus\mathfrak{a})$ is via the action of $SL(2,K)$ on $\mathcal{H}^{r_1}\times\mathbb{H}^{r_2}$. More precisely, K^2 admits natural embeddings $K^2\hookrightarrow \Big(\mathbb{R}^{r_1}\times\mathbb{C}^{r_2}\Big)^2\simeq (\mathbb{R}^2)^{r_1}\times(\mathbb{C}^2)^{r_2}$ so that $\mathcal{O}_K\oplus\mathfrak{a}$ naturally embeds into $(\mathbb{R}^2)^{r_1}\times(\mathbb{C}^2)^{r_2}$ as a rank two $\mathcal{O}_K$-lattice. As such, $SL(\mathcal{O}_K\oplus\mathfrak{a})$ acts on the image of $\mathcal{O}_K\oplus\mathfrak{a}$ in $(\mathbb{R}^2)^{r_1}\times(\mathbb{C}^2)^{r_2}$ as automorphisms. Our task here is to understand the cusps of this action of $SL(\mathcal{O}_K\oplus\mathfrak{a})$ on $\mathcal{H}^{r_1}\times\mathbb{H}^{r_2}$. For this, we go as follows.

First, the space $\mathcal{H}^{r_1}\times\mathbb{H}^{r_2}$ admits a natural boundary $\mathbb{R}^{r_1}\times\mathbb{C}^{r_2}$, in which the field K is imbedded via Archmidean places in S_∞: $K\hookrightarrow \mathbb{R}^{r_1}\times\mathbb{C}^{r_2}$. Consequently, $\mathbb{P}^1(K)\hookrightarrow\mathbb{P}^1(\mathbb{R})^{r_1}\times\mathbb{P}^1(\mathbb{C})^{r_2}$ with $\begin{bmatrix}1\\0\end{bmatrix}:= \infty\mapsto(\infty^{(r_1)},\infty^{(r_2)})$. As usual, via fractional linear transformations, $SL(2,\mathbb{R})$ acts on $\mathbb{P}^1(\mathbb{R})$, and $SL(2,\mathbb{C})$ acts on $\mathbb{P}^1(\mathbb{C})$, hence so does $SL(2,K)$ on

$$\mathbb{P}^1(K)\hookrightarrow\mathbb{P}^1(\mathbb{R})^{r_1}\times\mathbb{P}^1(\mathbb{C})^{r_2}.$$

Being a discrete subgroup of $SL(2,\mathbb{R})^{r_1}\times SL(2,\mathbb{C})^{r_2}$, for the action of $SL(\mathcal{O}_K\oplus\mathfrak{a})$ on $\mathbb{P}^1(K)$, we call the corresponding orbits (of $SL(\mathcal{O}_K\oplus\mathfrak{a})$ on $\mathbb{P}^1(K)$) the *cusps.*

With this, we have the following fundamental result rooted back to Maaβ.

Cusp and Ideal Class Correspondence. ([We7]) *There is a natural bijection between the ideal class group $CL(K)$ of K and the cusps $\mathcal{C}_\Gamma$ of $\Gamma=SL(\mathcal{O}_K\oplus\mathfrak{a})$ acting on $\mathcal{H}^{r_1}\times\mathbb{H}^{r_2}$ given by*

$$\mathcal{C}_\Gamma\to CL(K),\qquad \begin{bmatrix}\alpha\\\beta\end{bmatrix}\mapsto\Big[\mathcal{O}_K\,\alpha+\mathfrak{a}\,\beta\Big].$$

Easily, one checks that the inverse map Π^{-1} is given as follows: For $\mathfrak{b}$, choose $\alpha_\mathfrak{b},\beta_\mathfrak{b}\in K$ such that $\mathcal{O}_K\cdot\alpha_\mathfrak{b}+\mathfrak{a}\cdot\beta_\mathfrak{b}=\mathfrak{b}$; With this, then $\Pi^{-1}([\mathfrak{b}])$ is simply the class of the point $\begin{bmatrix}\alpha_\mathfrak{b}\\\beta_\mathfrak{b}\end{bmatrix}$ in $SL(2,\mathcal{O}_K\oplus$

$a)\backslash\mathbb{P}^1(K)$. Moreover, there always exists $M_{\begin{bmatrix}\alpha\\ \beta\end{bmatrix}} := \begin{pmatrix}\alpha & \alpha^*\\ \beta & \beta^*\end{pmatrix} \in$ $SL(2,K)$ such that $M_{\begin{bmatrix}\alpha\\ \beta\end{bmatrix}} \cdot \infty = \begin{bmatrix}\alpha\\ \beta\end{bmatrix}$.

C.4.2.5. Stablizer Groups of Cusps

Recall that under the Cusp-Ideal Class Correspondence, there are exactly h unequivalent cusps η_i, $i = 1, 2, \ldots, h$. Moreover, if we write the cusp $\eta := \eta_i = \begin{bmatrix}\alpha_i\\ \beta_i\end{bmatrix}$ for suitable α_i, $\beta_i \in K$, then the associated ideal class is exactly the one for the fractional ideal $\mathcal{O}_K\alpha_i + \mathfrak{a}\beta_i =: \mathfrak{b}_i$. Denote the stablizer group of η in $SL(\mathcal{O}_K \oplus \mathfrak{a})$ by Γ_η.

Lemma. ([We7]) *The associated 'lattice' for the cusp η is given by $\mathfrak{a}\mathfrak{b}^{-2}$. Moreover,*

$$A^{-1}\Gamma_\eta A = \left\{ \begin{pmatrix} u & z\\ 0 & u^{-1}\end{pmatrix} : u \in U_K, z \in \mathfrak{a}\mathfrak{b}^{-2}\right\}.$$

Set $\Gamma'_\eta := \left\{ A \begin{pmatrix} 1 & z\\ 0 & 1\end{pmatrix} A^{-1} : z \in \mathfrak{a}\mathfrak{b}^{-2}\right\}$, Then

$$\Gamma_\eta = \Gamma'_\eta \times \left\{ A \begin{pmatrix} u & 0\\ 0 & u^{-1}\end{pmatrix} A^{-1} : u \in U_K\right\}.$$

Note also that componentwise, $\begin{pmatrix} u & 0\\ 0 & u^{-1}\end{pmatrix} z = \frac{uz}{u^{-1}} = u^2 z$. So, in practice, what we really get is the following decomposition

$$\Gamma_\eta = \Gamma'_\eta \times U_K^2$$

with

$$U_K^2 \simeq \left\{ A \cdot \begin{pmatrix} u & 0\\ 0 & u^{-1}\end{pmatrix} \cdot A^{-1} : u \in U_K\right\} \simeq \left\{ A \begin{pmatrix} 1 & 0\\ 0 & u^2\end{pmatrix} A^{-1} : u \in U_K\right\}.$$

With this, we are ready to construct a fundamental domain for the action of $\Gamma_\eta \subset SL(\mathcal{O}_K \oplus \mathfrak{a})$ on $\mathcal{H}^{r_1} \times \mathbb{H}^{r_2}$. This is based on a construction of a fundamental domain for the action of Γ_∞ on $\mathcal{H}^{r_1} \times \mathbb{H}^{r_2}$. More precisely, with an element $A = \begin{pmatrix} \alpha & \alpha^* \\ \beta & \beta^* \end{pmatrix} \in SL(2, K)$ (always exists!), we have

i) $A \cdot \infty = \begin{bmatrix} \alpha \\ \beta \end{bmatrix}$; and

ii) The isotropy group of η in $A^{-1}SL(\mathcal{O}_K \oplus \mathfrak{a})A$ is generated by translations $\tau \mapsto \tau + z$ with $z \in \mathfrak{a}\mathfrak{b}^{-2}$ and by dilations $\tau \mapsto u\tau$ where u runs through the group U_K^2.

(Here, we use A, α, β, $\mathfrak{b}$ as running symbols for A_i, α_i, β_i, $\mathfrak{b}_i := \mathcal{O}_K\alpha_i + \mathfrak{a}\beta_i$.)

Consider then the map $\mathcal{H}^{r_1} \times \mathbb{H}^{r_2} \xrightarrow{\mathrm{ImJ}} \mathbb{R}_{>0}^{r_1+r_2}$ defined by

$$(z_1, \ldots, z_{r_1}; P_1, \ldots, P_{r_2}) \mapsto (\mathrm{Im}(z_1), \ldots, \mathrm{Im}(z_{r_1}); J(P_1), \ldots, J(P_{r_2}))$$

where if $z = x + iy \in \mathcal{H}$ resp. $P = z + rj \in \mathbb{H}$, we set $\mathrm{Im}(z) = y$ resp. $J(P) = r$. It induces a map

$$\left(A^{-1} \cdot \Gamma_\eta \cdot A\right)\Big\backslash\left(\mathcal{H}^{r_1} \times \mathbb{H}^{r_2}\right) \to U_K^2\Big\backslash\mathbb{R}_{>0}^{r_1+r_2},$$

which exhibits $\left(A^{-1} \cdot \Gamma_\eta \cdot A\right)\Big\backslash\left(\mathcal{H}^{r_1} \times \mathbb{H}^{r_2}\right)$ as a torus bundle over $U_K^2\Big\backslash\mathbb{R}_{>0}^{r_1+r_2}$ with fiber the $n = r_1 + 2r_2$ dimensional torus $\left(\mathbb{R}^{r_1} \times \mathbb{C}^{r_2}\right)\Big/\mathfrak{a}\mathfrak{b}^{-2}$. Having factored out the action of the translations, we only have to construct a fundamental domain for the action of U_K^2 on $\mathbb{R}_{>0}^{r_1+r_2}$. For this, we look first at the action of U_K^2 on the norm-one hypersurface $\mathbf{S} := \left\{y \in \mathbb{R}_{>0}^{r_1+r_2} : N(y) = 1\right\}$. By taking logarithms, it is transformed bijectively into a trace-zero hyperplane which is isomorphic to the space $\mathbb{R}^{r_1+r_2-1}$

$$\begin{aligned} \mathbf{S} \xrightarrow{\log} \mathbb{R}^{r_1+r_2-1} &:= \left\{(a_1, \ldots, a_{r_1+r_2}) \in \mathbb{R}^{r_1+r_2} : \sum a_i = 0\right\}, \\ y \mapsto \quad & \left(\log y_1, \ldots, \log y_{r_1+r_2}\right), \end{aligned}$$

where the action of U_K^2 on $\mathbf{S}$ is carried out over an action on $\mathbb{R}^{r_1+r_2-1}$ by the translations $a_i \mapsto a_i + \log \varepsilon^{(i)}$. By Dirichlet's Unit Theorem, the logarithm transforms U_K^2 into a lattice in $\mathbb{R}^{r_1+r_2-1}$. Accordingly, the exponential map transforms a fundamental domain, e.g., a fundamental parallelepiped, for this action back into a fundamental domain $\mathbf{S}_{U_K^2}$ for the action of U_K^2 on $\mathbf{S}$. The cone over $\mathbf{S}_{U_K^2}$, that is, $\mathbb{R}_{>0} \cdot \mathbf{S}_{U_K^2} \subset \mathbb{R}_{>0}^{r_1+r_2}$, is then a fundamental domain for the action of U_K^2 on $\mathbb{R}_{>0}^{r_1+r_2}$. Denote by $\mathcal{T}$ a fundamental domain for the action of the translations by elements of $\mathfrak{a}\mathfrak{b}^{-2}$ on $\mathbb{R}^{r_1} \times \mathbb{C}^{r_2}$, and set

$$\begin{aligned}&\mathrm{ReZ}\left(z_1, \ldots, z_{r_1}; P_1, \ldots, P_{r_2}\right)\\ &:= \left(\mathrm{Re}(z_1), \ldots, \mathrm{Re}(z_{r_1}); Z(P_1), \ldots, Z(P_{r_2})\right)\end{aligned}$$

with $\mathrm{Re}(z) := x$ resp. $Z(P) := z$ if $z = x + iy \in \mathcal{H}$ resp. $P = z + rj \in \mathbb{H}$, then what we have just said proves the following

Theorem. ([We7]) *A fundamental domain for the action of* $A^{-1}\Gamma_\eta A$ *on* $\mathcal{H}^{r_1} \times \mathbb{H}^{r_2}$ *is given by*

$$\mathbf{E} := \left\{\tau \in \mathcal{H}^{r_1} \times \mathbb{H}^{r_2} : \mathrm{ReZ}\,(\tau) \in \mathcal{T},\ \mathrm{ImJ}\,(\tau) \in \mathbb{R}_{>0} \cdot \mathbf{S}_{U_K^2}\right\}.$$

For later use, we also set $\mathcal{F}_\eta := A_\eta^{-1} \cdot \mathbf{E}$.

C.4.2.6. Fundamental Domain

Guided by Siegel's discussion on totally real fields [S] and the discussion above, we are now ready to construct fundamental domains for $SL(\mathcal{O}_K \oplus \mathfrak{a})\Big\backslash\left(\mathcal{H}^{r_1} \times \mathbb{H}^{r_2}\right)$.

As the first step, we generalize Siegel's 'distance to cusps'. For this, recall that for a cusp $\eta = \begin{bmatrix}\alpha\\ \beta\end{bmatrix} \in \mathbb{P}^1(K)$, by the Cusp-Ideal Class Correspondence, we obtain a natural ideal class associated to the fractional ideal $\mathfrak{b} := \mathcal{O}_K \cdot \alpha + \mathfrak{a} \cdot \beta$. Moreover, by assuming that α, β are all contained in $\mathcal{O}_K$, as we may, we know that the corresponding stablizer group Γ_η is given by

$$A^{-1} \cdot \Gamma_\eta \cdot A = \left\{\gamma = \begin{pmatrix} u & z\\ 0 & u^{-1}\end{pmatrix} \in \Gamma : u \in U_K, z \in \mathfrak{a}\mathfrak{b}^{-2}\right\},$$

where $A \in SL(2,K)$ satisfying $A\infty = \eta$ which may be further chosen in the form $A = \begin{pmatrix} \alpha & \alpha^* \\ \beta & \beta^* \end{pmatrix} \in SL(2,K)$ so that $\mathcal{O}_K\beta^* + \mathfrak{a}^{-1}\alpha^* = \mathfrak{b}^{-1}$.

Now for $\tau = (z_1, \ldots, z_{r_1}; P_1, \ldots, P_{r_2}) \in \mathcal{H}^{r_1} \times \mathbb{H}^{r_2}$, set

$$\begin{aligned} N(\tau) :=& N\Big(\mathrm{ImJ}(\tau)\Big) \\ =& \prod_{i=1}^{r_1} \mathrm{Im}(z_i) \cdot \prod_{j=1}^{r_2} J(P_j)^2 = \Big(y_1 \cdots y_{r_1}\Big) \cdot \Big(v_1 \cdots v_{r_2}\Big)^2. \end{aligned}$$

Then for all $\gamma = \begin{pmatrix} a & b \\ c & d \end{pmatrix} \in SL(2,K)$,

$$N\Big(\mathrm{ImJ}(\gamma \cdot \tau)\Big) = \frac{N(\mathrm{ImJ}(\tau))}{\|N(c\tau + d)\|^2}. \qquad (*)$$

(Note that here only the second row of γ appears.) Moreover, define the *reciprocal distance $\mu(\eta, \tau)$ from the point $\tau \in \mathcal{H}^{r_1} \times \mathbb{H}^{r_2}$ to the cusp $\eta = \begin{bmatrix} \alpha \\ \beta \end{bmatrix}$ in $\mathbb{P}^1(K)$* by

$$\begin{aligned} \mu(\eta, \tau) :=& N\Big(\mathfrak{a}^{-1} \cdot (\mathcal{O}_K\alpha + \mathfrak{a}\beta)^2\Big) \\ & \cdot \frac{\mathrm{Im}(z_1) \cdots \mathrm{Im}(z_{r_1}) \cdot J(P_1)^2 \cdots J(P_{r_2})^2}{\prod_{i=1}^{r_1} |(-\beta^{(i)} z_i + \alpha^{(i)})|^2 \prod_{j=1}^{r_2} \|(-\beta^{(j)} P_j + \alpha^{(j)})\|^2} \\ =& \frac{1}{N(\mathfrak{a}\mathfrak{b}^{-2})} \cdot \frac{N(\mathrm{ImJ}(\tau))}{\|N(-\beta\tau + \alpha)\|^2}. \end{aligned}$$

Lemma 1. ([We7]) (i) *μ is well-defined;*
(ii) *μ is invariant under the action of $SL(\mathcal{O}_K \oplus \mathfrak{a})$. That is to say,*

$$\mu(\gamma\eta, \gamma\tau) = \mu(\eta, \tau), \qquad \forall \gamma \in SL(\mathcal{O}_K \oplus \mathfrak{a}).$$

(iii) *There exists a positive constant C depending only on K and $\mathfrak{a}$ such that if $\mu(\eta, \tau) > C$ and $\mu(\eta', \tau) > C$ for $\tau \in \mathcal{H}^{r_1} \times \mathbb{H}^{r_2}$ and $\eta, \eta' \in \mathbb{P}^1(K)$, then $\eta = \eta'$.*

(iv) *There exists a positive real number $T := T(K)$ depending only on K such that for $\tau \in \mathcal{H}^{r_1} \times \mathbb{H}^{r_2}$, there exists a cusp η such that $\mu(\eta, \tau) > T$.*

Now for the cusp $\eta = \begin{bmatrix} \alpha \\ \beta \end{bmatrix} \in \mathbb{P}^1(K)$, define the 'sphere of influence' of η by

$$F_\eta := \Big\{ \tau \in \mathcal{H}^{r_1} \times \mathbb{H}^{r_2} : \mu(\eta, \tau) \geq \mu(\eta', \tau), \forall \eta' \in \mathbb{P}^1(K) \Big\}.$$

Lemma 2. ([We7]) *The action of $SL(\mathcal{O}_K \oplus \mathfrak{a})$ in the interior F_η^0 of F_η reduces to that of the isotropy group Γ_η of η, i.e., if τ and $\gamma\tau$ both belong to F_η^0, then $\gamma\tau = \tau$.*

Consequently, we arrive at the following way to decompose the orbit space $SL(\mathcal{O}_K \oplus \mathfrak{a}) \Big\backslash \Big(\mathcal{H}^{r_1} \times \mathbb{H}^{r_2} \Big)$ into h pieces glued in some way along pants of their boundary.

Theorem. (We]) *Let $i_\eta : \Gamma_\eta \Big\backslash F_\eta \hookrightarrow SL(\mathcal{O}_K \oplus \mathfrak{a}) \Big\backslash \Big(\mathcal{H}^{r_1} \times \mathbb{H}^{r_2} \Big)$ denote the natural map. Then*

$$SL(\mathcal{O}_K \oplus \mathfrak{a}) \Big\backslash \Big(\mathcal{H}^{r_1} \times \mathbb{H}^{r_2} \Big) = \cup_\eta i_\eta \Big(\Gamma_\eta \Big\backslash F_\eta \Big),$$

where the union is taken over a set of h cusps representing the ideal classes of K. Each piece corresponds to an ideal class of K.

Note that the action of Γ_η on $\mathcal{H}^{r_1} \times \mathbb{H}^{r_2}$ is free. Consequently, all fixed points of $SL(\mathcal{O}_K \oplus \mathfrak{a})$ on $\mathcal{H}^{r_1} \times \mathbb{H}^{r_2}$ lie on the boundaries of F_η.

We can give a more precise description of the fundamental domain, based on our understanding of that for stabilizer groups of cusps. To state it, denote by $\eta_1, \ldots, \eta_h$ inequivalent cusps for the action of $SL(\mathcal{O}_K \oplus \mathfrak{a})$ on $\mathcal{H}^{r_1} \times \mathbb{H}^{r_2}$. Choose $A_{\eta_i} \in SL(2, K)$ such that $A_{\eta_i}\infty = \eta_i$, $i = 1, 2, \ldots, h$. Write $\mathbf{S}$ for the norm-one hypersurface $\mathbf{S} := \Big\{ y \in \mathbb{R}_{>0}^{r_1+r_2} : N(y) = 1 \Big\}$, and $\mathbf{S}_{U_K^2}$ for the action of U_K^2 on $\mathbf{S}$. Denote by $\mathcal{T}$ a fundamental domain for the action of the translations by elements of $\mathfrak{a}\mathfrak{b}^{-2}$ on $\mathbb{R}^{r_1} \times \mathbb{C}^{r_2}$, and

$$\mathbf{E} := \left\{\tau \in \mathcal{H}^{r_1} \times \mathbb{H}^{r_2} : \mathrm{ReZ}\,(\tau) \in \mathcal{T},\ \mathrm{ImJ}\,(\tau) \in \mathbb{R}_{>0} \cdot \mathbf{S}_{U_K^2}\right\}$$

a fundamental domain for the action of $A_\eta^{-1}\Gamma_\eta A_\eta$ on $\mathcal{H}^{r_1} \times \mathbb{H}^{r_2}$. The intersections of $\mathbf{E}$ with $i_\eta(F_\eta)$ are connected. Consequently, we have the following

Theorem.′ ([We7]) (1) $D_\eta := A_\eta^{-1}\mathbf{E} \cap F_\eta$ *is a fundamental domain for the action of* Γ_η *on* F_η*;*
(2) *There exist* $\alpha_1, \ldots, \alpha_h \in SL(\mathcal{O}_K \oplus \mathfrak{a})$ *such that* $\cup_{i=1}^h \alpha(D_{\eta_i})$ *is connected and hence a fundamental domain for* $SL(\mathcal{O}_K \oplus \mathfrak{a})$.

That is to say, a fundamental domain may be given as $S_Y \cup \mathcal{F}_1(Y_1) \cup \cdots \cup \mathcal{F}_h(Y_h)$ with S_Y bounded, $\mathcal{F}_i(Y_i) = A_i \cdot \widetilde{\mathcal{F}}_i(Y_i)$ and

$$\widetilde{\mathcal{F}}_i(Y_i) := \left\{\tau \in \mathcal{H}^{r_1} \times \mathbb{H}^{r_2} : \mathrm{ReZ}(\tau) \in \Sigma,\ \mathrm{ImJ}(\tau) \in \mathbb{R}_{>T} \cdot \mathbf{S}_{U_K^2}\right\}.$$

Moreover, all $\mathcal{F}_i(Y_i)$'s are disjoint from each other when Y_i are sufficiently large.

C.4.2.7. Stability and Distances to Cusps

Define now the *distance of* τ *to the cusp* η by

$$d(\eta, \tau_\Lambda) := \frac{1}{\mu(\eta, \tau_\Lambda)} \geq 1.$$

Then we are ready to state the following fundamental result, which exposes a beautiful intrinsic relation between stability and the distance to cusps.

Theorem. ([We7]) *The lattice* Λ *is semi-stable if and only if the distances of corresponding point* $\tau_\Lambda \in \mathcal{H}^{r_1} \times \mathbb{H}^{r_2}$ *to all cusps are all bigger or equal to 1.*

C.4.2.8. Moduli of Rank Two Semi-Stable $\mathcal{O}_K$-Lattices

For a rank two $\mathcal{O}_K$-lattice Λ, denote by $\tau_\Lambda \in \mathcal{H}^{r_1} \times \mathbb{H}^{r_2}$ the corresponding point. Then, by the previous subsection, Λ is semi-stable if and only if for all cusps η, $d(\eta, \tau_\Lambda) := \frac{1}{\mu(\eta,\tau_\Lambda)}$ are bigger than or equal to 1. This then leads to the consideration of the following truncation of the fundamental domain $\mathcal{D}$ of $SL(\mathcal{O}_K \oplus \mathfrak{a})\Big\backslash\Big(\mathcal{H}^{r_1} \times \mathbb{H}^{r_2}\Big)$: For $T \geq 1$, denote by

$$\mathcal{D}_T := \Big\{\tau \in \mathcal{D} : d(\eta, \tau_\Lambda) \geq T^{-1},\ \forall \text{cusp } \eta\Big\}.$$

The space $\mathcal{D}_T$ may be precisely described in terms of $\mathcal{D}$ and certain neighborhood of cusps. To explain this, we first establish the following

Lemma. *For a cusp η, denote by*

$$X_\eta(T) := \Big\{\tau \in \mathcal{H}^{r_1} \times \mathbb{H}^{r_2} : d(\eta, \tau) < T^{-1}\Big\}.$$

Then for $T \geq 1$,

$$X_{\eta_1}(T) \cap X_{\eta_2}(T) \neq \emptyset \qquad \Leftrightarrow \eta_1 = \eta_2.$$

With this, we are ready to state the following

Theorem. ([We7]) *There is a natural identification between*
(a) *the moduli space of rank two semi-stable $\mathcal{O}_K$-lattices of volume $N(\mathfrak{a})\Delta_K$ with underlying projective module $\mathcal{O}_K \oplus \mathfrak{a}$;* and
(b) *the truncated compact domain $\mathcal{D}_1$ consisting of points in the fundamental domain $\mathcal{D}$ whose distances to all cusps are bigger than 1.*

In other words, the truncated compact domain $\mathcal{D}_1$ is obtained from the fundamental domain $\mathcal{D}$ of $SL(\mathcal{O}_K \oplus \mathfrak{a})\Big\backslash\Big(\mathcal{H}^{r_1} \times \mathbb{H}^{r_2}\Big)$ by delecting the disjoint open neighborhoods $\cup \cup_{i=1}^h \mathcal{F}_i(1)$ associated to inequivalent cusps $\eta_1, \eta_2, \ldots, \eta_h$, where $\mathcal{F}_i(T)$ denotes the neighborhood of η_i consisting of $\tau \in \mathcal{D}$ whose distance to η_i is strictly less than T^{-1}.

For later use, we set also

$$\mathcal{D}_T := \mathcal{D} \backslash \cup \cup_{i=1}^h \mathcal{F}_i(T), \qquad T \geq 1.$$

C.4.3. Epstein Zeta Functions and Their Fourier Expansions

C.4.3.1. Epstein Zeta Function and Eisenstein Series

For a fixed integer $r \geq 1$ and a fractional ideal $\mathfrak{a}$ of a number field K, define *the Epstein type zeta function* $\widehat{E}_{r,\mathfrak{a};\Lambda}(s)$ associated to an $\mathcal{O}_K$-lattice Λ with underlying projective module $P_{\mathfrak{a}} = \mathcal{O}_K^{(r-1)} \oplus \mathfrak{a}$ to be

$$\widehat{E}_{r,\mathfrak{a};\Lambda}(s) := \Big(\pi^{-\frac{rs}{2}}\Gamma(\frac{rs}{2})\Big)^{r_1}\Big(2\pi^{-rs}\Gamma(rs)\Big)^{r_2} \cdot \Big(N(\mathfrak{a})\Delta_K^{\frac{r}{2}}\Big)^s \cdot \sum_{\mathbf{x}\in\mathcal{O}_K^{(r-1)}\oplus\mathfrak{a}/U_{K,r}^+,\mathbf{x}\neq(0,\dots,0)} \frac{1}{\|\mathbf{x}\|_\Lambda^{rs}}$$

where $U_{K,r}^+ := \Big\{\varepsilon^r : \varepsilon \in U_K,\ \varepsilon^r \in U_K^+\Big\} = U_K^+ \cap U_K^r$. For example, note that in the case $r = 2$, $U_{K,2}^+ = U_K^2$, we have

$$\widehat{E}_{2,\mathfrak{a};\Lambda}(s) := \Big(\pi^{-s}\Gamma(s)\Big)^{r_1}\Big(2\pi^{-2s}\Gamma(2s)\Big)^{r_2} \cdot \Big(N(\mathfrak{a})\Delta_K\Big)^s \cdot \sum_{\mathbf{x}\in\mathcal{O}_K\oplus\mathfrak{a}/U_K^2,\mathbf{x}\neq(0,0)} \frac{1}{\|\mathbf{x}\|_\Lambda^{2s}}.$$

Set for $\mathrm{Re}(s) > 1$ also

$$\widehat{E}_{2,\mathfrak{a}}(\tau,s) := \Big(\pi^{-s}\Gamma(s)\Big)^{r_1}\Big(2\pi^{-2s}\Gamma(2s)\Big)^{r_2} \cdot \Big(N(\mathfrak{a})\Delta_K\Big)^s \cdot \sum_{(x,y)\in\mathcal{O}_K\oplus\mathfrak{a}/U_K,(x,y)\neq(0,0)} \left(\frac{N(\mathrm{ImJ}(\tau_\Lambda))}{\|x\cdot\tau+y\|^2}\right)^s.$$

Lemma. *For a rank two $\mathcal{O}_K$-lattice $\Lambda = (\mathcal{O}_K \oplus a, \rho_\Lambda)$, denote by τ_Λ the corresponding point in the moduli space $SL(\mathcal{O}_K \oplus a)\Big\backslash\Big(\mathcal{H}^{r_1} \times \mathbb{H}^{r_2}\Big)$. Then $\widehat{E}_{2,\mathfrak{a};\Lambda}(s) = \widehat{E}_{2,\mathfrak{a}}(\tau_\Lambda, s)$.*

C.4.3.2. Fourier Expansions

For simplicity, introduce the standard Eisenstein series by setting

$$E_{2,\mathfrak{a}}(\tau,s) := \sum_{(x,y)\in\mathcal{O}_K\oplus\mathfrak{a}/U_K,(x,y)\neq(0,0)} \left(\frac{N(\mathrm{ImJ}(\tau))}{\|x\cdot\tau+y\|^2}\right)^s, \qquad \mathrm{Re}(s) > 1.$$

Then the completed one becomes

$$\widehat{E}_{2,\mathfrak{a}}(\tau,s)=\Big(\pi^{-s}\Gamma(s)\Big)^{r_1}\Big(2\pi^{-2s}\Gamma(2s)\Big)^{r_2}\cdot\Big(N(\mathfrak{a})\Delta_K\Big)^{s}\cdot E_{2,\mathfrak{a}}(\tau,s). \tag{$*$}$$

As before, for the cusp $\eta=\begin{bmatrix}\alpha\\ \beta\end{bmatrix}$, choose a (normalized) matrix $A=\begin{pmatrix}\alpha & \alpha^*\\ \beta & \beta^*\end{pmatrix}\in SL(2,F)$ such that and that if $\mathfrak{b}=\mathcal{O}_K\alpha+\mathfrak{a}\beta$, then $\mathcal{O}_K\beta^*+\mathfrak{a}\alpha^*=\mathfrak{b}^{-1}$. Clearly, $A\infty=\eta$, and moreover,

$$A^{-1}\Gamma'_\eta A=\left\{\begin{pmatrix}1 & \omega\\ 0 & 1\end{pmatrix}:\omega\in\mathfrak{a}\mathfrak{b}^{-2}\right\}.$$

Since $\widehat{E}_{2,\mathfrak{a}}(\tau,s)$, and thus $E_{2,\mathfrak{a}}(\tau,s)$, is $SL(\mathcal{O}_K\oplus\mathfrak{a})$-invariant, $E_{2,\mathfrak{a}}(\tau,s)$ is $\Gamma'_\eta\subset SL(\mathcal{O}_K\oplus\mathfrak{a})$-invariant. Therefore $E_{2,\mathfrak{a}}(A\tau,s)$ is $\mathfrak{a}\mathfrak{b}^{-2}$-invariant, that is, $E_{2,\mathfrak{a}}(A\tau,s)$ is invariant under parallel transforms by elements of $\mathfrak{a}\mathfrak{b}^{-2}$. Consequently, we have the Fourier expansion

$$E_{2,\mathfrak{a}}(A\tau,s)=\sum_{\omega'\in(\mathfrak{a}\mathfrak{b}^{-2})^\vee}a_{\omega'}\Big(\mathrm{ImJ}(\tau),s\Big)\cdot e^{2\pi i\langle\omega',\mathrm{ReZ}(\tau)\rangle},$$

where $(\mathfrak{a}\mathfrak{b}^{-2})^\vee$ denotes the dual lattice of $\mathfrak{a}\mathfrak{b}^{-2}$. Thus, if we use Q to denote a fundamental parallolgram of $\mathfrak{a}\mathfrak{b}^{-2}$ in $\mathbb{R}^{r_1}\times\mathbb{C}^{r_2}$, then

$$a_{\omega'}(\mathrm{ImJ}(\tau),s):=\frac{1}{\mathrm{Vol}(\mathfrak{a}\mathfrak{b}^{-2})}\sum_{(c,d)\in(\mathcal{O}_K\oplus\mathfrak{a})A/U_K,(c,d)\neq(0,0)}$$
$$\int_Q\left(\frac{N(\mathrm{ImJ}(\tau))}{\|c\tau+d\|^2}\right)^s\cdot e^{-2\pi i\langle\omega',\mathrm{ReZ}(\tau)\rangle}\prod_{\sigma:\mathbb{R}}dx_\sigma\cdot\prod_{\tau:\mathbb{C}}dx_\tau dy_\tau.$$

Theorem. ([We7]) *We have the following Fourier expansion*

$$\begin{aligned}E_{2,\mathfrak{a}}(A\tau,s)=&\zeta([\mathfrak{a}^{-1}\mathfrak{b}],2s)\cdot N(\mathfrak{a}\mathfrak{b}^{-1})^{-2s}\cdot N(\mathrm{ImJ}(\tau))^s\\&+\frac{1}{\mathrm{Vol}(\mathfrak{a}\mathfrak{b}^{-2})}\sum_{\begin{pmatrix}* & *\\ c & c\omega+d\end{pmatrix}\in\mathcal{R}}\frac{1}{N(c)^{2s}}\cdot(\pi^{\frac12})^{r_1}\\&\cdot\left(\frac{\Gamma\Big(s-\frac12\Big)}{\Gamma(s)}\right)^{r_1}\left(\frac{\pi}{2s-1}\right)^{r_2}\cdot N(\mathrm{ImJ}(\tau))^{1-s}\end{aligned}$$

$$+\frac{1}{\mathrm{Vol}(\mathfrak{a}\mathfrak{b}^{-2})}\sum_{\begin{pmatrix} * & * \\ c & d\end{pmatrix}\in\mathcal{R}}\frac{e^{2\pi i\langle\omega',\frac{d}{c}\rangle}}{N(c)^{2s}}\cdot N(\mathrm{ImJ}(\tau))^{\frac{1}{2}}\cdot N(\omega')^{s-\frac{1}{2}}\left(\frac{2\pi^s}{\Gamma(s)}\right)^{r_1}$$

$$\times\prod_{\sigma:\mathbb{R}}K_{s-\frac{1}{2}}(2\pi|\omega'|_\sigma y_\sigma)\cdot\left(\frac{2\pi^{2s}|\omega'|^{2s-1}}{\Gamma(2s)}\right)^{r_2}\cdot\prod_{\tau:\mathbb{C}}K_{2s-1}(2\pi|\omega'|_\tau r_\tau).$$

C.4.4. Explicit Formula for Rank Two Zetas: Rankin-Selberg & Zagier Method

The original Rankin-Selberg method gives a way to express the Mellin transfrom of the constant term in the Fourier expansion of an automorphic function as the scalar product of the automorphic function with an Eisenstein series if the automorphic function is very small when approach to the cusp. In a paper by Zagier, this method is extended to a much broad type of automorphic functions. Here, we use a generalization of the Rankin-Selberg & Zagier method to give an explicit expression for rank two zeta functions of number fields in terms of Dedekind zeta functions.

In the sequel, T is assumed to be a positive real number ≥ 1. Our aim is to compute the integration

$$\iiint_{\mathcal{D}_T}\widehat{E}_{2,\mathfrak{a}}(\tau,s)d\mu(\tau).$$

Here $\mathcal{D}_T$ is the compact part obtained from the fundamental domain $\mathcal{D}$ for $SL(\mathcal{O}_K\oplus\mathfrak{a})\Big\backslash\Big(\mathcal{H}^{r_1}\times\mathbb{H}^{r_2}\Big)$ by cutting off the cusp neighborhoods defined by the conditions that the distance to cusps is less than T^{-1}. (Recall that, as such, $\mathcal{D}_1$ is simply the part corresponding to semi-stable lattices.)

For doing so, first formulate the integration

$$\iiint_{\mathcal{D}_T}\Big(\Delta_K\,\widehat{E}_{2,\mathfrak{a}}(\tau,s)\Big)\,d\mu(\tau)$$

where

$$\Delta_K:=\sum_{\sigma:\mathbb{R}}\Delta_\sigma+\sum_{\tau:\mathbb{C}}\Delta_\tau$$

with

$$\Delta_\sigma := y_\sigma^2\left(\frac{\partial^2}{\partial x_\sigma^2}+\frac{\partial^2}{\partial y_\sigma^2}\right),\quad \Delta_\tau := r_\tau^2\left(\frac{\partial^2}{\partial x_\tau^2}+\frac{\partial^2}{\partial y_\tau^2}+\frac{\partial^2}{\partial r_\tau^2}\right)-r\frac{\partial}{\partial r_\tau}.$$

(For the time being, by an abuse of notation, we use Δ_K to denote the hyperbolic Laplace operator for the space $\mathcal{H}^{r_1}\times\mathbb{H}^{r_2}$, not the absolute value of the discriminant of K which accordingly is changed to D_K.)

Note that

$$\Delta_\sigma\left(y_\sigma^s\right)=s(s-1)\cdot y_\sigma^s,\qquad\text{and}\qquad \Delta_\tau\left(r_\tau^{2s}\right)=2s(2s-2)\cdot r_\tau^{2s}$$

by the SL-invariance of the metrics, we conclude that

$$\Delta_K\left(\widehat{E}_{2,\mathfrak{a}}(\tau,s)\right)=\left(r_1\cdot\left(s(s-1)\right)+r_2\cdot\left(2s(2s-2)\right)\right)\cdot\widehat{E}_{2,\mathfrak{a}}(\tau,s).$$

Hence

$$\iiint_{\mathcal{D}_T}\widehat{E}_{2,\mathfrak{a}}(\tau,s)\,d\mu(tau)=\frac{r_1+4r_2}{s(s-1)}\iiint_{\mathcal{D}_T}\Delta_K\widehat{E}_{2,\mathfrak{a}}(\tau,s)\,d\mu(\tau).$$

On the other hand, using Stokes' Formula, we have

$$\begin{aligned}&\iiint_{\mathcal{D}_T}\Delta_K\widehat{E}_{2,\mathfrak{a}}(\tau,s)d\mu(\tau)\\ &=\iiint_{\mathcal{D}_T}\left(\Delta_K\widehat{E}_{2,\mathfrak{a}}(\tau,s)\right)\cdot 1d\mu(\tau)\\ &\qquad-\iiint_{\mathcal{D}_T}\widehat{E}_{2,\mathfrak{a}}(\tau,s)\cdot\left(\Delta_K 1\right)d\mu(\tau)\\ &=\iiint_{\mathcal{D}_T}\left(\left(\Delta_K\widehat{E}_{2,\mathfrak{a}}(\tau,s)\right)\cdot 1-\widehat{E}_{2,\mathfrak{a}}(\tau,s)\cdot\left(\Delta_K 1\right)\right)d\mu(\tau)\\ &=\iint_{\partial D(T)}\left(\frac{\partial\widehat{E}_{2,\mathfrak{a}}(\tau,s)}{\partial\nu}\cdot 1-\widehat{E}_{2,\mathfrak{a}}(\tau,s)\cdot\frac{\partial 1}{\partial\nu}\right)d\mu\\ &=\iint_{\partial D(T)}\frac{\partial\widehat{E}_{2,\mathfrak{a}}(\tau,s)}{\partial\nu}d\mu\end{aligned}$$

where $\frac{\partial}{\partial \nu}$ is the outer normal derivative, $d\mu$ is the volume element of the boundary $\partial \mathcal{D}_T$.

Theorem. ([We7]) *Up to a constant factor depending only on* K,

$$\iiint_{\mathcal{D}_T} \widehat{E}_{2,\mathfrak{a}}(\tau, s) d\mu(\tau) = \frac{\xi_K(2s)}{s-1}\Big(\Delta_K \cdot T\Big)^{s-1} - \frac{\xi_K(2-2s)}{s}\Big(\Delta_K \cdot T\Big)^{-s}.$$

Consequently, we have the following

Theorem′. ([We7]) *The rank two non-abelian zeta function* $\xi_{K,2}(s)$ *of* K *is given by*

$$\xi_{K,2}(s) = \frac{\xi_K(2s)}{s-1}\Delta_K^{s-1} - \frac{\xi_K(2s-1)}{s}\Delta_K^{-s} \qquad \mathrm{Re}(s) > 1.$$

C.4.5. Zeros of Rank Two Non-Abelian Zeta Functions for Number Fields

Let us first start with Suzuki's weak result and then give Lagarias' unconditional result [LS]. Meanwhile, for an independent parallel work, please go to Ki's beautiful paper [Ki].

Proposition. (Suzuki) *If the Riemann Hypothesis for the Riemann zeta function holds, then all zeros of* $\xi_{\mathbb{Q},2}(s)$ *lie on the critical line* $\mathrm{Re}(s) = \frac{1}{2}$.

Proof. Let

$$F(z) = -Z\left(\frac{1}{2} + 2iz\right) \qquad \text{with} \qquad Z(s) = s(1-s)\xi(s).$$

Lemma. (1) $F(z+\frac{i}{4}) - F(z-\frac{i}{4}) = iz(1+4z^2)\,\xi_{\mathbb{Q},2}(\frac{1}{2}+iz)$.
(2) *Assume the RH, then all zeros of* $F(z+\frac{i}{4}) - F(z-\frac{i}{4})$ *are real. In particular, then the RH implies that* $\xi_{\mathbb{Q},2}(\frac{1}{2}+zi)$ *admits only real zeros.*

Proof. (1) Simple calculation. (2) Clearly, $F(z)$ is an entire function of order 1, so there are constants A, B such that

$$F(z) = e^{A+Bz} \cdot \prod_{\rho: F(\rho)=0} \left(1 - \frac{z}{\rho}\right) \cdot \exp\left(\frac{z}{\rho}\right).$$

Note that essentially, ρ are zeros of the completed Riemann zeta but transformed from z to $\frac{1}{2}+2iz$. Hence, *by the RH*, all ρ are real. Moreover, since for $x \in \mathbb{R}$, $F(x)$ takes only real values. Hence, constants A and B are both real.

Now let $z_0 = x_0 + iy_0$ be a zero of

$$F\left(z+\frac{i}{4}\right) - F\left(z-\frac{i}{4}\right) = iz(1+4z^2)\,\xi_{\mathbb{Q},2}\left(\frac{1}{2}+iz\right).$$

Then $z_0 = 0$ and/or z_0 is a zero of $\xi_{\mathbb{Q},2}(\frac{1}{2}+iz)$ since $\xi_{\mathbb{Q},2}(\frac{1}{2}+iz)$ admits simple poles at $z = \pm\frac{1}{2}i$.

In any case, $F(z_0+\frac{i}{4}) = F(z_0-\frac{i}{4})$. By taking absolute values on both sides,

$$\left|e^{A+B(z_0+\frac{i}{4})}\cdot\prod\left(1-\frac{z_0+\frac{i}{4}}{\rho}\right)\cdot\exp\left(\frac{z_0+\frac{i}{4}}{\rho}\right)\right| = \left|e^{A+B(z_0-\frac{i}{4})}\cdot\prod\left(1-\frac{z_0-\frac{i}{4}}{\rho}\right)\cdot\exp\left(\frac{z_0-\frac{i}{4}}{\rho}\right)\right|.$$

Since $B \in \mathbb{R}$ and $\rho_n \in \mathbb{R}$ (which is obtained by the RH as said above), we get

$$1 = \prod_{n=1}^{\infty}\frac{(x_0-\rho_n)^2+(y_0-\frac{1}{4})^2}{(x_0-\rho_n)^2+(y_0+\frac{1}{4})^2}.$$

Thus if $y_0 > 0$, then the right hand side is < 1, while if $y_0 < 0$, then the right hand side is > 1. Contradiction. This leads then $y_0 = 0$, hence completes the proof.

With this in mind, note that in the proof above, the RH was used to ensure that ρ are real, which have the effect that then in the calculation for the exponential factor $\exp\left(\frac{z_0+\frac{i}{4}}{\rho}\right)$, the ratio $\frac{\left|\exp\left(\frac{z_0+\frac{i}{4}}{\rho}\right)\right|}{\left|\exp\left(\frac{z_0-\frac{i}{4}}{\rho}\right)\right|}$ gives us the exact value 1. That is to say, this factor of ratio of exp's does not contribute.

Lagarias then offers his own unconditional result as a part of his understanding of de Branges's work. The trick is to use the functional equation. So instead of working on individual ρ_n in the product, we may equally use the functional equation to pair ρ and $1-\rho$ for the zeros of the completed Riemann zeta function, or even to group ρ, $1-\rho$, $\overline{\rho}$ and $1-\overline{\rho}$ together. Consequently, the exponential factor appeared inside the infinite product may be totally omitted. That is to say, from the very beginning, we may simply assume that the Hadamard product involved takes the form

$$F(z) = e^{A+Bz} \cdot \prod_{\rho:F(\rho)=0}' \left(1 - \frac{z}{\rho}\right)$$

where $\prod'$ means that ρ's are paired or grouped as above. Form here, it is an easy exercise to deduce the following result of (Suzuki and) Lagarias.

Theorem. ([LS]) *All zeros of* $\xi_{\mathbb{Q},2}(s)$ *lie on the line* $\mathrm{Re}(s) = \frac{1}{2}$.
Proof. Alternatively, as above, we have

$$\begin{aligned}&\left|e^{A+B(z_0+\frac{i}{4})} \cdot \prod \left(1 - \frac{z_0+\frac{i}{4}}{\rho}\right) \cdot \exp\left(\frac{z_0+\frac{i}{4}}{\rho}\right)\right| \\ &\quad = \left|e^{A+B(z_0-\frac{i}{4})} \cdot \prod \left(1 - \frac{z_0-\frac{i}{4}}{\rho}\right) \cdot \exp\left(\frac{z_0-\frac{i}{4}}{\rho}\right)\right|.\end{aligned}$$

Since $B \in \mathbb{R}$, so if we can take care of the factors $\exp\left(\frac{z_0+\frac{i}{4}}{\rho}\right)$ and $\exp\left(\frac{z_0-\frac{i}{4}}{\rho}\right)$ in a nice way, we are done. For this, as said above, let us group ρ, $\bar{\rho}$, $1-\rho$, $1-\bar{\rho}$ together, we see that $\frac{1}{\rho} + \frac{1}{\bar{\rho}} = \frac{2\,\mathrm{Re}(\rho)}{|\rho|^2}$ and $\frac{1}{1-\rho} + \frac{1}{1-\bar{\rho}} = \frac{2-2\,\mathrm{Re}(\rho)}{|1-\rho|^2}$ are all reals, hence, the same prove as above works.

As observed in [We7], the above method works for the functions $\xi_{\mathbb{Q},2}^T(s)$ as well, provided that $T \geq 1$. In particular, we then obtain the following

RH in Rank Two. ([We7]) *All zeros of rank two non-abelian zeta functions for number fields are on the critical line* $\mathrm{Re}(s) = \frac{1}{2}$.

References

[Am] S.A. Amitsur, Groups with representations of bounded degree II, Illinois J. Math. 5 (1961), 198-205

[An] A.N. Andrianov, Euler products that correspond to Siegel's modular forms of genus 2, Russian Math. Surveys **29**:3(1974), 45-116

[Ar] S. Arakelov, Intersection theory of divisors on an arithmetic surface, Izv. Akad. Nauk SSSR Ser. Mat., 38 No. **6**, (1974)

[Ar1] J. Arthur, Eisenstein series and the trace formula, Proc. Sympos. Pure Math., XXXIII, 253-274, AMS, Providence, R.I., 1979.

[Ar2] J. Arthur, A trace formula for reductive groups. I. Terms associated to classes in $G(\mathbb{Q})$. Duke Math. J. **45** (1978), no. 4, 911–952

[Ar3] J. Arthur, A trace formula for reductive groups. II. Applications of a truncation operator. Compositio Math. **40** (1980), no. 1, 87–121.

[Ar4] J. Arthur, A measure on the unipotent variety, Canad. J. Math **37**, (1985) pp. 1237–1274

[A] E. Artin, Quadratische Körper im Gebiete der höheren Kongruenzen, I,II, *Math. Zeit*, **19** 153-246 (1924) (See also *Collected Papers*, pp. 1-94, Addison-Wesley 1965)

[AT] E. Artin & J. Tate, *Class Field Theory*, Benjamin Inc, 1968

[At] M. Atiyah, Vector Bundles over an elliptic curve, *Proc. LMS, VII*, 414-452 (1957) (See also *Collected Works*, Vol. 1, pp. 105-143, Oxford Science Publications, 1988)

[AB] M. Atiyah & R. Bott, The Yang-Mills equations over Riemann surfaces. Philos. Trans. Roy. Soc. London Ser. A 258 (1983), no. 1505, 523–615.

[Be1] A. Beilinson, Higher regulators and values of L-functions, J. Soviet Math. **30** (1985), 2036-2070

[Be2] A. Beilinson, Height pairings between algebraic cycles, in *Current Trends in Arithmetical Algebraic Geometry*, Contemporary Math. **67** (1987), 1-24

[BR] U. Bhosle & A. Ramanathan, Moduli of parabolic G-bundles on curves. Math. Z. 202 (1989), no. 2, 161–180.

[Bo1] A. Borel, Some finiteness properties of adele groups over number fields, Publ. Math., IHES, **16** (1963) 5-30

[Bo2] A. Borel, *Introduction aux groupes arithmetictiques*, Hermann, 1969

[Bor] A. Borisov, Convolution structures and arithmetic cohomology, Compositio Math. 136 (2003), no. 3, 237–254

[B-PGN] L. Brambila-Paz, I. Grzegorczyk, & P.E. Newstead, Geography of Brill-Noether Loci for Small slops, J. Alg. Geo. **6** 645-669 (1997)

[Bu] J.-F. Burnol, Weierstrass points on arithmetic surfaces. Invent. Math. 107 (1992), no. 2, 421–432.

[CF] J.W.S. Cassels & A. Fröhlich, *Algebraic Number Theory*, Academic Press, 1967

[Co] A. Connes, Trace formula in noncommutative geometry and the zeros of the Riemann zeta function. Selecta Math. (N.S.) 5 (1999), no. 1, 29–106.

[C] H. Cramér, Studien über die Nullstellen der Riemannschen Zeta funktion, Math. Z. **4** (1919), 104-130

[D] P. Deligne, Catégories Tannakiennes, *Grothendieck Festschrift*, Vol. I, Progress in Math. 86 (1990), 111-195

[DM] P. Deligne & J.S. Milne, Tannakian categories, in *Hodge Cycles, Motives and Shimura Varieties*, LNM **900**, (1982), 101-228

[De1] Ch. Deninger, On the Γ-factors attached to motives. Invent. Math. **104** (1991), no. 2, 245–261.

[De2] Ch. Deninger, Local L-factors of motives and regularized determinants. Invent. Math. **107** (1992), no. 1, 135–150.

[De3] Ch. Deninger, Motivic L-functions and regularized determinants, Motives (Seattle, WA, 1991), 707-743, Proc. Sympos. Pure Math, **55** Part 1, AMS, Providence, RI, 1994

[De4] Ch. Deninger, Lefschetz trace formulas and explicit formulas in analytic number theory. J. Reine Angew. Math. **441** (1993), 1–15.

[De5] Ch. Deninger, Some analogies between number theory and dynamical systems on foliated spaces. Proceedings of the International Congress of Mathematicians, Vol. I (Berlin, 1998). Doc. Math. 1998, Extra Vol. I, 163–186

[DS] Ch. Deninger & M. Schröter, A distribution-theoretic proof of Guinand's functional equation for Cramér's V-function and generalizations. J. London Math. Soc. (2) **52** (1995), no. 1, 48–60.

[DR] U.V. Desale & S. Ramanan, Poincaré polynomials of the variety of stable bundles, Maeh. Ann **216**, 233-244 (1975)

[FW] G. Faltings & G. Wüstholz, Diophantine approximations on projective spaces. Invent. Math. 116 (1994), no. 1-3, 109–138.

[Fo] G.B. Folland, *A course in abstract harmonic analysis*, Studies in advanced mathematics, CRC Press, 1995

[Fo1] J.-M. Fontaine, Repre'sentations p-adiques semi-stables. Asterisque No. 223 (1994), 113–184.

[Fo2] J.-M. Fontaine, Representations l-adiques potentiellement semi-stables. Ast No. 223 (1994), 321–347.

[FV] M.D. Fried & H. Völklein, The embedding problem over a Hilbertian PAC-field. Ann. of Math. (2) **135** (1992), no. 3, 469–481

[Ge] G. van der Geer, *Hilbert Modular Surfaces*, Ergebnisse der Mathematik und ihrer Grenzgebiete, 3. Folge Bd.16, Springer (1988)

[GS] G. van der Geer & R. Schoof, Effectivity of Arakelov Divisors and the Theta Divisor of a Number Field, Sel. Math., New ser. **6** (2000), 377-398

[Gr1] D.R. Grayson, Reduction theory using semistability. Comment. Math. Helv. **59** (1984), no. 4, 600–634

[Gr2] D.R. Grayson, Reduction theory using semistability. II. Comment. Math. Helv. **61** (1986), no. 4, 661–676.

[Gu1] R.C. Gunning, *Lectures on Riemann surfaces*, Princeton Math. Notes, 1966

[Gu2] R.C. Gunning, *Lectures on vector bundles over Riemann surfaces*, Princeton Math. Notes, 1967

[Ha1] Sh. Haran, Index theory, potential theory, and the Riemann hypothesis. *L-functions and arithmetic* (Durham, 1989), 257–270, London Math. Soc. Lecture Note Ser., 153, Cambridge Univ. Press, 1991.

[Ha2] Sh. Haran, Riesz potentials and explicit sums in arithmetic. Invent. Math. 101 (1990), no. 3, 697–703.

[HN] G. Harder & M.S. Narasimhan, On the cohomology groups of moduli spaces of vector bundles over curves, Math Ann. **212**, (1975) 215-248

[Har] R. Hartshorne, *Algebraic Geometry*, GTM 52, Springer-Verlag, 1977

[H] H. Hasse, *Mathematische Abhandlungen*, Walter de Gruyter, 1975.

[Hi] H. Hida, *Modular Forms and Galois Cohomology*, Cambridge Studies in Advanced Mathematics 69, 2000

[Hil] D. Hilbert, Über die Theorie der relativ-Abelschen Zahlkörper, Nachr. der K. Ges. der Wiss. Göttingen, 377-399 (1898)

[Iw1] K. Iwasawa, *Algebraic theory of algebraic functions*, Lecture Notes at Princeton Univ. noted by T. Kimura, 1975

[Iw2] K. Iwasawa, Letter to Dieudonné, April 8, 1952, in N.Kurokawa and T. Sunuda: *Zeta Functions in Geometry*, Advanced Studies in Pure Math. **21** (1992), 445-450

[JLR] H. Jacquet, E. Lapid & J. Rogawski, Periods of automorphic forms. J. Amer. Math. Soc. 12 (1999), no. 1, 173–240

[JL1] J. Jorgenson & S. Lang, On Crame'r's theorem for general Euler products with functional equation. Math. Ann. 297 (1993), no. 3, 383–416.

[JL2] J. Jorgenson & S. Lang, *Basic analysis of regularized series and products.* Lecture Notes in Mathematics, 1564. Springer-Verlag, Berlin, 1993

[JL3] J. Jorgenson & S. Lang, *Explicit formulas.* Lecture Notes in Mathematics, 1593. Springer-Verlag, Berlin, 1994.

[Ka] Y. Kawada, Class formulations. VI. Restriction to a subfamily. J. Fac. Sci. Univ. Tokyo Sect. I 8 1960 229–262 (1960).

[KT] Y. Kawada & J. Tate, On the Galois cohomology of unramified extensions of function fields in one variable. Amer. J. Math. 77, (1955). 197–217.

[Ke] G.R. Kempf, Instability in invariant theory, Ann. Math. **108** (1978), 299-316

[Ki] H. Ki, All but finitely many non-trivial zeros of the approximations of the Epstein zeta function are simple and on the critical line. Proc. London Math. Soc. (3) 90 (2005), no. 2, 321–344.

[KW] H.H. Kim & L. Weng, Volume of truncated fundamental domains, 2005

[Laf] L. Lafforgue, *Chtoucas de Drinfeld et conjecture de Ramanujan-Petersson.* Asterisque No. 243 (1997)

[LS] J. Lagarias & M. Suzuki, The Riemann Hypothesis for certain integrals of Eisenstein series, 2004

[L1] S. Lang, *Algebraic Number Theory*, Springer-Verlag, 1986

[L2] S. Lang, *Fundamentals on Diophantine Geometry*, Springer-Verlag, 1983

[L3] S. Lang, *Introduction to Arakelov theory*, Springer Verlag, 1988

[La1] R. Langlands, *On the functional equations satisfied by Eisenstein series*, Springer LNM **544**, 1976

[La2] R. Langlands, the volume of the fundamental domain for some arithmetical subgroups of Chevalley groups, in *Algebraic Groups and Discontinuous Subgroups,* Proc. Sympos. Pure Math. 9, AMS (1966) pp.143–148

[La3] R. Langlands, Automorphic representations, Shimura varieties, and motives. Ein Marchen. *Automorphic forms, representations and L-functions* pp. 205–246, Proc. Sympos. Pure Math., XXXIII, AMS, 1979.

[Li] X. Li, A note on the Riemann-Roch theorem for function fields. *Analytic number theory*, Vol. **2**, 567–570, Progr. Math., **139**, 1996

[Ma] B. Mazur, Deforming Galois representations, in *Galois groups over* **Q**, MSRI Publ. **16**, (1989), 385-437

[MS] V.B. Mehta & C.S. Seshadri, Moduli of vector bundles on curves with parabolic structures. Math. Ann. 248 (1980), no. 3, 205–239.

[Mi] H. Minkowski, *Geometrie der Zahlen*, Leipzig and Berlin, 1896

[MW] C. Moeglin & J.-L. Waldspurger, *Spectral decomposition and Eisenstein series.* Cambridge Tracts in Math, **113**. Cambridge University Press, 1995

[Mo] C. Moreno, *Algebraic curves over finite fields.* Cambridge Tracts in Mathematics, **97**, Cambridge University Press, 1991

[M] D. Mumford, *Geometric Invariant Theory*, Springer-Verlag, (1965)

[Na] M. Namba, *Branched coverings and algebraic functions.* Pitman Research Notes in Mathematics Series, 161. John Wiley & Sons, Inc., 1987. viii+201 pp.

[NR] M.S. Narasimhan & S. Ramanan, Moduli of vector bundles on a compact Riemann surfaces, Ann. of Math. **89** 14-51 (1969)

[NS] M.S. Narasimhan & C.S. Seshadri, Stable and unitary vector bundles on a compact Riemann surface. Ann. of Math. (2) 82 1965

[Ne] J. Neukirch, *Algebraic Number Theory*, Grundlehren der Math. Wissenschaften, Vol. **322**, Springer-Verlag, 1999

[Pa] A.N. Parshin, On the arithmetic of two-dimensional schemes. I. Distributions and residues. (Russian) Izv. Akad. Nauk SSSR Ser. Mat. 40 (1976), no. 4, 736–773, 949.

[RR] S. Ramanan & A. Ramanathan, Some remarks on the instability flag. Tohoku Math. J. (2) 36 (1984), no. 2, 269–291.

[Ra1] A. Ramanathan, Stable principal bundles on a compact Riemann surface. Math. Ann. 213 (1975), 129–152.

[Ra2] A. Ramanathan, Moduli for principal bundles over algebraic curves. I. Proc. Indian Acad. Sci. Math. Sci. 106 (1996), no. 3, 251–328.

[Ra3] A. Ramanathan, Moduli for principal bundles over algebraic curves. I. Proc. Indian Acad. Sci. Math. Sci. 106 (1996), no. 4, 421–449.

[RSS] M. Rapoport, N. Schappacher, & P. Schneider, *Beilinson's conjectures on special values of L-functions*, Perspectives in Math, Vol. **4.** (1988)

[RZ] M. Rapoport & Th. Zink, *Period spaces for p-divisible groups.* Annals of Mathematics Studies, 141. Princeton University Press, Princeton, NJ, 1996.

[Sch] A.J. Scholl, An introduction to Kato's Euler systems, in *Galois Representations in arithmetic algebraic geometry*, London Math. Soc. Lecture Note Series **254**, (1998) 379-460

[S] J.-P. Serre, *Algebraic Groups and Class Fields*, GTM 117, Springer (1988)

[Se1] C.S. Seshadri, Space of unitary vector bundles on a compact Riemann surface. Ann. of Math. (2) 85 (1967) 253–336.

[Se2] C. S. Seshadri, *Fibrés vectoriels sur les courbes algébriques*, Asterisque **96**, 1982

[Sh] I. Shafarevich, *Lectures on minimal models and birational transformations of two dimensional schemes.* Notes by C. P. Ramanujam. Tata Institute of Fundamental Research Lectures on Mathematics and Physics, No. **37** (1966)

[Shi] G. Shimura, *Introduction to the arithmetic theory of automorphic functions*, Iwanami, (1971)

[Sie] C.L. Siegel, *Lectures on the geometry of numbers*, notes by B. Friedman, rewritten by K. Chandrasekharan with the assistance of R. Suter, Springer-Verlag, 1989.

[St1] U. Stuhler, Eine Bemerkung zur Reduktionstheorie quadratischer Formen, Arch. Math. (Basel) **27** (1976), no. 6, 604–610

[St2] U. Stuhler, Zur Reduktionstheorie der positiven quadratischen Formen. II, Arch. Math. (Basel) **28** (1977), no. 6, 611–619

[T] J. Tate, Fourier analysis in number fields and Hecke's zeta functions, Thesis, Princeton University, 1950

[Ti] J. Tilouine, *Deformations of Galois representations and Hecke algebras*, Published for The Mehta Research Institute of Mathematics and Mathematical Physics, 1996.

[To] B. Totaro, Tensor products in p-adic Hodge theory. Duke Math. J. 83 (1996), no. 1, 79–104.

[W1] A. Weil, Généralisation des fonctions abéliennes, J. Math Pures et Appl, **17**, (1938) 47-87

[W2] A. Weil, *Sur les courbes algébriques et les variétés qui s'en déduisent*, Herman, Paris (1948)

[W3] A. Weil, Sur les formules explicites de la the'orie des nombres. (French) Izv. Akad. Nauk SSSR Ser. Mat. 36 (1972), 3–18.

[W4] A. Weil, *Basic Number Theory*, Springer-Verlag, 1973

[W5] A. Weil, *Adeles and algebraic groups.* With appendices by M. Demazure and Takashi Ono. Progress in Math., 23. 1982.

[We1] L. Weng, Ω-admissible theory II: Deligne pairings over moduli spaces of punctured Riemann surfaces, Math. Ann **320** (2001), 239-283

[We2] L. Weng, Stability and New Non-Abelian Zeta Functions, *Number theoretic methods* (Iizuka, 2001), 405–419, Dev. Math., 8, Kluwer Acad. Publ., 2002.

[We3] L. Weng, Refined Brill-Noether Locus and Non-Abelian Zeta Functions for Elliptic Curves, *Algebraic geometry in East Asia* (Kyoto, 2001), 245–262, World Sci. 2002

[We4] L. Weng, Non-abelian zeta functions for function fields, Amer. J. Math. 127 (2005), no. 5, 973–1017

[We5] L. Weng, Non-abelian class field theory for Riemann surfaces, Appendix to this paper, this volume

[We6] L. Weng, Non-Abelian L-Functions for number fields, available at http://xxx.lanl.gov/abs/math.NT/0412008, to appear

[We7] L. Weng, Rank Two Non-Abelian Zeta and its Zeros, available at http://xxx.lanl.gov/abs/math.NT/0412009, submitted

[We8] L. Weng, Automorphic Forms, Eisenstein Series and Spectral Decompositions, this volume

[We9] L. Weng, Non-Abelian L-Functions for number fields II, in preperation

[Z] Zagier, D. The Rankin-Selberg method for automorphic functions which are not of rapid decay. J. Fac. Sci. Univ. Tokyo Sect. IA Math. 28(3), 415–437 (1982)

Lin WENG
Graduate School of Mathematics
Kyushu University
Fukuoka 812-8581, Japan
Email: weng@math.kyushu-u.ac.jp

Appendix:

Non-Abelian CFT for Function Fields over C

In this appendix, using what we call a micro reciprocity law, we complete Weil's program [W] for non-abelian class field theory of Riemann surfaces.

1. Refined Structures for Tannakian Categories

Let $\mathbb{T}$ be a Tannakian category with a fiber functor $\omega : \mathbb{T} \to \mathrm{Ver}_{\mathbb{C}}$, where $\mathrm{Ver}_{\mathbb{C}}$ denotes the category of finite dimensional $\mathbb{C}$-vector spaces. An object $t \in \mathbb{T}$ is called *decomposable* if there exist non-zero objects $x, y \in \mathbb{T}$ such that $t = x \oplus y$. An object is called *indecomposable* if it is not decomposable. If moreover every object x of $\mathbb{T}$ can be written uniquely as a sum of irreducible objects $x = x_1 \oplus x_2 \oplus \cdots \oplus x_n$, then $\mathbb{T}$ is called a Tannakian category with *unique factorization*. Usually, we call x_i's the *irreducible components* of x.

A Tannakian subcategory $\mathbb{S}$ of a unique factorization Tannakian category $\mathbb{T}$ is called *completed* if for $x \in \mathbb{S}$, all its irreducible components x_i's in $\mathbb{T}$ are also in $\mathbb{S}$. $\mathbb{S}$ is called *finitely generated* if as an abelian category, it is generated by finitely many objects. Moreover, $\mathbb{S}$ is called *finitely completed* if (a) $\mathbb{S}$ is finitely generated; (b) $\mathbb{S}$ is completed; and (c) $\mathrm{Aut}^{\otimes}\omega\big|_{\mathbb{S}}$ is a finite group.

2. An Example

Let M be a compact Riemann surface of genus g. Fix an effective divisor $D = \sum_{i=1}^{N} e_i P_i$ with $e_i \in \mathbb{Z}_{\geq 2}$ once for all. For simplicity, in this note we always assume that $(M; D) \neq (\mathbb{P}^1; e_1 P_1)$, or $(\mathbb{P}^1; e_1 P_1 + e_2 P_2)$ with $e_1 \neq e_2$. (These cases may be easily treated.)

By definition a parabolic semi-stable bundle

$$\Sigma := (E =: E(\Sigma); P_1, \dots, P_N; Fil(E|P_1), \dots, Fil(E|_{P_N}); \\ a_{11}, \dots, a_{1r_1}; \dots; a_{N1}, \dots, a_{Nr_N})$$

of parabolic degree 0 is called a *GA bundle* over M along D if

(i) the parabolic weights are all rational, i.e., $a_{ij} \in \mathbb{Q} \cap [0, 1)$;

(ii) there exist $\alpha_{ij} \in \mathbb{Z}, \beta_{ij} \in \mathbb{Z}_F > 0$ such that

(a) $(\alpha_{ij}, \beta_{ij}) = 1$;

(b) $a_{ij} = \alpha_{ij}/\beta_{ij}$; and

(c) $\beta_{ij}|e_i$, for all i, j.

Denote by $[\Sigma]$ the Seshadri equivalence class associated with Σ. Moreover, for $[\Sigma]$, define $\omega_D([\Sigma])$ as $E(\mathrm{Gr}(\Sigma))|_P$, i.e., the fiber of the bundles associated with Jordan-Hölder graded parabolic bundle of Σ at a fixed $P \in M^0 := M\backslash|D| = M\backslash\{P_1, \ldots, P_N\}$. Put $\mathcal{M}(M; D) := \{[\Sigma] : \Sigma \text{ is a GA bundle over } M \text{ along } D\}$.

Proposition. *$\mathcal{M}(M; D)$ is a unique factorization Tannakian category and $\omega_D : \mathcal{M}(M; D) \to \mathrm{Vec}_{\mathbb{C}}$ is a fiber functor.*

Proof. (1) By a result of Mehta-Seshadri [MS, Prop. 1.15], $\mathcal{M}(M; D)$ is an abelian category. Then from the unitary representation interpretation of a GA bundle, a fundamental result due to Seshadri, (see [MS, Thm 4.1], also in Step 3 of Section 5 below,) $\mathcal{M}(M; D)$ is closed under tensor product. The rigidity may be checked directly.

(2) Since the Jordan-Hölder graded bundle is a direct sum of stable and hence irreducible objects and is unique, (see [MS, Rm 1.16],) so, $\mathcal{M}(M; D)$ is a unique factorization category.

(3) By definition, we know that the functor ω is exact and tensor. So we should check whether it is faithful. This then is a direct consequence of the fact that $\mathcal{M}(M; D)$ is a unique factorization category and that any morphism between two irreducible objects is either zero or a constant multiple of the identity map.

3. Reciprocity Map

In $[\Sigma]$, choose its associated Jordan-Hölder graded bundle $\mathrm{Gr}(\Sigma)$ as a representative. Then by the above mentioned fundamental result of Seshadri, $\mathrm{Gr}(\Sigma)$ corresponds to a unitary representation $\rho_{\mathrm{Gr}(\Sigma)} : \pi_1(M^0) \to U(r_\Sigma)$, where r_Σ denotes the rank of $E(\Sigma) := E$.

For each element $g \in \pi_1(M^0)$, we then obtain a $\mathbb{C}$-isomorphism of $E(\mathrm{Gr}(\Sigma))|_P$. Thus, in particular, we get a natural morphism

$$W : \pi_1(M^0) \to \mathrm{Aut}^{\otimes}\omega_D.$$

Now note that $\pi_1(M^0)$ is generated by $2g$ hyperbolic transformations $A_1, B_1, \ldots, A_g, B_g$ and N parabolic transformations $S_1, \ldots, S_N$

satisfying a single relation

$$A_1B_1A_1^{-1}B_1^{-1}\cdots A_gB_gA_g^{-1}B_g^{-1}S_1\cdots S_N=1,$$

and that $\rho_{\mathrm{Gr}(\Sigma)}(S_i^{e_i})=1$ for all $i=1,\ldots,N$. ([MS, §1].) Denote by $J(D)$ the normal subgroup of $\pi_1(M^0)$ generated by $S_1^{e_1},\ldots,S_N^{e_N}$. Then naturally we obtain the following *reciprocity map*

$$W(D):\pi_1(M^0)/J(D)\to \mathrm{Aut}^{\otimes}\omega_D.$$

4. Main Theorem

As usual, a Galois covering $\pi:M'\to M$ is called branched at most at D if (1) π is branched at $P_1,\ldots,P_N$; and (2) the ramification index e_i' of points over P_i divides e_i for all $i=1,\ldots,N$. Clearly, by changing D, we get all finite Galois coverings of M.

Non-Abelian CFT. (1) (Existence and Conductor Theorem) *There is a natural one-to-one correspondence w_D between*

$$\Big\{\mathbb{S}:\text{finitely completed Tannakian subcategory of }\mathcal{M}(M;D)\Big\}$$

and

$$\Big\{\pi:M'\to M:\text{finite Galois covering branched at most at }D\Big\};$$

(2) (Reciprocity Law) *The reciprocity map induces a natural group isomorphism*

$$\mathrm{Aut}^{\otimes}(\omega_D\big|_{\mathbb{S}})\simeq \mathrm{Gal}\,(w_D(\mathbb{S})).$$

5. Proof

Step 1: Galois Theory. By a result of Bungaard, Nielsen, Fox, M. Kato, and Namba, (see e.g., [Na, Thms 1.2.15 and 1.3.9],) we know that the assignment $(\pi:M'\to M)\mapsto \pi_*(\pi_1(M'\backslash\pi^{-1}\{P_1,\ldots,P_N\}))$ gives a one-to-one correspondence between isomorphism classes of finite Galois coverings $\pi:M'\to M$ branched at most at D and finite index (closed) normal subgroups $K=K(\pi)$ of $\pi_1(M^0)$ containing $J(D)$. Moreover, we have a natural isomorphism $\mathrm{Gal}(\pi)\simeq \pi_1(M^0)/K(\pi)\Big(\simeq\big(\pi_1(M^0)/J(D)\big)\Big/\big(K(\pi)/J(D)\big)\Big)$. Thus the problem is transformed to the one for finite index normal subgroups of

$\pi_1(M^0)$ which contain $J(D)$, or the same, finite index normal subgroups of $G(D) := \pi_1(M^0)/J(D)$.

Step 2: Tannakian Category Theory: Geometric Side. Consider now the category $\mathbb{T}(D)$ of equivalence classes of unitary representations of $G(D)$. Clearly, $\mathbb{T}(D)$ forms a unique factorization Tannakian category, whose fiber functor $\omega(D)$ may be defined to be the forget functor. Now fixed once for all a representative $\rho_\Sigma : G(D) \to U(r_\Sigma)$ for each equivalence classes $[\Sigma]$. (The choice of the representative will not change the essentials below as the resulting groups are isomorphic to each other.)

Let $\mathbb{S}$ be a finitely completed Tannakian subcategory of $\mathbb{T}(D)$. Then as in the definition of reciprocity map above, we have a natural morphism $G(D) \overset{\omega_\mathbb{S}}{\to} \mathrm{Aut}^\otimes\omega\Big|_\mathbb{S}$. Denote its kernel by $K(\mathbb{S})$. Then, by definition, $G(D)/K(\mathbb{S})$ is a finite group, and $K(\mathbb{S}) = \cap_{[\Sigma]\in\mathbb{S}}\mathrm{ker}\rho_\Sigma$.

Since $\mathbb{S}$ is finitely completed, there exists a finite subset of $\mathbb{S}$, denoted by $S = \{[\Sigma_1], [\Sigma_2], \ldots, [\Sigma_t]\}$, which generates $\mathbb{S}$ as a completed Tannakian subcategory. Set $[\Sigma_0] := \oplus_{i=1}^t[\Sigma_i]$. Then for any $[\Sigma] \in \mathbb{S}$, $\mathrm{ker}(\rho_{\Sigma_0}) \subset \mathrm{ker}(\rho_{[\Sigma]})$, since $\mathbb{S}$ is generated by S. Also, by definition, $\mathrm{ker}(\rho_{\Sigma_0}) = \cap_{i=0}^t\mathrm{ker}(\rho_{\Sigma_i})$. Thus $K(\mathbb{S}) = \mathrm{ker}(\rho_{\Sigma_0})$.

Therefore, for any $[\Sigma] \in \mathbb{S}$, $\rho_\Sigma = \tilde{\rho}_\Sigma \circ \Pi(D;\mathbb{S})$ where $\Pi(D;\mathbb{S}) : G(D) \to G(D)/K(\mathbb{S})$ denotes the natural quotient map and $\tilde{\rho}_\Sigma$ is a suitable unitary representation of $G(D)/K(\mathbb{S})$.

Now set $\tilde{\mathbb{S}} := \{[\tilde{\rho}_\Sigma] : [\Sigma] \in \mathbb{S}\}$. $\tilde{\mathbb{S}}$ is a finitely completed Tannakian subcategory in $\mathcal{U}ni(G(D)/K(\mathbb{S}))$, the category of equivalence classes of unitary representations of $G(D)/K(\mathbb{S})$.

In particular, since the unitary representation $\tilde{\rho}_{\Sigma_0}$ of $G(D)/K(\mathbb{S})$ maps $G(D)/K(\mathbb{S})$ injectively into its image, for any two elements $g_1, g_2 \in G(D)/K(\mathbb{S})$, $\tilde{\rho}_{\Sigma_0}(g_1) \neq \tilde{\rho}_{\Sigma_0}(g_2)$.

With this, by applying the van Kampen Completeness Theorem ([Ka]), which claims that for any compact group G, if Z is a subset of the category $\mathcal{U}ni(G)$ such that for any two elements g_1, g_2, there exists a representation ρ_{g_1,g_2} in Z such that $\rho_{g_1,g_2}(g_1) \neq \rho_{g_1,g_2}(g_2)$, then the completed Tannakian subcategory generated by Z is the whole category $\mathcal{U}ni(G)$ itself, we conclude that $\tilde{\mathbb{S}} = \mathcal{U}ni(G(D)/K(\mathbb{S}))$. But as categories, $\mathbb{S}$ is equivalent to $\tilde{\mathbb{S}}$, thus by the Tannaka duality, (see e.g., [DM] or [Ta]) we obtain a natural isomorphism $\mathrm{Aut}^\otimes\omega|_\mathbb{S} \simeq$

$G(D)/K(\mathbb{S})$.

On the other hand, if K is a finite index (closed) normal subgroup of $G(D)$. Set $\tilde{\mathbb{S}} := \mathcal{U}ni(G(D)/K)$ with the fiber functor $\omega_{\tilde{\mathbb{S}}}$. Compositing with the natural quotient map $\Pi : G(D) \to G(D)/K$ we then obtain an equivalent category $\mathbb{S}$ consisting of corresponding unitary representations of $G(D)$. $\mathbb{S}$ may also be viewed as a Tannakian subcategory of $\mathcal{U}ni(G(D))$. We next show that indeed such an $\mathbb{S}$ is a finitely completed Tannakian subcategory.

From definition, $\mathrm{Aut}^{\otimes}\omega|_{\mathbb{S}} \simeq \mathrm{Aut}^{\otimes}\omega_{\tilde{\mathbb{S}}}$ which by the Tannaka duality theorem is isomorphic to $G(D)/K$. So it then suffices to show that $\mathbb{S}$ is finitely generated. But this is then a direct consequence of the fact that for any finite group there always exists a unitary representation such that the group is injectively mapped into the unitary group.

Step 3: A Micro Reciprocity Law. With Steps 1 and 2, the proof of the Main Theorem is then completed by the following

Weil-Narasimhan-Seshadri Correspondence ([MS, Thm 4.1]) *There is a natural one-to-one correspondence between isomorphism classes of unitary representations of fundamental groups of M^0 and Seshadri equivalence classes of semi-stable parabolic bundles over M^0 of parabolic degree zero.*

Indeed, with this theorem, the Seshadri equivalence classes of GA bundles over M along D correspond naturally in one-to-one to the equivalence classes of unitary representations of the group $\pi_1(M^0)/J(D)$. Thus by Step 2, the finitely completed Tannakian subcategories of $\mathcal{M}(M;D)$ are in one-to-one correspondence to the finite index closed normal subgroup of $\pi_1(M^0)/J(D)$, which by Step 1 in one-to-one correspondence to the finite Galois coverings of M branched at most at D. This proves the theorem.

6. An Application to Inverse Galois Problem

As it stands, we may use our main theorem to see whether a finite group can be realized as a Galois group of certain coverings by constructing suitable finitely completed Tannakian subcategory of $\mathcal{M}(M;D)$. In particular, to relate with the so-called Inverse Galois Problem, or better, the Regular Inverse Galois Problem, via a fundamental result of Belyi, we only need to consider the finitely completed Tannakian subcategories of $\mathcal{M}(\mathbb{P}^1;\Delta)$, for some effective divisors $\Delta = e_\infty \cdot [\infty] + e_1 \cdot [1] + e_0 \cdot [0]$ supported on $\{\infty, 1, 0\} \subset \mathbb{P}^1$.

An Example. By definition, a collection of objects of a finitely completed Tannakian subcategory in $\mathcal{M}(M;D)$ is called primitive generators if it is a smallest collection which generates the subcategory as an *abelian category.*

Since ramification points are all fixed to be $\{\infty, 1, 0\}$, to simplify the notation, set

$$\Sigma_{11} := \Big(\mathcal{O};0;0;0\Big);$$
$$\Sigma_{12} := \left(\mathcal{O}(-1);0;\frac{1}{2};\frac{1}{2}\right)$$
$$\Sigma_{21} := \left(\mathcal{O}(-1)\oplus\mathcal{O}(-1);\frac{2}{3},\frac{1}{3};\frac{1}{2},0;\frac{1}{2},0\right).$$

Then we may view Σ_{ij} as parabolic vector bundles over $\mathbb{P}^1$ with parabolic points $\infty, 1$ and 0. That is to say, the bundle part is given by the line bundles or rank two vector bundle over $\mathbb{P}^1$, while the parabolic weights are given by $(0;0;0), (0;\frac{1}{2};\frac{1}{2})$ and $(\frac{2}{3},\frac{1}{3};\frac{1}{2},0;\frac{1}{2},0)$ at points $(\infty;1;0)$. Here for rank two cases, the filtration is choosen to be the one such that $E_P \supset \mathbb{Q}(e_1+e_2) \supset \{0\}$ with the i-th factor gives $\mathcal{O}(-1)_P = \mathbb{C}\cdot e_i$.

Proposition. *Let* $\mathcal{R}(\Sigma_{11},\Sigma_{12},\Sigma_{21})$ *be the Tannakian subcategory generated by parabolic vector bundles* Σ_{11},Σ_{12} *and* Σ_{21}. *Then the category* $\mathcal{R}(\Sigma_{11},\Sigma_{12},\Sigma_{21})$ *is generated by* $\{\Sigma_{11},\Sigma_{12},\Sigma_{21}\}$. *Moreover,*

$$\mathrm{Aut}^{\otimes}\omega\big|_{\mathcal{R}(\Sigma_{11},\Sigma_{12},\Sigma_{21})} \simeq S_3.$$

Remark. This proposition, via our non-abelian reciprocity law, reproves a well-known result that S_3 may be realized as a Galois group of a branched covering of $\mathbb{P}^1$ ramified at $\infty, 1, 0$ with ramification index $3, 2, 2$ respectively.

Proof. We first need to show that all tensor products could be realized as extensions of Σ_{ij}. For this, we check case by case as follows.
(1) Clearly, $\Sigma_{11}\otimes\Sigma_{ij} = \Sigma_{ij}$;

(2) $\Sigma_{12} \otimes \Sigma_{12} = \Sigma_{11}$. Indeed,

$$\begin{aligned}
&\Sigma_{12} \otimes \Sigma_{12} \\
&= \left(\mathcal{O}(-2); 0+0; \frac{1}{2}+\frac{1}{2}; \frac{1}{2}+\frac{1}{2}\right) = (\mathcal{O}(-2); 0; 1; 1) \\
&= (\mathcal{O}(-2+2\cdot 1); 0; 1-1; 1-1) = (\mathcal{O}; 0; 0; 0) \\
&= \Sigma_{11};
\end{aligned}$$

(3) $\Sigma_{12} \otimes \Sigma_{21} = \Sigma_{21}$. Indeed,

$$\begin{aligned}
&\Sigma_{12} \otimes \Sigma_{21} \\
&= \left(\mathcal{O}(-2) \oplus \mathcal{O}(-2); \frac{2}{3}+0, \frac{1}{3}+0; \frac{1}{2}+\frac{1}{2}, \frac{1}{2}+0; \frac{1}{2}+\frac{1}{2}, \frac{1}{2}+0\right) \\
&= \left(\mathcal{O}(-2) \oplus \mathcal{O}(-2); \frac{2}{3}, \frac{1}{3}; 1, \frac{1}{2}; 1, \frac{1}{2}\right) \\
&= \left(\mathcal{O}(-2+1) \oplus \mathcal{O}(-2+1); \frac{2}{3}, \frac{1}{3}; \frac{1}{2}; 0; \frac{1}{2}; 0\right) \\
&= \Sigma_{21};
\end{aligned}$$

(4) $\Sigma_{21} \otimes \Sigma_{21} = \Sigma_{21} \oplus \Sigma_{12} \oplus \Sigma_{11}$. Indeed,

$$\begin{aligned}
&\Sigma_{21} \otimes \Sigma_{21} \\
&= \Big(\mathcal{O}(-2)^{\oplus 4}; \frac{2}{3}+\frac{2}{3}, \frac{2}{3}+\frac{1}{3}, \frac{1}{3}+\frac{2}{3}, \frac{1}{3}+\frac{1}{3}; \frac{1}{2}+\frac{1}{2}, \frac{1}{2}+0, \\
&\qquad\qquad 0+\frac{1}{2}, 0+0; \frac{1}{2}+\frac{1}{2}, \frac{1}{2}+0, 0+\frac{1}{2}, 0+0\Big) \\
&= \left(\mathcal{O}(-2)^{\oplus 4}; 1+\frac{1}{3}, 1, 1, \frac{2}{3}; 1, \frac{1}{2}, \frac{1}{2}, 0; 1, \frac{1}{2}, \frac{1}{2}, 0\right) \\
&= \Big(\mathcal{O}(-2+1)^{\oplus 2} \oplus \mathcal{O}(-2+1) \oplus \mathcal{O}(-2+2); 1+\frac{1}{3}-1, \\
&\qquad\qquad 1-1, 1-1, \frac{2}{3}; 1-1, \frac{1}{2}, \frac{1}{2}, 0; 1-1, \frac{1}{2}, \frac{1}{2}, 0\Big) \\
&= \left(\mathcal{O}(-1)^{\oplus 2} \oplus \mathcal{O}(-1) \oplus \mathcal{O}(0); \frac{2}{3}, \frac{1}{3}, 0, 0; \frac{1}{2}, \frac{1}{2}, 0, 0; \frac{1}{2}, \frac{1}{2}, 0, 0\right) \\
&= \Sigma_{21} \oplus \Sigma_{12} \oplus \Sigma_{11}.
\end{aligned}$$

Now note that the above decompositions for the tensor products are exactly the same as that for the irreducible representations of S_3. So $\mathcal{R}(\Sigma_{11}, \Sigma_{12}, \Sigma_{21})$ is equivalent to the category of finite dimensional complex representations of S_3. As a direct consequence, we conclude that the structure group $\mathrm{Aut}^{\otimes}\omega|_{\mathcal{R}(\Sigma_{11},\Sigma_{12},\Sigma_{21})}$ is simply S_3.

This example shows that how the approach works. First we use the category of representations of a finite group to find how the tensor products of irreducible representations decompose. Then we find whether there exist corresponding semi-stable parabolic vector bundles which give an equivalent category. As such, the problem becomes a combinatorial one, since on $\mathbb{P}^1$, all vector bundles are direct sums of $\mathcal{O}(n)$ for $n \in \mathbb{Z}$. Further examples with all dihedral groups may be easily constructed as well, since their irreducible representations are at most of rank 2. We leave this to the reader.

Up to this point, naturally, we may ask to which extend this approach to the Inverse Galois Problem work over number fields K. To answer this, we come back to moduli spaces of parabolic vector bundles constructed by Narasimhan and Seshadri. Unlike existing approaches to this problem, we have enough rational points ready to use: the anti-canonical line bundles of these moduli spaces are (roughly speaking) positive. Therefore, for us, the essential problem is whether the categories generated by K-rational parabolic bundles correspond to coverings defined over K.

References

[DM] P. Deligne & J.S. Milne, Tannakian categories, in *Hodge Cycles, Motives and Shimura Varieties*, LNM **900**, (1982), 101-228

[Ka] E. van Kampen, Almost periodic functions and compact groups, Ann. of Math. **37** (1936), 78-91

[MS] V.B. Mehta & C.S. Seshadri, Moduli of vector bundles on curves with parabolic structures. Math. Ann. **248** (1980), no. 3, 205–239.

[MFK] D. Mumford, J. Fogarty & F. Kirwan, *Geometric Invariant Theory*, Springer-Verlag, 1994

[Na] M. Namba, *Branched coverings and algebraic functions*. Pitman Research Notes in Mathematics Series **161**, Longman Scientific & Technical, 1987

[NS] M.S. Narasimhan & C.S. Seshadri, Stable and unitary vector bundles on a compact Riemann surface. Ann. of Math. **82** (1965), 540-567

[Se1] C.S. Seshadri, Moduli of π-vector bundles over an algebraic curve. *Questions on Algebraic Varieties*, C.I.M.E., III, (1969) 139–260

[Se2] C. S. Seshadri, *Fibrés vectoriels sur les courbes algébriques*, Asterisque **96**, 1982

[Ta] T. Tannaka, *Theory of topological groups*, Iwanami, 1949 (in Japanese)

[W] A. Weil, Généralisation des fonctions abéliennes, J. Math Pures et Appl, **17**, (1938) 47-87

Lin WENG
Institute for Fundamental Research
The *L* Academy